Polymers
Properties and Applications

2

H. H. Kausch

Polymer Fracture

With 180 Figures

Springer-Verlag
Berlin Heidelberg New York 1978

CHEMISTRY

Professor Dr. rer. nat.
HANS-HENNING KAUSCH-BLECKEN VON SCHMELING

Laboratoire des Polymères, Département des Matériaux,
École Polytechnique Fédérale de Lausanne, Suisse

This volume continues the series *Chemie, Physik und Technologie der Kunststoffe in Einzeldarstellungen,* which is now entitled *Polymers/ Properties and Applications.*

ISBN 3-540-08786-9 Springer-Verlag Berlin Heidelberg New York

ISBN 0-387-08786-9 Springer-Verlag New York Heidelberg Berlin

Library of Congress Cataloging in Publication Data:
Kausch, H. H. Polymer fracture. (Polymers ; v. 2). Bibliography: p. Includes index.
1. Polymers and polymerization – – Fracture. I. Title.
II. Series. Ta 455.P 58 K 38   620.1'92   78-14205

Typesetting and printing: Schwetzinger Verlagsdruckerei GmbH, 6830 Schwetzingen. Bookbinding: Konrad Triltsch, Graphischer Betrieb, 8700 Würzburg

2152/3020–543210

Preface

This book on "Polymer Fracture" might as well have been called "Kinetic Theory of Polymer Fracture". The term "kinetic theory", however, needs some definition or, at least, some explanation. A kinetic theory deals with and particularly considers the effect of the existence and discrete size, of the motion and of the physical properties of molecules on the macroscopic behavior of an ensemble, gaseous or other. A kinetic theory *of strength* does have to consider additional aspects such as elastic and anelastic deformations, chemical and physical reactions, and the sequence and distribution of different disintegration steps.

In the last fifteen years considerable progress has been made in the latter domains. The deformation and rupture of molecular chains, crystals, and morphological structures have been intensively investigated. The understanding of the effect of those processes on the strength of polymeric materials has especially been furthered by the development and application of *spectroscopical methods* (ESR, IR) and of the *tools of fracture mechanics.* It is the aim of this book to relate the conventional and successful statistical, parametrical, and continuum mechanical treatment of fracture phenomena to new results on the behavior of highly stressed molecular chains.

The ultimate deformation of chains, the kinetics of the mechanical formation of free radicals and of their reactions, the initiation and propagation of crazes and cracks, and the interpretation of fracture toughness and energy release rates in terms of molecular parameters will be discussed. Although the prime interest had been given to these subjects it was not possible to cover the literature completely. The author apologizes in advance for all intentional and unintentional omissions that will have occurred. In any event this book draws on the existing literature on morphology, viscoelasticity, deformation and fracture of high polymers. It is hoped that it forms a useful supplement of such monographs in view of a molecular interpretation of polymer fracture.

Lausanne, September 1978   H. H. Kausch

Acknowledgements

The author would like to thank his friends, colleagues, and collaborators for their encouragement and help. Their multiple efforts are highly esteemed. E. H. Andrews, W. Döll, H. Hahn, D. Langbein, J. A. Sauer, W. O. Statton, J. Sohma, I. M. Ward, and J. G. Williams have been kind enough to read appropriate chapters of this manuscript and to make well appreciated comments. Many colleagues have assisted the timely preparation of this book by making available copies of their relevant manuscripts prior to publication. Thanks are also due to the many authors and publishers for their permissions to reproduce graphic material and to K. L. DeVries, and W. O. Statton, Salt Lake City, W. Döll, Freiburg, E. W. Fischer, Mainz, K. Friedrich, Bochum, E. Gaube, Hoechst AG, S. Gogolewski, Lodz, J. W. S. Hearle, Manchester, H. Hendus, BASF AG, D. Hull, Liverpool, G. Kämpf, Bayer AG, P. Stockmayer, Stuttgart, and J. Wristers, Baytown, for providing valuable photographs or electron micrographs.

The author acknowledges the financial assistance of the Deutsche Forschungsgemeinschaft during some of the studies carried out; he thanks the Battelle Institut e. V., Frankfurt, which had made available its technical facilities during the early stages of preparing the material; in organizing the manuscript the author has been greatly assisted by his collaborators, Ing. EPFL T. Pesek and Dr. S. Sikka, and by the speedy and reliable typing of Frau Weinhold and Mme Steireif. All this help is gratefully acknowledged.

Table of Contents

Chapter 1

Deformation and Fracture of High Polymers, Definition and Scope of Treatment

The importance of a thorough understanding of the deformation behavior and the strength of polymeric engineering materials need not be emphasized. It is obvious to anyone who wants to use polymers as load bearing, weather-resisting, or deformable components or who wants to grind or degrade them. "Strength" and "fracture" of a sample are the positive and negative aspects of one and the same phenomenon, namely that of stress-biased material disintegration. The final step of such disintegration manifests itself as macroscopic failure of the component under use, be it a water pipe, a glass fiber reinforced oil tank, or a plastic grocery bag. The preceding intermediate steps — nonlinear deformation, environmental attack, and crack initiation and growth — are often less obvious, although they cause and/or constitute the damage developed within a loaded sample.

Efforts of fracture research are directed, therefore, towards three goals:

1. To understand the cause and development of a failure and to assess the importance of any one process contributing to a failure or delaying it

2. To see whether, how, and with what effect the most damaging process can be controlled or even eliminated, and how the strength-determining material properties can be improved

3. With the first two goals only partly attainable, one would like to have knowledge of the endurance limit of a part in terms of stress or strain level, environment, number of loading cycles, or of loading time.

The first goal is equivalent to the ability to predict the material response under all kinds of foreseeable conditions of mechanical, thermal, or environmental attack, to characterize the constantly changing state of the material, and to know at any moment which conditions of attack are critical for the state the material just attained. To achieve this goal in all generality is hardly possible, in view of the complexity involved in characterizing polymeric structures, let alone their deformation. A conclusive discussion of those two subjects and of their interrelation with and dependency on other aspects of polymer science is perhaps achieved in the framework of polymer science encyclopedias [1, 2] and partially in specialized textbooks on the viscoelastic [3, 4], mechanical [5], and physical [6—8] behavior of polymers. Although the first goal cannot be achieved in all generality, the literature on fracture (cf. textbooks [9—12]) deals extensively and successfully with interpretations of the fracture event which focus on just those processes which account for the most obvious changes of sample morphology during fracture development.

The previous statements on the goals of fracture research are well illustrated by the fracture patterns of rigid polyvinylchloride (PVC) shown in Figures 1.1–1.3. The samples, water pipes internally pressurized, fractured in a brittle manner when the circumferential stress level was high, in a partially ductile manner and only after prolonged times at medium stresses, and through thermal crack growth (creep crazing) after very long times at low stress levels. The three processes leading to pipe failure in these three cases are respectively the rapid extension of a flaw, the flow of matter, and the thermally activated growth of a flaw. In all three cases the volume element within which the fracture is initiated is finite, consequently nonhomogeneous deformation must have occurred locally. We will deal with the nature and interpretation of such heterogeneous deformation of a supposedly homogeneous material later.

Fig. 1.1. Brittle fracture of an internally pressurized water pipe (polyvinylchloride, PVC), taken from Ref. [13].

Fig. 1.2. Ductile failure of a PVC pipe at medium stress levels (rupture as a result of the reduction of wall thickness at a defect at the indicated position), taken from Ref. [13].

Fig. 1.3. Long-time failure of a PVC pipe through development of a creep craze (from [13]).

We also understand the importance of the first goal with respect to achieving the second. In the described example of pressurized PVC pipes it was known that in the medium time range the performance of a pipe was limited by its creep behavior. We would wish, therefore, to decrease the creep rate by increasing the rigidity of the material; in doeing so, however, we would facilitate crack propagation and, possibly, craze initiation and thus reduce the permissible stress levels at short and very long times of application.

The third goal, the knowledge of the endurance limit of a part in terms of stress or strain level, is generally directly attainable within a certain experimental region. Figure 1.4 summarizes, for instance, the test results on PVC pipes at temperatures between 20 and 60 °C, circumferential stresses between 10 and 50 MN/m^2 and a range of times to fracture of 0.3 to 10^4 h. Such representations, which in nongraphi-

Fig. 1.4. Times-to-failure of internally pressurized PVC water pipes at different stresses and temperatures (after [13]).

cal form can be extended to interrelations between more than three parameters, form an important basis of our understanding of material behavior within the multi-dimensional space covered by the parameters investigated. Material selection and optimization of its properties through parameter variation become possible to the extent the mathematical framework of parameter-property coefficients holds. With regard to fracture this means that one would like to define a fracture criterion in terms of the relevant structural and environmental parameters.

A good example for the occurrence of unpredicted events is offered by the behavior of polyethylene (PE) pipes — and of other semicrystalline materials — under constant circumferential stress; they showed a rapid drop of sustained stresses after long loading times (Fig. 1.5). With PE — as with PVC — an initial time range is observed where the times to failure are only a weak function of stress. Depending on the temperature, the failure mode is either brittle (Fig. 1.1) or ductile (Fig. 1.2 and 1.6). Both materials are also comparable insofar as the thermally activated growth of a creep crack (Fig. 1.3, 1.7, and 1.8) can be held responsible for the fail-

3

$$\sigma = p \, \frac{d_m}{2s}$$

Fig. 1.5. Times-to-failure of high-density polyethylene (HDPE) water pipes under internal pressure p at different stresses and temperatures. d_m: average diameter, s: wall thickness; A: ductile failure, B: creep crazing (specimens are shown in Figs. 1.6–1.8; after [14]).

Fig. 1.6. Ductile failure of a HDPE water pipe under the conditions shown as point A in Fig. 1.5: $\sigma_v = 7$ MN/m^2, $T = 80$ °C (after [14])

Fig. 1.7. Surface of a creep craze formed in HDPE under conditions shown as point B in Fig. 1.5: $\sigma_v = 6$ MN/m^2, $T = 80$ °C (after [14]).

Fig. 1.8. Detail of the fracture surface close to the upper center of the mirror zone, the point of craze initiation (after [14]).

4

ure of a pipe after long times of service. Both materials are different insofar, as one finds in PE — but practically not in PVC — that the kinetics of creep crack development differs markedly from that of ductile failure (Fig. 1.5). This finding only underlines that it is necessary to deal directly with the physical nature of material defect development in order to predict reliably material behavior — especially for new applications — and/or improve properties through additional components or modified production.

In this monograph we wish to discuss the physical nature of defect development, and we will focus on linear thermoplastic and elastomeric polymers (Table 1.1.). We recognize that these materials cover a wide range of properties, although they are composed of molecules which exhibit a number of similarities: they are predominantly linear, flexible, highly anisotropic (nonextended) chains having molecular weights between 20 000 and more than 1 000 000. Figure 1.9 gives a represen-

molecule

chain end-to-end distance

segment

bond vectors

Fig. 1.9. Representation of linear flexible chain molecule (polyamide 6), (after [21]).

tation of a (polyamide 6) chain showing the nonextended *conformation*, an arbitrary division into *segments*, and — in the insert — the primary *bonds* forming the *backbone chain*. The relative sizes of the atoms and the proportion of volume occupied by them within the chain is illustrated by the Stuart model of a polyamide segment (Fig. 1.10). In actual size this segment has an extended length of 1.97 nm. If such a segment could be stressed in the chain axis direction, the bending and stretching of primary bonds would result in a chain stiffness of about 200 GN/m² [15], whereas the intermolecular attraction of segments through the much weaker van der Waal's forces entails a stiffness of "only" 3 to 8 GN/m² in a direction perpendicular to the chain axis. *The fact that polymer molecules are long, anisotropic, and flexible chains accounts for the characteristic properties of polymeric solids,*

5

Table 1.1. Mechanical Properties of Chain Molecules and Polymeric Solids*)

Polymer	Abbrev. (ISO/R 1043–1969)	Chain molecule			Oriented Fiber [17]		Isotropic Solid		Critical stress σ_c (MN/m²) (Ref. [18])
		Monomer molecular weight M_0 (g/mol)	Monomer length $L(10^{-10}$m) (from unit cell)	Young's modulus E_c (MN/m²) (Refs. [15, 16])	Young's modulus E (MN/m²) [19]	Strength σ_b (MN/m²)	Young's modulus E (MN/m²)	Strength σ_b (MN/m²)	
Semicrystalline Polymers									
1. Polyethylene high density	PE	28	2.534	235,000	8000–50,000 [19]	480	1000	27	160
low density			2.54				200	13	
2. Polypropylene	PP	42	6.5/3	41,000	6000–9000		1250–2400	33	98
3. Polytetrafluorethylene	PTFE	100	2.60	153,000			400–980	23	117
4. Poly(oxymethylene)	POM	30	1.92	53,000		1200–2550	3500	68	216
5. Poly(ethylene oxide)	(PEO)	44	19.3/9	9800			530		
6. Poly(propylene oxide)		58	7.15/2				133		
7. Polyethylene terephthalate	PETP	192	10.77	74,000–137,000	20,000–25,000	800–1400	3000	53	155
8. Polyamide									
polycaprolactam	PA 6	226	17.2	24,500	5000–8500	390–870	1250	61	
polyhexamethylene adipamide	PA 66	226	17.3			420–930	1500–4100	80	179
aromatic	(Kevlar 49)	119	6.6	180,000	144,000	3000			
Amorphous Polymers									
9. Polystyrene	PS	104	2.21	9000–11,800			3250	50	
10. Polyvinyl chloride	PVC	62.5	2.55	(160,000–230,000)			2500–3000	55	
11. Polyvinyl alcohol	PVAL	44	2.52	250,000	35,000–50,000	1200–2800		28	142

Table 1.1. (continued)

Polymer	Abbrev. (ISO/R 1043--1969)	Chain molecule			Oriented Fiber [17]		Isotropic Solid		
		Monomer molecular weight M_0 (g/mol)	Monomer length $L(10^{-10}$m) (from unit cell	Young's modulus E_c(MN/m²) (Refs. [15, 16])	Young's modulus E (MN/m²)	Strength σ_b (MN/m²)	Young's modulus E (MN/m²)	Strength σ_b (MN/m²)	Critical stress σ_c (MN/m²) (Ref. [18])
Amorphous Polymers									
12. Polyvinyl acetate	PVAC	86					1300--2800	29--49	
13. Polymethyl methacrylate	PMMA	100	2.11				3000--4000	70	68
14. Polycarbonate (bisphenol A)	PC	254	10.75				2350	53	145
(Cross-linked Rubbers, for comparison)									
Natural rubber	NR	136	8.1				1--2	17--25	
Butyl rubber	IIR	56	4.7				0.7--1.5	18--21	
Polyurethane	PUR						0.50	20--30	
Silicon rubber	SIR						0.15	2--8	

*) Reference values of technical polymers at room temperature (critical stresses refer to temperature of ductile-brittle transition); data taken from D. W. van Krevelen [7] and Polymer Handbook [20] unless otherwise indicated.

Fig. 1.10. Arbitrary segment of polyamide 6 chain (extended length 1.97 nm, cross section 0.176 nm^2).

namely the anisotropy of macroscopic properties, the microscopic heterogeneity and the nonlinearity and strong time-dependency of polymer response to mechanical excitation.

All solid plastics, from rubbery networks to rigid glasslike bodies, are aggregates of such molecular chains, which cohere through physical (temporary) or chemical (permanent) "cross-links". The wide variety of the macroscopic mechanical properties of these materials is caused by differences in aggregation (crystallinity or super-structure) and in molecular parameters, as for instance by differences in chain structure, intermolecular segmental attraction, segment mobility, or cross-link density. Unfortunately these parameters are not fixed values attached to a molecular species or a polymeric body, but depend on polymer processing and on environmental conditions. In fact simply by changing the time or temperature scale of the mechanical excitation of one and the same polymer a wide range of responses will be obtained: from that of a brittle glass to that of a highly viscous liquid. This general behavior is depicted in a time-temperature diagram, given by Retting [22] for hard PVC (Fig. 1.11). Within the relatively small temperature interval from 0 to 100 °C (rapid loading) or −100 to +60 °C (creep) the indicated changes from glass to rubber, from − macroscopically − brittle to highly ductile defor-

Fig. 1.11. Time-temperature diagram of the mechanical properties of PVC (after Retting [22]).

mation and fracture are observed. The changes in the type of response of a given aggregate of chains clearly indicate that different deformation processes must be dominant in different regions of the time-temperature regime.

In discussing the physical nature of defect development it is necessary to specify which form of local deformation eventually leads to a disintegration of the polymeric solid or to a localized heterogeneous deformation and to crack initiation. In previously proposed kinetic theories quite different processes were held responsible for the development of macroscopic failure. *Viscous flow* (within cotton fibers) was suggested by Busse and his colleagues [23] to determine the fatigue life of these fibers. The *"slippage" of secondary bonds* and the net decrease of the number of bonds was considered by Tobolsky and Eyring [24] for the two cases where repair of bonds under stress is possible or impossible. A model involving two independent networks was discussed by Stuart and Anderson [25] who argued that only *fracture* within the first network with unsymmetrical potential barriers between broken and unbroken states contributes to failure of glasses. A theory of correlated *Brownian motion* of groups of atoms (*Vielfachplatzwechsel*) proposed by Holzmüller [26, 27] is based on Boltzmann statistics of phonon energy distribution where the phonons are the independent statistical units.

The idea that the *breaking of primary chemical bonds* plays a decisive role in fracture was independently forwarded by Zhurkov [28, 29] and Bueche [30, 31]. Bueche, and later Halpin [32, 33] and Smith [34, 35], had been concerned with the strength of rubbery polymers, and found that the time-temperature dependence of the ultimate properties could be explained by the deformation behavior and the *finite extensibility of molecular strands.* Prevorsek and Lyons [36, 37] emphasized that the random thermal motion of chain segments leads to *nucleation and growth of* voids without breakage of any load-carrying bonds. The lifetime of a sample then is determined by the time it takes one flaw to grow to a critical size. Niklas and Kausch [13] discussed the effect of the *dissociation of dipole-dipole associates* on the strength of PVC.

To the molecular processes specified above one has to add the unspecified ones: internal destruction, damage, or crack formation probability. In analogy to the molecular description of deformation used by Blasenbrey and Pechhold [38] all of these molecular processes can be referred to the four physical rearrangements that can occur between neighboring chain segments with parallel chain axes: change of conformation (segment rotation, *gauche-trans* transformation), cavitation, slip, and chain scission. In Figure 1.12 these rearrangements of segment pairs are shown. Of these chain scission and — to some extent — cavitation and slip are potentially detrimental to the load-bearing capacity of a polymer network. Conformational changes on the other hand appear to be "conservative" processes, which by themselves will modify and delay but never generate any progress in the development of a fracture process.

Here it seems to be necessary to say a word about the meaning of the terms fracture, rupture, and failure as they will be used within this text. Andrews [9] says, "Fracture is the creation of new surfaces within a body . . . The term rupture may be regarded as truly synonymous but not the term failure since the latter indicates only that some change has taken place in the body (generally an engineering com-

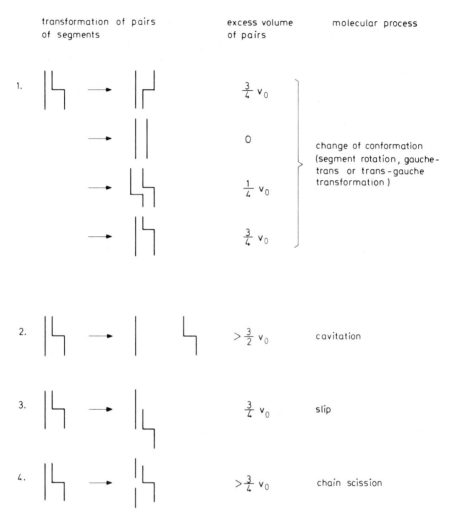

Fig. 1.12. Four basic transformations of pairs of linear chain segments. Excess volume of pairs refers to PE, with v_0 being the volume of a CH_2-group [38].

ponent of some kind) which renders it unsuitable for its intended use. Fracture is only one of many possible modes of failure". We will supplement this definition by saying that as a result of the creation of new surfaces either holes have to open up or formerly connected parts of the fractured specimen have to become detached from each other.

In mathematical terms we may define a fracture process as one by which the sum of the degrees of connectedness of the parts of a sample or a structural element is increased by at least one. This definition includes the piercing of films and tubes, but excludes minute damages as caused by crazing or chain scission.

The term fracture process is meant to describe the whole development of a fracture from its initiation through crack growth and propagation to completion of the fracture. In contrast to thermal or environmental degradation and failure

10

the fracture process is understood as stress-biased material disintegration. It carries on if and only if a driving force exists, which has literally to be a force. Forces, however, will give rise to deformations. This means that fracture initiation will always be preceeded by specimen deformation.

From Figure 1.12 it could be concluded that, in the absence of flow, chain scission and chain strength determine the mechanical properties of polymers. In fact, there are a number of observations which support this assumption:

— with increasing number n of chain atoms, a saturated linear hydrocarbon will turn from gas to liquid ($n \geqslant 5$) and to solid ($n \geqslant 18$); the solid will attain technically useful strength (10 MN/m^2) if n is around 2000; with increasing molecular weight and chain orientation that strength may still be increased 50-fold [19, 20];
— the mechanical loading and rupture of virtually all natural and synthetic polymers leads to the scission of molecular chains and the formation of free radicals [39];
— the strength and the hardness of solids increase with volume concentration of primary bonds [40];
— in a number of uniaxial tensile fracture experiments the energy of activation of sample fracture coincided with the activation energy for main chain bond scission [39].

The observation that sample strength increases with molecular weight and chain orientation is illustrated in more detail by Figure 1.13. In low molecular weight material chain slip occurs readily, so that the sample strength depends solely on the strength of intermolecular attraction. Noticeable macroscopic strength is only obtained once the molecular weight is sufficient to permit physical cross-links through entanglements or chain folds between several chains. In the range of molecular weights between 1.5 and 3 x 10^4 g/mol, fiber strength increases with molecular weight, whereas at still higher molecular weights, the effect of the increasing number of defects introduced in processing such a fiber, tends to outweigh the effect of the decreasing number of chain ends (Fig. 1.13, Curve I). It is also indicated in Figure 1.13 (Curves I–IV) that high strength with a high degree of orientation (e.g., obtained at higher drawing temperatures T_d) requires high molecular weights [41]. The effects of crystal structure and molecular weight on macroscopic tensile strength are shown by curves V and VI [43].

On the other hand, the data of Table I do not support a simple relation between the strength values of individual chains and of the solids formed thereof. Young's moduli of bulk specimens are one to two orders of magnitude smaller than the corresponding chain moduli, bulk strengths at room temperature are two to three orders of magnitude smaller than chain strengths. The critical stress, which is the fracture stress at that temperature where fracture changes from ductile to brittle, i.e., at a temperature well below the glass transition temperature, is also one to two orders of magnitude smaller than the corresponding chain strengths. The tensile strengths observed at different morphological levels are indicated in Figure 1.14.

In the cited monographs [2–10] very little has been said about the details of how molecular destruction processes effect ultimate properties. In the last 10 years, however, considerable progress has been made in precisely this area, in the investigation of the deformation, strength, and scission of chain molecules. Mainly dealing

Fig. 1.13. Effect of molecular weight and temperature T_d of the fiber orientation process (drawing) on the tensile strength of polyamide 6. I: low-dernier filaments, $T_d = 170\,°C$ [41], II–IV: effect of drawing temperature, $T_d = 150\,°C$, $140\,°C$, $20\,°C$ [42], V, VI: effect of crystal structure [42].

with the notable contributions of Russian scientists working in this area, the work by Regel, Slutsker, and Tomashevskii [11] has just become available in Russian.

With these words of introduction we are in a position better to define the scope of this book. The central theme is the behavior of chain molecules under conditions of extreme mechanical excitation and the likely role of chain scission in the initiation and development of macroscopic fracture. We will study the structure and deformation of solid polymers (Chapter 2) in an attempt to define the conditions under which elastic straining of chains occurs. We will then discuss existing non-morphologic fracture theories (Chapter 3) in view of our central theme. The molecular description of fracture begins with a brief account of the strength of primary bonds (Chapter 4) and of the thermomechanical excitation and scission of chain segments (Chapter 5). The study of free radicals formed as a consequence of stress-induced chain scission (Chapters 6–8) constitutes the main part of this volume. It is the only part of this book where a comprehensive coverage of the existing literature has been attempted. In the final chapter, 9, the role of molecular chains in heterogeneous fracture will be analyzed. For this purpose a brief introduction is given into the mechanics of fracture (stress analysis of cracks in elastic and elastic-plastic materials, slow crack propagation). The fracture mechanical functions critical energy release rate G_{Ic} and fracture toughness K_{Ic} are interpreted from a non-

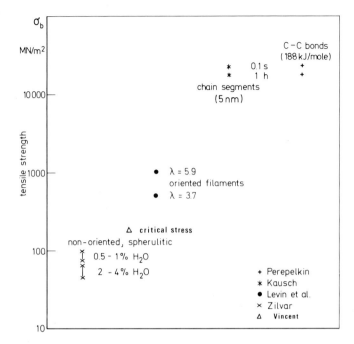

Fig. 1.14. Tensile strengths of polyamide 6 at different levels of morphologic organization.

continuum viewpoint. The same approach is applied to the phenomenon of crazing. A discussion of molecular and morphological aspects of crack propagation and of possible failure criteria concludes this monograph.

As mentioned in the introduction, a kinetic theory of fracture deals with and particularly considers the existence and discrete size, the motion, and the response of molecules or their parts in a stressed sample. It thus requires a molecular description of polymer structure and deformation. Such a description is available in the cited books [1–12], [20] or papers [15, 38]. In Chapter 2 only a very brief summary is offered which is not meant to be self-supporting or complete.

References for Chapter 1

1. H. A. Stuart (ed.): Physik der Hochpolymeren. Berlin – Heidelberg: Springer-Verlag 1952.
2. H. F. Mark, N. G. Gaylord, N. M. Bikales (eds.): Encyclopedia of Polymer Science and Technology. New York: Interscience 1964.
3. J. D. Ferry: Viscoelastic Properties of Polymers (2nd ed.). New York – London: Wiley 1970.
4. N. G. McCrum, B. E. Read, G. Williams: Anelastic and Dielectric Effects in Polymeric Solids. London – New York: Wiley 1967.

5. I. M. Ward: Mechanical Properties of Solid Polymers. London: Wiley-Interscience 1971.
6. A. D. Jenkins (ed.): Polymer Science. Amsterdam − London: North Holland 1972.
7. D. W. van Krevelen: Properties of Polymers, Correlations with Chemical Structures. Amsterdam − London: Elsevier 1972.
8. J. A. Manson, L. H. Sperling: Polymer Blends and Composites. New York − London: Plenum 1976.
9. E. H. Andrews: Fracture in Polymers. Edinburgh London: American Elsevier 1968.
10. G. M. Bartenev, Yu. S. Zuyev: Strength and Failure of Viscoelastic Materials. Oxford − London: Pergamon 1968.
11. V. R. Regel, A. I. Slutsker, E. E. Tomashevskii: Kinetic Nature of the Strength of Solids (in Russian). Moscow: Nauka 1974.
12. R. Hertzberg: Deformation and Fracture Mechanics of Engineering Materials. New York: Wiley 1976.
13. H. Niklas, H. H. Kausch: Kunststoffe *53*, 839 and 886 (1963).
14. E. Gaube, H. H. Kausch: Kunststoffe *63*, 391−397 (1973).
15. J. A. Sauer, A. E. Woodward, In: Polymer Thermal Analysis. Vol. II. P. E. Slade, L. T. Jenkins (eds.). New York: Marcel Dekker 1970. pp. 107−224.
16. I. Sakurada, T. Ito, K. Nakamae: J. Polym. Sci. *15*, 75 (1966).
17. K. E. Perepelkin: Angew. Makrom. Chemie *22*, 181−204 (1972).
18. P. I. Vincent: Polymer *13*, 558 (1972).
19. N. J. Capiati, R. S. Porter: J. Polym. Sci., Polym. Phys. Ed. *13*, 1177−1186 (1975).
20. J. Brandrup, E. H. Immergut: Polymer Handbook (2nd ed.). New York: Wiley 1975.
21. H. H. Kausch: J. Polymer Sci. *C32*, 1−44 (1971).
22. W. Retting: Kolloid-Z *213*, 69 (1966).
23. W. F. Busse, E. T. Lessing, D. L. Loughborough, L. Larrick: J. Appl. Phys. *13*, 715 (1942).
24. A. Tobolsky, H. Eyring: J. Chem. Phys. *11*, 125 (1943).
25. O. L. Anderson, D. A. Stuart: Ind. Eng. Chem. *46*, 154 (1954).
26. W. Holzmüller: Z. Phys. Chemie (Leipzig) *202*, 440 (1954) and *203*, 163 (1954).
27. W. Holzmüller: Kolloid-Z *203*, 7 (1965) and *205*, 24 (1965).
28. S. N. Zhurkov, B. N. Narzulaev: J. Techn. Phys. (USSR) *23*, 1677 (1953).
29. S. N. Zhurkov: Z. Phys. Chemie (Leipzig) *213*, 183 (1960).
30. F. Bueche: J. Appl. Phys. *26*, 1133 (1955) and *28*, 784 (1957).
31. F. Bueche, J. C. Halpin: J. Appl. Phys. *35*, 36 (1964).
32. J. C. Halpin: Rubber Chem. Technol. *38*, 1007 (1965).
33. J. C. Halpin, H. W. Polley: J. Composite Materials *1*, 64 (1967).
34. T. L. Smith: J. Polym. Sci. *32*, 99 (1958).
35. T. L. Smith: J. Polym. Sci. A, *1*, 3597 (1963).
36. D. C. Prevorsek, W. J. C. Withwell: Textiles Res. J. *33*, 963 (1963).
37. D. C. Prevorsek, M. L. Brooks: J. Appl. Polym. Sci. *11*, 925 (1967).
38. W. Pechhold, S. Blasenbrey: Kolloid-Z. Z. Polymere *241*, 955 (1970).
39. S. N. Zhurkov: Int. J. of Fracture Mechanics *1*, 311−323 (1965).
40. L. Holliday, W. A. Holmes-Walker: J. Appl. Polym. Sci. *16*, 139−155 (1972).
41. D. Prevorsek: J. Polym. Sci. *C32*, 343−375 (1971).
42. V. M. Poludennaja, N. S. Volkova, D. N. Archangelskij, A. G. Zigockij, A. A. Konkin: Chim. Volokna *5*, 6−7 (1969)
43. H. Kraessig: Textilveredlung *4*, 26−37 (1969)
44. B. Ya. Levin, A. V. Savitskii, A. Ya. Savostin, E. Ye. Tomashevskii: Polymer Science USSR *13*, 1061−1068 (1971).
45. V. Zilvar, I. Boukal, J. Hell: Trans. J. Plastics Inst. *35*, 403−408 (1967).

Chapter 2

Structure and Deformation

I. Elements of the Superstructure of Solid Polymers

A. Amorphous Regions

The basic structural elements of high polymer solids are the chain molecules. The variety of their structure and flexibility permits different modes of organization and mechanical interaction. At this point the characteristic elements of structure and superstructure of amorphous and semicrystalline polymers will be introduced. The interrelations between chain parameters (structure and regularity), crystal or superstructural parameters (degree of crystallinity, lattice structure, nucleation and growth kinetics, defects), and environmental parameters are extensively discussed in the literature and are not the object of this book [1–3].

   If no regular organizatzion or no colloidal structure within a solid aggregate of chains becomes apparent one speaks of an amorphous state. In recent years the interpretation of the non-ordered or amorphous state of solid high polymers has been subject to lively discussions and careful studies. Until about 1960 the general view prevailed that the chain molecules in isotropic, non-crystalline polymers as in many rubber, glassy polymers such as PS*), PVC, PMMA, PC, or quenched "semi-crystalline" polymers such as PCTFE, PTFE, PETP have a random conformation and that the random coil or spaghetti model gives an adequate description of the "structure" of these polymers. In the years following 1960 the concept of a *close range order* of the chain molecules within X-ray amorphous polymers gained grow-

---

*   In the annex (Table A.1) a list of standardized abbreviations of those polymer names is given
    which are used in this monograph. The abbreviations are in accordance with the international
    standards proposed by ASTM, DIN, and ISO.

ing support. The evidence for a close range order was thought to derive from a comparison of segment volume and amorphous density, from electron microscopic observation of structural elements, from calorimetric investigations, from a discussion of crystallisation kinetics, and from a study of network orientation. After 1970 Kirste, Schelten, Fischer, Benoit, and their associates [4—6, 91] were applying neutron scattering techniques to amorphous polymers in addition to employing light and small angle x-ray scattering and spectroscopic methods. They concluded that their experiments and the previous experiments of other authors have given no positive clue either against the random coil hypothesis or in favor of the concept of an existing close range order in PS, PMMA, PC, SI or the amorphous regions of PE. Flory [7], [8] also summarizes his extended observations on polymers in solution, rubbery networks, and amorphous one- and two-phase materials in favor of the random coil structure of chains in bulk. These authors [4—8], [9], [100—104] state in fact that the radius of gyration of a polymer molecule in the bulk agrees well with the unperturbed dimensions measured in the melt or in dilute solution.

Hosemann, Pechhold, and Yeh [9—12] propose that amorphous states and defect-rich, originally ordered states are commensurable. They thus retain elements

Fig. 2.1 a—e. Model representations of the amorphous state. (a) interpenetrating coils [4, 7], (b) and (c) honeycomb and meander model [10, 11], (d) folded chain fringed micellar grains [12], (e) fringed micellar domain structure [13].

of close range order. Their amorphous state is not principally but only gradually different from a well ordered state. Figure 2.1 shows model representations of the various interpretations of the nature of the amorphous state:

a) interpenetrating random coils [4], [7]

b) and c) honeycomb and meander model [10], [11]

d) folded chain fringed micellar grains [12]

e) fringed micellar domain structure [13].

In agreement with the considerations above the term amorphous is used to indicate the absence of crystallographic reflections. If we wish to place importance on the absence or presence of a close range order within amorphous regions we will explicitly say so. In the following discussion ordered structural elements are to be characterized. In doing so one should keep in mind that for the interpretation of ordered structures in real polymer solids one finds a situation somewhat inverse to the interpretation of the amorphous state, i.e. one should keep in mind that in any real polymer well ordered regions are of limited size and finite degree of perfection.

B. Crystallites

From the beginning of polymer science it had been known that native as well as synthetic high polymers crystallize [14a]. Analysis of x-ray diffraction patterns has revealed the lattice structure and unit cell dimensions of high polymer crystallites. Up to 1957 it was believed that the crystallites are of the fringe micelle type. A typical micelle was supposed to bundle some tens of hundreds of different molecules which — after leaving the micelle and passing through amorphous regions — would at random join other micelles. In 1957 Fischer [15], Keller [16], and Till [17] independently from each other discovered and proposed that high polymers form single crystal lamellae within which chain folding occurs. Figure 2.2 shows an electron micrograph of a stack of PE single crystals [18] grown from dilute solution and Figure 2.3 the arrangement of the chain molecules in such lamellar crystals. Here

Fig. 2.2. Polyethylene single crystals grown from dilute solution (Courtesy E. W. Fischer, Mainz).

17

Fig. 2.3. Chain folding in polyethylene single crystals.

the PE chain is folded — with adjacent reentry of the chain after each fold — within the (110) plane of the ortho-rhombic PE crystal. The unit cell dimensions have been determined [19] to be

a = 0.74 nm
b = 0.493 nm
c = 0.353 nm (chain axis direction).

The largest of the lozenge-shaped single crystals shown has an edge length of 4700 nm and a thickness of about 10 to 15 nm (originally the crystal seems to have been a hollow pyramid which became flattened during preparation). Lamellar crystals — including extended chain crystals — of various degrees of perfection and size are the product of crystallization from solution or from the melt under appropriate conditions. The extreme mechanical anisotropy of the chain molecules is reflected by that of their crystallites. For example the elastic response of PE single crystals in the three axis directions differs by two orders of magnitude. The elastic modulus is found [20] to be:

$E_a =$ 5.7 GN/m^2 GPa
$E_b =$ 2.1 GN/m^2
$E_c =$ 180–340 GN/m^2

C. Superstructure

At this point the superstructure or morphology of polymers is reviewed with a desire to identify and define elements of their heterogeneity. The most important heterogeneities derive from the tendency of many polymers to (partially) crystallize. More or less well defined crystalline lamellae are found in form of single crystals as described above, stacked and/or grown upon each other in the form of axialites or sheaving layer structures, as twisted aggregates in spherulites, and in the form of sandwich-like structures in highly oriented fibrils [1–3]. The radially symmetric growth of twisting lamellae (Fig. 2.4) from many nuclei which leads to the spherulitic structure, is shown in Figures 2.5. This is the largest and dominant feature of melt-crystallized specimens.

Fig. 2.4. Detail of spherulite in polyethylene showing stacks of crystal lamellae (Courtesy E. W. Fischer, Mainz).

Fig. 2.5. Spherulitic structure of melt crystallized polyoxymethylene.

Presently it is discussed [91, 92, 100] whether and to what extent the lamellar crystals of a solid semicrystalline polymer show the same regularly folded structure as solution grown single crystals. Schelten et al. [91] concluded that the results of the small-angle neutron scattering of deuterated PE molecules (PED) in a protonated PE host (PEH) are incompatible with a regularly folded morphology. Yoon and Flory [92] supported Scheltens conclusions. They compared the intensity distributions calculated for several morphological models with the experimental scattering data [91]. The best fit could be obtained for a model where 30% of "the chains emanating from a crystalline layer escape therefrom instead of re-entering the layer at the adjacent lattice site". After extended analysis of these and additional own data, Fischer [100] stated that the radii of gyration (of quenched PE, of PEO crystallized by slow cooling, and of isothermally crystallized isotactic PP) were only slightly affected by crystallisation or annealing even though drastic changes of the long spacing (12—25 nm) had occurred. With regard to the problem of chain folding and adjacent reentry, the majority of the results seemed to indicate that the adjacent reentry model is not compatible with neutron scattering behavior. On this basis Fischer [100] advanced a "solidification-model of crystallization" which excludes large scale diffusional motion of chain segments during crystallization.

The boundaries between spherulites resemble grain boundaries; these grain boundary zones are enriched in low molecular weight material, impurities, chain ends, and defects. The deformation and strength of such a "composite" structure naturally depend on the compliances of all of its components. In such a composite large compliances (small constants of elasticity) have to be assigned to the cohesion across grain boundaries and across lamellar fold surfaces. The cohesion between chains within a crystal lamella is much stronger than the intercrystalline interaction; this renders a certain stability to the lamellar elements in sample deformation. The deformation behavior of such a non-oriented semicrystalline polymer will, therefore, depend much more on the nature of the secondary forces between the structural elements than on the length or strength of the chain molecules.

The different forms of superstructure are related to the structure and composition of the molecular species involved, and to the conditions of polymerization, processing, and environmental treatment, i.e. on nearly every parameter involved in producing a polymeric solid. For illustration of the size and shape of supermolecular structures formed during the initial phase of polymerization of a homopolymer attention is called to the fibrous or globular structures reported by Wristers [21] for polyolefines. These structures are obtained after polymerization in gaseous or liquid environment and at low or high efficiency of various Ti-, V-, Cr-, or Al-based catalysts. In Figures 2.6 to 2.8 electron micrographs of such nascent polymer structures are reproduced [21]. Polypropylene at low catalyst efficiency is formed in globules of 0.5 μm diameter (Fig. 2.6), at high catalyst efficiency fibers several μm long are observed (Fig. 2.7); the fiber diameter is closely correlated with the lateral face dimension of the primary catalyst crystal and varies between 0.37 and 2 μm for $TiCl_3$-crystal breadths of between 5 and 50 nm. Polyethylene samples prepared with $TiCl_4$-Al(i-Bu)$_3$ or other colloidal catalysts show less regular surface structures (Fig. 2.8).

Fig. 2.6. Nascent polymer morphology: polypropylene prepared at low catalyst efficiency (taken from Ref. [21]).

Fig. 2.7. Polypropylene prepared at high catalyst efficiency (taken from Ref. [21]).

In the course of the bulk polymerization of lactams it was observed [22] that the resultant polymers precipitated in the form of spherulites. Other examples are the fibrous, globular, or star-shaped aggregates existing in PVC [23–26] which have sizes of between 10 and about 500 nm. In good solvents the largest aggregates disintegrate into single molecules but only at temperatures larger than 200 °C [25].

Fig. 2.8. Polypropylene prepared with TiCl$_4$–Al(i-Bu)$_3$ (taken from Ref. [21]).

A large group of multi-phase polymeric systems is formed by the block co-polymers, graft copolymers, random copolymers, and poly-blends. This group of polymers has received increasing attention (e.g. 27–29) because of the controllable, "tailor-made" properties of these engineering materials. Here the superstructure is determined by the relative weight, sequential order, and limited miscibility or even incompatibility

- of the covalently connected molecular segments of hard (glassy, thermoplastic) and soft (elastomeric, rubbery) chains as within the (two- or three-) block copoly-mers and random copolymers
- of the (two) phases grafted onto eachother within graft-copolymers or
- of the molecules mixed together in poly-blends.

The variety of ensuing multi-phase structures cannot be discussed here in any detail; it ranges from the extremely ordered hexagonal patterns of spherical domains of one phase embedded within the continuous phase (mostly of the dominant component) to the partial phase separation of solution-cast poly-blends occurring only after heat treatment [27–29]. In Figure 2.9 an electron micrograph of the morphology of a butadiene styrene diblock copolymer is reproduced [31]. At 17% by weight the styrene forms the discrete phase of spherical or cylindrical aggregates in a continuous butadien phase. Owing to the homogeneity of block lengths the aggregates are arranged in highly regular lattices. The sizes of the aggregates correspond very well to the molecular conditions. They must be larger than one diameter of the coil formed by the PS segment and must be smaller than two such diameters to permit phase separation while maintaining material continuity [28–30].

For a study of the effect of chain length and strength quite naturally structures which contain a high degree of oriented chains are more promising than others. One of the most widely investigated ones are the semicrystalline fibers. The characteristic, microfibrillar and sandwichlike structure of highly oriented fibers will be discussed in connection with the electron spin resonance investigations of chain scission especially observed in these fibers.

Still higher degrees of chain orientation are found in the extended chain crystallites, row structures, and highly lamellar "hard-elastic" or "springy" forms of crystallizable polymers which recently have become of considerable interest. Follow-

21

Fig. 2.9. Thin film of butadien-styrene diblock copolymer cast from petrol ether solution and annealed 1 h at 100 °C. At 17% by weight the styrene aggregates in the form of spheres and of cylinders (Courtesy G. Kämpf, Uerdingen).

ing early investigations of Keller [32], Pennings et al. [33] and Kawai et al. [34] much work has been devoted to the *row structures* or *shish-kebabs* which consist of ribbon-like backbones overgrown by folded chain platelets. Each backbone fiber again seems to be composed of ultimate fibrils (microfibers) less than 10 nm in diameter [35]. These structures do form in crystallization under molecular orientation, in sheared dilute solution as well as in extrusion from the melt [36]. Comprehensive reviews on these microfibers have recently been given by Tucker and George [35–36]. According to them the transverse dimension of these microfibers is of the order of 0.8 to 20 nm in various celluloses, and between 7 and 50 nm in most thermoplastic fibers; microfiber lengths are given as 18 to over 1000 nm [35]. At the time of this writing Zwijnenburg and Pennings [37] report on a technique to obtain PE macrocrystals of virtually infinite length. There, a seed crystal and – subsequently – the growing crystalline PE fiber is pulled under steady-state conditions from a flowing solution. The diameter of the obtained fiber, an agglomeration of shish-kebab fibrils, varied between 10 and 40 μm [37].

In the more recent years the solid state extrusion of thermoplastic polymers involving very high pressure (up to 5 kbar = 0.5 GN/m^2), temperatures between 30 and about 250 °C, and extensional flow has been studied [38–44]. In the case of PE such a treatment led to a new high-pressure, high-temperature phase of hexagonal symmetry at a high level of chain extension. In analogy to the extended-chain crystals observed by e.g. Anderson [45] in the fracture surfaces of low-molecular weight PE the term *extended-chain* crystals is presently being used for pressure crystallized PE also. Weeks and Porter found that highly oriented strands of such material (M_w = 58 000) have an exceptional stiffness of 70 GN/m^2 at room temperature which is comparable to that of E-glass [40]. In addition a good tensile strength of 500 MN/m^2 is reported [41]. Lupton and Regester [39] worked with ultrahigh-molecular-weight (UHMW) HDPE with M_w between 2 and $3 \cdot 10^6$ g/mol and obtained a stiffness of 2 to 3 GN/m^2 and a strength of 33–39 MN/m^2, but an excellent tensile-impact resistance of 300 to 600 kJ/m^2. Capaccio and Ward [93–98] prepared ultra oriented polymers (PE, PP, POM) by means of cold drawing and hydrostatic extrusion. A fibrillar structure seemed to be retained in those samples. With increasing draw ratio the morphology would be characterized primarily by a continuum of oriented material in which a statistical distribution of chain ends would constitute the only discontinuities. Thus an increasingly smaller fraction of the material would show the conventional morphology comprising crystals and oriented amorphous chains, including tie molecules, in series.

In the last few years it has been discovered that by controlled drawing and annealing treatment of a number of fiber forming thermoplastics a highly crystalline morphology with "springy" or hard elastic response could be obtained [46–51, 105]. Springy behavior, the characteristics of which will be discussed in the next section, has been found so far in PP, POM, PE, PMP, PES, and even in PA 66 [47, 105]. The morphology is proposed to consist of stacks of flat, highly regular, c-axis oriented folded chain lamellae with the lamellae interconnected at points about 100 nm apart [46–51, 105].

D. Characterization

The experimental methods available to determine or characterize the structure of polymer chains or of their aggregates are discussed in the general references of Chapter 1. In addition information on X-ray diffraction [3], neutron scattering [4–6], electron- and light-scattering [4, 52, 53], optical and electron microscopy [3, 14b], thermal [3, 54] and viscoelastic behavior [14c, 55–57], and nuclear magnetic resonance (NMR) technique [3] can be obtained from the references of this Chapter. Infrared absorption (IR) and electron spin resonance (ESR) techniques will be introduced in Chapters 5 and 6 respectively. Except for the microscopic methods all of the mentioned techniques provide an average information, an information to which all structural elements within the active volume of an investigated sample contribute according to their individual response, concentration, and position. Any macroscopic information which is not solely taken as such but is used to characterize an inhomogeneous solid generally involves elemental response functions and orientation distributions which are not completely known. For this reason a certain ambiguity is introduced into the deconvolution of such experimental data. This leaves some room for individual interpretations as to the concentration and nature of structural elements. This ambiguity has been inherent in the discussion of the existence or non-existence of close range order in amorphous polymers and of the nature and distribution of defects within crystallites. Of even more concern for the structural interpretation of mechanical properties is the fact that the nature of the connecting members between structural elements is much less well known than the nature of the latter. This is true on all levels: for the arrangement and interaction of non-ordered chain segments, for the fold surfaces of lamellar crystals and the connections between them, for the amorphous regions between crystallites in fibers or unoriented polycrystalline samples, and for the boundaries of microfibrils or spherulites. This will be further illustrated in the next section.

II. Deformation

A. Phenomenology

The deformation of the complex structures described in the last section involves a number of different – and characteristic – phenomena ranging from a linear elastic to a highly viscous response. Figure 2.10 gives examples of the different types of quasi-static uniaxial-deformation behavior of thermoplastic polymers in the non-rubbery temperature region. The stress-strain curve a) is that of a brittle polymer (polystyrene at room temperature) characterized by its very limited extensibility and the steep and monotonous increase of stress with strain. Curve b) represents a hard elastic polymer, lamellar polypropylene film [58], which combines strong elastic behavior with good extensibility at a high stress level and almost complete recovery (within a period of days). Curve c) is that of a ductile polymer. The initial monotonous increase of the engineering stress generally is less steep than that of a brittle polymer, i.e. the secant modulus is smaller. The engineering stress σ reaches a maximum at the yield point (Streckgrenze) which marks the beginning of the so-called cold drawing, a notable reduction in cross-section of the strained specimen. If the stress-strain curve extends considerably beyond the yield point one speaks of a tough polymer; the shape of the curve depends on the strain hardening behavior of the polymer and on the tendency to neck formation. Curve d) is characteristic for soft polymers, e.g. for thermoplastics at a temperature close to their glass transition temperature or plasticised polymers, which do not yet show rubber elastic behavior.

As indicated earlier the different stress-strain curves are not characteristic for particular, chemically defined species of polymers but for the physical state of a polymeric solid. If the environmental parameters are chosen accordingly transitions from one type of behavior (e.g. brittle, curve a) to another (e.g. ductile, curve c) will be observed. These phenomenological aspects of polymer deformation are discussed in detail in [14], [52–53], [55–57], and in the general references of Chapter 1. A decrease of rate of strain or an increase of temperature generally tend to increase the ductility and to shift the type of response from that of curve a) towards that of curves c) and d). At small strains (between zero and about one per cent) the uniaxial stress σ and the strain ϵ are linearly related (Hooke's law):

$$\sigma = E \cdot \epsilon \tag{2.1}$$

Even at these small deformations the apparent Young's modulus E is a function of the rate of strain. This shows that E is not solely determined by the energy elastic deformation of bond angles, bond lengths, and intermolecular distances but also involves *time-dependent* displacements of atoms and small atom-groups. In the following region (1 to about 5% of strain) stress and strain are no longer proportional to each other, structural and conformational rearrangements occur that are mechanically although not thermodynamically reversible, one speaks of *anelastic* (viscoelastic in the narrow sense) or *paraelastic* behavior. Beyond the yield point large scale reorientation of chains and lamellar crystals begins, a process generally termed plastic deformation. A *true plastic deformation* can be understood as the transformation from one equilibrium state to another equilibrium state without the creation of frozen-in tensions. The latter point is particularly important in view of the fact that the average post-yield deformation is brought about largely by mechanically reversible anelastic conformational changes of the molecules and not by flow of molecules past each other. Unless a state of equilibrium had been established through proper annealing treatment highly drawn samples may retract considerably. Within the context of this book emphasis should be placed, however, not so much on the processes leading to or accompanying molecular reorientation – which is basically a reinforcing effect – but on the damaging processes of chain scission, void formation, and flow. The latter processes gradually occur in the strain region from just before the yield point up to final fracture. Among the damaging processes one has also to include the phenomenon of normal stress yielding or crazing of glassy polymers which will be discussed in Chapter 9.

Not represented in Figure 2.10 is the deformation behavior of cross-linked networks of highly flexible elastomeric chains. The characteristic tensile deformation of elastomers is not based on energy elasticity but rather on the change of entropy accompanying the deformation and orientation of randomly coiled chain molecules

Fig. 2.10. Characteristic uniaxial deformation behavior of polymers. a. brittle (PS at room-temperature), b. hard elastic (lamellar PP film), c. ductile (PVC at room-temperature), d. soft (PTFE at 100 °C).

[59]. To a first approximation stress, uniaxial extension ratio λ, and shear modulus G are related by

$$\sigma = G\,(\lambda - 1/\lambda^2). \tag{2.2}$$

The shear modulus G is proportional to the absolute temperature T, Boltzmann constant k, and the number N of subchains between cross-links:

$$G = NkT = \rho RT\,(1 - 2\,M_c/M)/M_c \tag{2.3}$$

where M and M_c are the number-average molecular weights of the uncrosslinked polymer and of the subchains respectively, ρ is polymer density.

The *pure* rubber-elastic deformation is mechanically completely reversible and does not involve chain scission or creep. A *real* rubber, however, like any viscoelastic solid, shows on a molecular scale energy and entropy elastic deformation accompanied by viscous flow. Thus it shows stress relaxation at constant strain, creep at constant load, and energy dissipation in dynamic excitation. Representation of the macroscopic mechanical response of viscoelastic solids — even in a strain region where no large-scale reorientation occurs — requires, therefore, the use of damped elastic elements consisting of springs (modulus G) and rate sensitive damping elements (dashpots characterized by a viscosity η). The simplest elements are the Maxwell element with spring (G) and dashpot (η) in series and the Voigt (Kelvin) element with spring G and dashpot η in parallel. The Maxwell element has a relaxation time $\tau = \eta/G$, the Voigt-Kelvin element has the same relaxation time which here more precisely is called retardation time. The phenomenological theory of viscoelasticity [55] describes the mechanical response of a body in terms of a distribution of basic viscoelastic elements primarily characterized by their relaxation times τ_i. If the spectra of the molecular relaxation times, $H\,(\ln\tau)$, are known, then the viscoelastic modulus functions are, in principle, derivable therefrom [14c, 14d, 55]. The time-dependent shear or stress relaxation modulus G(t) becomes in terms of a continuous spectrum of relaxation times:

$$G(t) = G_e + \int_{-\infty}^{+\infty} H\,\exp(-t/\tau)\,\mathrm{d}\ln\tau \tag{2.4}$$

where G_e is the modulus at infinite times which — in the case of a solid — is supposedly larger than zero. The modulus functions are conveniently treated as complex quantities consisting of a (recoverable) storage and an (irrecoverable) loss component; the ratio of loss (e.g. G'') and storage component (G') is called the loss tangent or loss factor, i.e. $\tan\delta = G''/G'$. The phenomenological theory of viscoelasticity is based on the Boltzmann superposition principle, i.e. on the linear additivity of the effects of mechanical history. It should be emphasized that the aim of that theory is not so much to derive the form of the relaxation time spectra from detailed structural considerations but rather to utilize the response functions obtained in one experiment, e.g. creep, to predict the response of the same material under different excitations, e.g. dynamic loading [14, 55]. The general application of this approach is restricted to the linear range of response. The theory has been extended into the non-linear range [14d], it has also been adapted to account for limited changes in the structure and/or of the orientation of the stressed system [14e]. In the context of this book use will be made of the established results of the phenomenological theory as well as of the "molecular theory" of viscoelasticity [55–57], particularly of the molecular friction coefficient introduced in it.

B. Molecular Description

A kinetic theory of fracture intends to interrelate the motion and response of molecules to the ultimate properties of a stressed sample. A kinetic theory, therefore, entails a molecular description of the deformation of the microscopically heterogeneous and anisotropic aggregates of chains to such an extent that critical deformation processes can be identified. The macroscopic deformation of any aggregate potentially involves the deformation, displacement, and/or reorientation of so different substructural elements as bond vectors, chain segments or crystal lamellae. The molecular origin of observed deformation mechanisms have been elucidated by various spectroscopic methods — including mechanical relaxation spectroscopy — and by the previously listed tools which characterize the morphology. They are supplemented by dynamic scattering and diffraction techniques [52, 53].

The following molecular description of the uniaxial straining of unoriented semicrystalline polyethylene characterizes the ductile deformation of fiber-forming spherulitic thermoplastics and may serve to illustrate the large variety of deformation mechanisms involved. At small strains of less than one per cent one observes the anisotropic elastic response of the (orthorhombic) PE crystallites [57] and of the amorphous material [53]. At the same time those anelastic deformations of CH_2-groups and chain segments occur that characterize the low temperature β-, γ-, and δ-relaxation mechanisms [10, 56]. At larger strains (1 to 5 per cent) additional chain segments change their relative position and conformational changes are initiated (bond rotation). A detailed study on the behavior of the chains within amorphous regions had been carried out by Petraccone et al. [53]. In crystalline regions subjected to strains of this order, dislocations and dislocation networks are generated (in lamellar crystals observed by means of moiré patterns). Depending on environmental parameters and the type of the dislocations their movement leads to crystal plastic deformation through twinning or slip or to a phase transformation from the orthorhombic to a monoclinic unit cell. An extensive review on the deformation behavior of polymer single crystals has recently been given by Sauer et al. [57] and in the book of Wunderlich [3]; a detailed account of the contribution of the various structural elements and defects to the deformation of semicrystalline polymers is found in the literature in a large number of papers of which only a few have been cited here [47–62]. Although the above mentioned effects lead to a non-linear stress-strain response, the initially present substructure is still maintained. Such a deformation is called homogeneous.

Still larger stresses cause a destruction of the substructure, involving reorientation of chain segments and of lamellar crystals (crystal rotation, chain tilt and slip), opening-up of voids, and the first breakages of chains take place. These processes account for the ductile deformation. At this stage the mechanical energy input is largely dissipated as heat as shown in later Chapters. Since the deformation proceeds at an almost constant level of engineering stress it has generally been termed plastic deformation – although on a molecular basis it does not correspond to the plastic deformation of metals. The initial post-yield deformation of PE is predominantly not brought about by mechanically irreversible flow. In a strain region of up to about 50 per cent the lamellar crystals slip and/or orient their a-axes perpendicular to the draw direction. The c-axes (chain axes) resume a preferred angle of 35 to 40° with respect to the draw direction. The latter orientation, however, is reversible and changes into a random distribution if external stresses are removed [61]. In a strain region of between 100 and 400 per cent an increasing c-axis orientation in the draw direction is observed. In this strain region the polymer undergoes the important transition from a spherulitic to a fibrillar structure. According to Peterlin [62] micro-fibrils with lateral dimensions of 20 to 40 nm are formed which contain nearly un-modified blocks of folded chains broken away from crystal lamellae but still inter-connected by unfolded sections of tie molecules. A model of the fibrillar structure of a highly drawn fiber is shown in Fig. 2.11. The originally random distribution of

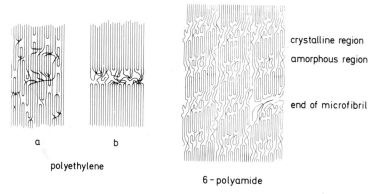

crystalline region

amorphous region

end of microfibril

a   b

polyethylene

6 – polyamide

Fig. 2.11. Models of the structure of drawn semicrystalline polymers, polyethylene: (a) as drawn [18], (b) annealed [18] 6-polyamide, as drawn.

chain segments and crystal blocks has become highly oriented. The load carrying capability of such a structure is – as will be studied in considerable detail – many times larger than that of the unoriented, spherulitic sample.

The described molecular deformation processes affect the macroscopic fracture behavior of a given high polymer. The relative magnitude of the deformation completely determines:
– the level of stress and strain to be attained in "homogeneous" deformation
– the volume density of elastically stored energy
– the rates of energy dissipation and local heating
– network reorientation and possible strain hardening
– chain scission, void formation, and other forms of structural weakening.

III. Model Representation of Deformation

For the interpretation of the complex mechanical response of a highly anisotropic polymeric network it is indispensable to have a simplifying model representation of the arrangement and interaction of the structural elements and of their deformation. Such model representations will be useful in the further investigations which necessarily focus on some aspects while averaging or neglecting a large number of others. In this section one would like to characterize the assumptions made on the geometry and mode of interaction of structural elements in basic theories of the deformation of a composite body. These theories had been developed to account for the behavior at small strains. They can be extended into theories of strength if and only if failure criteria are introduced which become effective within the range of validity of the particular deformation theory.

There is a large body of literature on the purely *elastic* interaction of non-prismatic structural elements within a deformed "composite" solid (e.g. 63–70); this literature also forms part of the theoretical foundation of the behavior of *viscoelastic* composite solids. While not generally discussing these foundations three groups of model descriptions of multi-phase materials will be singled out:

– those of Voigt [63], Reuss [64], or Takayanagi [71] which assume a homogeneous distribution of strain *or* stress,
– the calculations of Kerner [65], Hashin [66], van der Poel [67], or Uemura et al. [72] which minimize the elastic energy of a system containing assumedly spherical inclusions, and
– the empirically modified models of Eilers [68], Guth [69], Mooney [70], and Nielsen [73] which are based on the original calculations of Einstein [74], where the disturbances of an elastic field in the vicinity of – and caused by – completely rigid, spherical inclusions is considered.

It is to be investigated which *chain* properties are effectively recognized by these groups of model descriptions of polymer deformation. One knows that the semicrystalline samples discussed previously are polycrystalline solids containing dispersed amorphous regions with frequently ill-defined boundaries and equally ill-defined interaction between amorphous and crystalline regions. In a direct approach Takayanagi [63] represented semicrystalline polymers by "diagrams" (Fig. 2.12) involving straight boundaries and conditions of plane stress or strain. He obtained a

Takayanagi diagram

Fig. 2.12. Takayanagi diagram of semicrystalline polymer; content of amorphous material corresponds to $\lambda \cdot \varphi$.

29

meaningful interpretation of the relative contributions of crystalline and amorphous phases to the complex moduli of these materials. The Takayanagi approach can well accomodate information on the deformation behavior of individual crystalline [57] or amorphous regions [53] that becomes available from other sources. In the analysis of e.g. the response of a fiber to dynamic loading through a Takayanagi diagram the nature of chain segment orientation is considered in detail but likely distributions of chain length and axial chain stresses do not enter the model.

The two-shell model of Kerner [65] conforms to the conditions of the second group of models. The dilatation of a spherical inclusion surrounded by a homogeneous medium is derived subject to the condition that displacements and tractions at the surface of the inclusion are continuous. The homogeneous medium is supposed to have the elastic properties of the composite as a whole. The model interrelates shear (G_i) and compressive (K_i) moduli (or Poisson's ratios ν_i) of an arbitrary number of isotropic elements with the macroscopic moduli G_c and K_c.

In the case of isotropic inclusions of an elastic shear modulus G_f comparable to that of the matrix, G_m, Eq. (2.5) describes the variation of the complex shear modulus, G_c^*, of the composite as a function of the volume content V_f of the discrete phase:

$$\frac{G_c^*}{G_m^*} = \frac{(1 - V_f)\, G_m^* + (\alpha + V_f)\, G_f^*}{(1 + \alpha V_f)\, G_m^* + \alpha(1 - V_f)\, G_f^*} \tag{2.5}$$

The quantity α is derived from the Poisson's ratio ν_m as:

$$\alpha = \frac{(8 - 10\,\nu_m)}{(7 - 5\,\nu_m)} \tag{2.6}$$

As will be noted no molecular anisotropies and no effects due to size and size distribution of the particles of the discrete phase are recognized. Through $E^* = 2(1 + \nu)G^*$ Eq. (2.5) can be used to predict also the complex tensile modulus. A good example for the applicability of Eq. (2.5) is furnished by the experimental data obtained by Dickie et al. [75]. For the dynamic tensile modulus of a physical mixture (polymer blend) of 75% by weight polymethylmethacrylate (PMMA, continuous phase) and 25% butylacrylate (PBA, discrete phase) within experimental error correspondence of calculated and measured data was obtained (Fig. 2.13,

Fig. 2.13. Complex dynamic tensile modulus of blends (solid line) and graft copolymers (circles) respectively of polymethyl-methacrylate (75% by weight) and polybutylacrylate (25% by weight) at 110 Hz (after [75, 76]).

solid curves). A copolymer of equal volumetric composition, where 25% by volume of the elastomeric butylacrylate had been grafted onto PMMA, yielded the experimental data points shown. In the region between the glass transition temperatures of the PBA ($-20\,^\circ$C) and of the PMMA (at these conditions $160\,^\circ$C) a difference between the response of the mixture and of the graft copolymer can be observed. This difference can be related to the existence of the primary bonds which connect matrix and inclusion and which lead to frozen-in stresses.

A three-shell model of a two-phase solid was proposed by van der Poel ([67], Fig. 2.14). An interior sphere representing a hard component (G_f, K_f, V_f) is surrounded by a soft shell (G_a, K_a, V_a), and by a continuum, for which the elastic

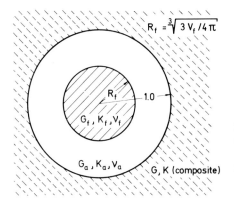

Fig. 2.14. Arrangement of phases in the three-shell model of Van der Poel [67]; bulk and shear moduli (K, G) and Poisson's ratios v of filler (index f) and matrix (index a) and volume concentration V_f enter calculated composite moduli K and G (after [76]).

moduli G and K are being calculated in terms of the mentioned elastic constants and of the volume concentration (V_f) of the hard component. The condition of mechanical equilibrium under shear and hydrostatic pressure leads to an explicit expression for K and to a complicated determinant for G which has to be solved numerically and for each individual system. The van der Poel approach had yielded the best results in describing the elasticity of elastomers highly filled (up to 50 vol.%) with mineral fillers (see e.g. [77]). Again no orientation distributions or physical anisotropies of the constituent particles enter the model.

As mentioned frequently the mechanical and optical response of molecules — and of their crystallites — is highly anisotropic. Depending on the property under consideration the carriers of the molecular anisotropy are the bond vectors (infrared dichroism), chain segments (optical and mechanical anisotropy), or the end-to-end vectors of chains (rubber elastic properties). For the representation of the ensuing macroscopic anisotropies one has to recognize, therefore, the molecular anisotropy and the orientation distribution of the anisotropic molecular units (Fig. 1.9.). Since these are essentially one-dimensional elements their distribution and orientation behavior can be treated as that of rods; such a model had been used successfully to explain the optical anisotropy [78], and the anisotropies of thermal conductivity [79], thermal expansion or linear compressibility [80], and Young's modulus [59, 65, 80, 81] of rubbers and oriented thermoplastics. The rod models focus on the large anisotropies which exhibit properties such as the polarizability or the retractive elastic forces; only two components of such a property, namely that in chain axis direction and that perpendicular to it enter the model as independent parameters.

For example the macroscopic elastic response of an aggregate of randomly, partly, or well oriented chains can be represented by that of a corresponding distribution of one-dimensional finite elements (Fig. 2.15) if the following conditions are fulfilled:
- external forces P cause a homogeneous local strain which determines the axial stress $\psi\,(\vartheta,\,\varphi,\,t)$ of a chain segment
- forces between chains are not transferred by shear or bending moments
- the state of stress at any arbitrary point A is given as the integral over all the traction forces exerted by the chain segments intersecting the surface of a volume element enclosing the point A [82].

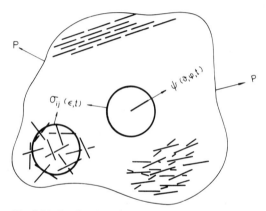

Fig. 2.15. Application of the condition of homogeneous strain ϵ (P, t) to a partially oriented system of finite, one-dimensional elements. The stress tensor σ_{ij} is obtained by space averaging over the traction vectors $\Psi\,(\vartheta,\,\varphi,\,t)$.

The components of stress of such a homogeneously strained, partly oriented aggregate of elastic elements with axial Modulus E_k, length L, volume concentration λ, and orientation distribution ρ was derived by Hsiao [82] as a special case of the Voigt-model employing a homogeneous distribution of strain:

$$\sigma_{ij} = \int E_k L^2 \,\lambda\rho\,\epsilon_{mn} s_m\, s_n\, s_i s_j d\omega. \tag{2.7}$$

Here the summation convention of repeated indices is used. The s_i are the components of the unit vector in the principal axis directions and ω the solid angle. Into this model only one molecular elastic component enters: the axial chain modulus. Any interactions by shear or in a direction perpendicular to the chain axes cannot be accounted for. It should only be applied, therefore, if these interactions can be meaningfully neglected.

A much more general representation of either different or differently oriented molecular domains can be achieved, of course, through three-dimensional elements. In the case of transverse symmetry the molecular elements must be characterized

32

by 5 elastic constants (compliances), the orientation of one or two axes, and the condition of stress and strain at the boundary of an element. Voigt [63] based his calculations on the assumption that there be no discontinous change of strain at any boundary; Reuss [64] made the assumption of homogeneous stress. Following either Voigt or Reuss the space averaging over the elastic constants c_{ijmn} or elastic compliances s_{ijmn} of the molecular domains leads to the upper or lower bound respectively of the macroscopic moduli [83]. For an affinely deforming aggregate of such elements Ward [84] and later Kausch [85] have calculated the macroscopic elastic moduli as a function of domain orientation. The calculated curves for the change of the elastic moduli with draw ratio are particularly characterized by the rate of change and by the extent of change ultimately attained. If only reorientation of otherwise unchanged molecular domains occurs in drawing, the properties of a "completely" oriented sample must correspond to the properties of the molecular domains. In Figure 2.16 the calculated changes of Young's modulus in the direc-

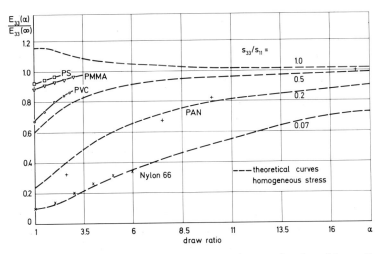

Fig. 2.16. Change of Young's modulus $E_{33} = \sigma_{33}/\epsilon_{33}$ as a function of draw ratio α and of the mechanical anisotropy $s_{33}/s_{11} = s_{3333}/s_{1111}$ of the molecular domains (after [13], [85]).

tion of draw as a function of draw ratio and domain anisotropy are represented in comparison to experimental data [13, 85]. The findings of Ward and Kausch may be summarized by saying:

— the orientation behavior of amorphous thermoplastics (PS, PMMA, PVC) could be represented by the affine deformation of oriented domains

— the observed data for PS, PMMA, and PVC were in agreement with the assumption of homogeneous stress at the boundaries of the orienting domains, the data for PC were better in agreement with a condition of homogeneous strain

— the orienting molecular domains have comparatively small elastic anisotropies (s_{3333}/s_{1111}-values ranging from 0.5 to 1), the axial compliances being more than an order of magnitude larger than those of the chains; this indicates that the domains are not equivalent to chains, micro-crystallites, or fringe micelles; if such domains do have a physical significance in amorphous polymers then they must be considered to be aggregates of chain segments with a stiffness about twice as large as that of the sample as a whole [13]

— the concept of *affinely* orienting lamellae in semicrystalline polymers is only a crude approximation [13, 87–89, 106].

The mathematical representation of the elastic behavior of oriented hetero-geneous solids can be somewhat improved through a more appropriate choice of the boundary conditions such as proposed by Hashin and Shtrikman [66] and Stern-stein and Lederle [86]. In the case of lamellar polymers the formalisms developed for reinforced materials are quite useful [87–88]. An extensive review on the experimental characterization of the anisotropic and non-linear viscoelastic behavior of solid polymers and of their model interpretation had been given by Hadley and Ward [89]. New descriptions of polymer structure and deformation derive from the concept of paracrystalline domains particularly proposed by Hosemann [9, 90] and Bonart [90], from a thermodynamic treatment of defect concentrations in bundles of chains according to the kink and meander model of Pechhold [10–11], and from the continuum mechanical analysis developed by Anthony and Kröner [14g, 99].

The theory of paracrystals recognized that chain molecules form three-dimensional lattices and establish a certain long range order between members of the same lattice while postulating at the same time that in a real solid the lattice will be distorted. The degree of lattice distortion is measured by the variation in length that the three distance vectors a_i between corresponding lattice points encounter if moved in the three lattice directions. If the dimensionless relative mean fluctuations g_{ik} of the distance vectors a_i are all zero the structure is crystalline and if all $g_{ik} > 0.1$ it is amorphous. The g_{ik} provide a quantitative measure of the colloidal structure of microheterogeneous solids. If for instance g_{13} and g_{23} are large compared with the other g_{ik} we deal with the nematic state (segments parallel but at random distances), if g_{31} and g_{32} are large compared with the other g_{ik} we have a smectic state where the segments arrange in layers [9]. The relative paracrystalline distance fluctuation was shown to be inversely related to the maximum number of network planes within one microdomain [9]. The g_{ik} are obtained from line profile measurements in small angle X-ray scattering (SAXS).

Hosemann's schematic representation of a two-dimensional paracrystalline lattice is given in Figure 2.17. If we interpret a superstructure (e.g. the fibrillar structure shown in Figure 2.11) in terms of paracrystals rather than crystals we arrive at the very same size and orientation distribution of the scattering elements — be they crystallites or paracrystals. The differences in interpretation become apparent, however, if intercrystalline rearrangements in annealing, during deformation, or after irradiation are discussed [9].

The general concept that there are only gradual differences between the more or less ordered regions of polymers is also employed by Pechhold and his co-workers in their quantitative microstructural theory (*Kinkenmodell*) of deformation [10, 11, 14f]. Pechhold recognizes that chain molecules exist in energetically different states of rotational isomers between which — cooperative and mostly rapid — transitions can be accomplished. He bases his quantitative thermodynamical calculations

Fig. 2.17. Schematic paracrystalline (a) and molecular representation of different two-dimensional lattices and superstructures (after [9]). a_1 = crystal, a_2 = ideal paracrystal, a_3 = real paracrystal, a_4 = amorphous state, a_5 = micro-paracrystallites (function of "micellar cross-links"), b = molecular model of linear polyethylene.

on kink isomers, which show only a small departure from the extended planar chain and are most likely to form bundles ([11], Fig. 2.18a). The neighbouring chains within such bundle form various pairs of segments (cf. Fig. 1.12) which have different volumes, axial lengths and interaction energies. The concentration of kinks determines the axial extension of the bundle. In the partition function of a bundle of chains all these quantities enter; the partition function relates, therefore, the kink

35

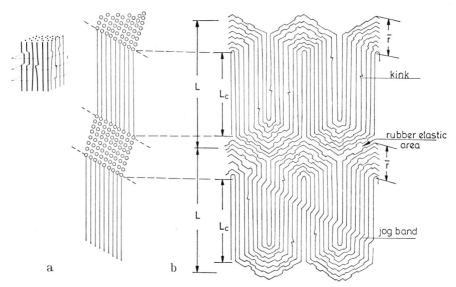

Fig. 2.18. Kink-model representation of (a) bundles of chains and (b) oriented fibers with sandwich-like structure (after [10, 11]).

concentration with bundle geometry and internal and free energy of the isomers. Consequently the elastic compliances of a transversely isotropic bundle of chains could be determined as a function of kink concentration, kink block size, width of kink steps, and temperature [10, 11, 14f]. The anelastic and plastic deformation may be interpreted in terms of successive dislocation motion and slip of bundle surface areas. The kink model offers in every respect a molecular description of polymer structure. Again there are only gradual differences between ordered and non-ordered regions. Pechhold estimates that an apparently perfect PE-crystallite may contain up to 4 kinks per 1000 CH_2-groups whereas in a melt-like structure this concentration is about 200 in 1000. Although this concentration is so large as to annihilate all close and long range order some logical principles should govern the space filling through chain molecules. Pechhold generated appropriate patterns, the honeycomb- and the meander model (Fig. 2.1c). He feels that the latter structure is the more probable one and may exist in semi-crystalline fibers (Fig. 2.18b) as well as in rubbers [11,14f]. The previously mentioned α, β and γ relaxations are interpreted in terms of the above model as the motion of kink-blocks, as freezing-in of segment rotation because of lack of free volume, and as the steps and jumps of kinks in the amorphous and crystalline regions respectively [11]. Although meanwhile the neutron scattering experiments [100–104] strongly disprove the existence of a distinct meander arrangement of chains, Pechhold's considerations have greatly fertilized the efforts to study the structure of amorphous regions.

The close geometrical description of the arrangement of chain molecules within bundles offered by the kink-model have been used by Kröner and Anthony [14g, 99] to develop a quantitative non-linear deformation theory based on the structure defect "disclination". The meander model, then, is reduced to a particular arrangement of disclinations.

With regard to the elucidation of phenomena occurring during crack propagation in polymers it is generally not necessary to consider intracrystalline processes in detail. If, however, the intracrystalline segments of tie molecules are concerned then one has to be aware of the fact that the intermolecular interaction determines what forces can be transmitted onto these chain segments and whether or not these forces are large enough to cause chain scission.

Thus far model representations of polymer deformation have been discussed. Each of those could be converted into a model representing the fracture behavior if it were possible to formulate an adequate fracture criterion within the range of validity of these models. Having dealt with deformation the fracture criteria to be formulated would have to involve finite extensibility, critical load, or limited volume concentration of stored or dissipated energy. Dealing with fracture one will find that strain, stress, and energy are not sufficient as variables and that one will have to add at least two new dimensions: time and structural discontinuity. This will be explained in the following Chapter.

References for Chapter 2

1. P. H. Geil: Polymer Single Crystals, New York: Interscience 1963.
2. A. Keller: Polymer crystals, Rep. Progr. Phys. *31*, 623–704 (1968).
3. B. Wunderlich· Macromolecular Physics: Crystal Structure, Morphology, Defects (Vol. 1), Crystal Nucleation, Growth, Annealing (Vol. 2), New York and London: Academic Press 1973.
4. E. W. Fischer: Structure of amorphous organic polymers in bulk, Proc. Conf. Non-Crystalline Solids, Clausthal-Zellerfeld, Sept. 1976.
5. R. G. Kirste, W. A. Kruse, J. Schelten: Makromol. Chem. *162*, 299 (1972).
6. J. P. Cotton, D. Decker, H. Benoit, B. Farnoux, J. Higgins, G. Jannink, R. Ober, C. Picot, J. des Cloizeaux: Macromolecules *7*, 863 (1974).
7. P. J. Flory: Statistical Mechanics of Chain Molecules, New York: Wiley 1969.
8. P. J. Flory: Spatial configuration of macromolecular chains, The 1974 Nobel Lecture, Brit. Polymer J. *8*, 1–10 (1976).
9. R. Hosemann: Makromol. Chem., Suppl. *1*, 559–577 (1975). R. Hosemann: Ber. Bunsenges. phys. Chem. *74*, 755–767 (1970).
10. W. Pechhold: Kolloid-Z. Z. Polymere *228*, 1 (1968).
11. W. Pechhold: J. Polymer Sci. *C32*, 123–148 (1971).
12. G. S. Y. Yeh: Polymer Prepr. *14/2*, 718 (1973).
13. H. H. Kausch: Kolloid-Z. Z. Polymere *234*, 1148–1149 (1969), *237*, 251–266 (1970) and J. Polymer Sci. *C32*, 1–44 (1971).
14. a) H. F. Mark: p. xiii. b) H. Gleiter, R. Hornbogen, J. Petermann: p. 149. c) J. D. Ferry: p. 27. d) R. S. Rivlin: p. 71. e) R. F. Landel, R. F. Fedors: p. 131. f) W. Pechhold: p. 301. g) K. H. Anthony, E. Kröner: p. 429. Deformation and Fracture of Polymers: H. H. Kausch, J. A. Hassel, R. I. Jaffee, eds., New York/London: Plenum Press 1973.
15. E. W. Fischer: Z. Naturforsch. *12a*, 753 (1957).
16. A. Keller: Phil. Mag. *2*, 1171 (1957).
17. P. H. Till: J. Polymer Sci. *24*, 301 (1957).
18. Courtesy of E. W. Fischer: Mainz 1974.
19. C. W. Bunn: Trans. Faraday Soc. *35*, 482 (1939).
20. T. Miyazawa, T. Kitagawa: J. Polymer Sci. *B2* (1964).
21. J. Wristers: J. Polymer Sci., Polym. Physics Ed. *11*, 1601–1617 (1973).
22. T. Komoto, M. Iguchi, H. Kanetsuna, T. Kawai: Makromol. Chem. *135*, 145 (1975).
23. K. C. Tsou, H. P. Geil: Int. J. Polym. Mater. *1*, 223 (1972).

24. T. Hattori, K. Tanaka, M. Matsuo: Polym. Eng. Sci. *12*, 199 (1972).
25. A. H. Abdel-Alim, A. E. Hamielec: J. Appl. Polym. Sci. *17*, 3033–3047 (1973).
26. A. H. Abdel-Alim: J. Appl. Polym. Sci. *19*, 2179–2185 (1975).
27. N. A. J. Platzer: Copolymers, Polyblends, and Composites, Adv. Chem. *142* (1975).
28. J. A. Manson, L. H. Sperling: Polymer Blends and Composites, New York and London: Plenum Press 1976.
29. Mehrphasensysteme, Spring Meeting of the German Chemical and Physical Societies, Bad Nauheim, March 29 – April 2, 1976, papers published in Angew. Makromol. Chem. *58/59* and *60/61* (1977).
30. G. Kämpf, M. Hoffmann, H. Krömer: Ber. Bunsen-Ges. *74*, 851 (1970).
31. Courtesy to G. Kämpf: Uerdingen, 1976.
32. A. Keller: J. Polymer Sci. *15*, 32–49 (1955).
33. A. J. Pennings, A. M. Kiel: Kolloid-Z. Z. Polymere *205*, 160–162 (1965).
34. T. Kawai, T. Matsumoto, M. Kato, H. Maeda: Kolloid-Z. Z. Polymere *222*, 1–10 (1968).
35. P. Tucker, W. George: Polym. Eng. Sci. *12*, 364–377 (1972).
36. P. Tucker, W. George: Text. Res. J. *44*, 56–70 (1974).
37. A. Zwijnenburg, A. J. Pennings: Kolloid-Z. Z. Polymere *254*, 868–881 (1976).
38. M. Takayanagi, In: Deformation and Fracture of High Polymers, H. H. Kausch, J. A. Hassel, R. I. Jaffee, eds., New York: Plenum Press 1973, p. 353–376.
39. J. M. Lupton, J. W. Regester: J. Appl. Polymer Sci. *18*, 2407–2425 (1974).
40. N. E. Weeks, R. S. Porter: J. Polymer Sci., Polymer Physics Ed. *12*, 635–643 (1974).
41. N. J. Capiati, R. S. Porter: ibid. *13*, 1177–1186 (1975).
42. R. B. Morris, D. C. Bassett: ibid. *13*, 1501–1509 (1975).
43. N. E. Weeks, R. S. Porter: ibid. *13*, 2031–2048 (1975).
44. N. E. Weeks, R. S. Porter: ibid. *13*, 2049–2065 (1975).
45. F. R. Anderson: J. Polymer Sci. *C3*, 123 (1963).
46. E. S. Clark, C. A. Garber: Int. J. Polym. Mater. *1*, 31 (1971).
47. W. O. Statton: Report UTEC MSE 73–100, University of Utah, Salt Lake City, 1973.
48. H. D. Noether, W. Whitney: Kolloid-Z. Z. Polymere *251*, 991 (1973).
49. I. K. Park, H. D. Noether: Colloid and Polymer Sci. *253*, 824–839 (1975).
50. D. Göritz: Colloid and Polymer Sci. *253*, 844–851 (1975).
51. B. Cayrol, J. Petermann: J. Polymer Sci., Polymer Phys. Ed. *12*, 2169–2172 (1974).
52. R. S. Stein: J. Polymer Sci. *C32*, 45–68 (1971).
53. V. Petraccone, I. C. Sanchez, R. S. Stein: J. Polymer Sci., Polymer Phys. Ed. *13*, 1991–2029 (1975).
54. B. Wunderlich, H. Baur: Adv. Polymer Sci. *7*, 151–368 (1970).
55. J. D. Ferry: Viscoelastic Properties of Polymers, New York: Wiley 1970.
56. J. A. Sauer, A. E. Woodward: Polymer Thermal Analysis, II. P. E. Slade, Jr., L. T. Jenkins, eds., New York: Dekker, Inc. 1970, p. 107.
57. J. A. Sauer, G. C. Richardson, D. R. Morrow: J. Macromol. Sci. – Revs. Macromol. Chem. *C9 (2)*, 149–267 (1973).
58. H. D. Noether: personal communication.
59. L. R. G. Treloar: The Physics of Rubber Elasticity, London: Oxford Univ. Press 1958.
60. M. Maeda, S. Hibi, F. Itoh, S. Nomura, T. Kawaguchi, H. Kawai: J. Polymer Sci., A-2, *8*, 1303–1322 (1970).
    S. Nomura, A. Asanuma, S. Suehiro, H. Kawai: ibid. *9*, 1991–2007 (1971).
    Rheo-Optical Studies of High Polymers, Onogi Laboratory, Kyoto University, Kyoto, Japan, 1971.
61. N. Kasai, M. Kakudo: J. Polymer Sci. A-2, 1955 (1964).
62. A. Peterlin: Adv. Polymer Science and Engineering, K. D. Pae, D. R. Morrow, and Yo Chen, eds., New York: Plenum Press 1972, p. 1.
63. W. Voigt: Lehrbuch der Kristallphysik, B. G. Teubner Leipzig (1910).
64. A. Reuss: Z. angew. Math. Mech. *9*, 49 (1929).
65. E. H. Kerner: Proc. Phys. Soc. B *69*, 802 (1956), Proc. Phys. Soc. B *69*, 808 (1956).
66. Z. Hashin: J. Mech. Phys. Solids *10*, 335 (1962). Z. Hashin, S. Shtrikman: J. Mech. Phys. Solids *10*, 343 (1962).

67. C. van der Poel: Rheol. Acta *1*, 198 (1958).
68. H. Eilers: Kolloid-Z. *97*, 313 (1941).
69. E. Guth: J. Appl. Phys. *16*, 20 (1954). E. Guth, O. Gold: Phys. Rev. *53*, 322 (1938).
70. M. Mooney: J. Colloid Sci. *6*, 162–170 (1951).
71. M. Takayanagi, K. Imada, T. Kajiyama: J. Polymer Sci., C *15*, 263–281 (1966).
72. S. Uemura, M. Takayanagi: J. Appl. Polymer Sci. *10*, 113 (1966).
73. L. E. Nielsen: Mechanical Properties of Polymers and Composites, New York: Marcel Dekker, 1974, Vol. 2, p. 387 ff.
74. A. Einstein: Annalen Phys. *19*, 289 (1906), Annalen Phys. *34*, 591 (1911).
75. R. A. Dickie, M. Cheung: J. Appl. Polymer Sci. *17*, 79–94 (1973).
76. H. H. Kausch: Angew. Makromol. Chem. *60/61*, 139 (1977).
77. F. R. Schwarzl, H. W. Bree, C. J. Nederveen: Proc. IV. Int. Congr. Rheology (Providence 1963). E. H. Lee, ed., New York: Interscience, 1965, Vol. 3, p. 241–263.
78. W. Kuhn: Kolloid-Z. *87*, 3 (1939). W. Kuhn, F. Grün: Kolloid-Z. *1*, 248 (1942).
79. K. Eiermann: Kunststoffe *51*, 512 (1961). K. Eiermann, K. H. Hellwege: J. Polymer Sci. *1*, 99 (1962).
80. J. Hennig: Kunststoffe *57*, 385 (1967). J. Hennig: Kolloid-Z. Z. Polymere *202*, 127 (1965).
81. W. Hellmuth, H. G. Kilian, F. H. Müller: Kolloid-Z. Z. Polymere *218*, 10 (1967). F. H. Müller: J. Polymer Sci. C *20*, 61 (1967).
82. C. C. Hsiao: J. Polymer Sci. *44*, 71 (1960).
83. J. Bishop, R. Hill: Phil. Mag. *42*, 414 and 1298 (1951).
84. I. M. Ward: Proc. Phys. Soc. *80*, 1176 (1962).
85. H. H. Kausch: J. Appl. Phys. *38*, 4213 (1967).
86. S. S. Sternstein, G. M. Lederle, In: Polymer Networks: Structure and Mechanical Properties, Newman and Chompff, eds., New York: Plenum Press, 1971.
87. J. C. Halpin, J. C. Kardos: J. Appl. Phys. *43*, 2235 (1972).
88. E. H. Andrews: Pure Appl. Chem. *39*, 179–194 (1974).
89. D. W. Hadley, I. M. Ward: Rep. Progr. Physics, *38*, 1143–1215 (1975).
90. R. Bonart, R. Hosemann: Kolloid-Z. Z. Polymere *186*, 16 (1962).
91. J. Schelten, D. G. H. Ballard, G. D. Wignall, G. W. Longman, W. Schmatz: Polymer *17*, 751 (1976).
92. D. Y. Yoon, P. J. Flory: Polymer *18*, 509–513 (1976).
93. G. Capaccio, I. M. Ward: Polymer *16*, 239 (1975).
94. G. Capaccio, T. A. Crompton, I. M. Ward: J. Polymer Sci. A *214*, 1641 (1976).
95. G. Capaccio, T. J. Chapman, I. M. Ward: Polymer *16*, 469 (1975).
96. G. Capaccio, T. A. Crompton, I. M. Ward: Polymer *17*, 645 (1976).
97. I. M. Ward: Seminar on Ultra-High Polymers, St. Margherita Ligure, May 1977.
98. G. Capaccio: Seminar on Ultra-High Polymers, St. Margherita Ligure, May 1977.
99. K. Anthony: Habilitationsschrift, Universität Stuttgart, November 1974.
100. E. W. Fischer: Europhysics Conf. Abstracts 2 E, 71–79 (1977).
101. J. Schelten, G. D. Wignall, D. G. H. Ballard, G. W. Longmann: Polymer *18*/11, 1111 (1977).
102. H. Benoit, D. Decker, R. Duplessix, C. Picot, P. Rempp, J. P. Cotton, B. Farnoux, G. Jannink, R. Ober: J. Polym. Sci., Polym. Phys. Ed. *14*, 2119–2128 (1976).
103. C. Picot, R. Duplessix, D. Decker, H. Benoit, F. Boue, J. P. Cotton, M. Daoud, B. Farnoux, G. Jannink, M. Nierlich, A. J. deVries, P. Pincus: Macromolecules *10*, 436–442 (1977).
104. Europhysics Conf. Abstracts *2F* (1978).
105. S. L. Cannon, G. B. McKenna, W. O. Statton: J. Polymer Sci., Macromol. Rev. *11*, 209–275 (1976).
106. H. G. Kilian, M. Pietralla (Polymer *19*, 664–672, 1978) derive from the anisotropy A of the thermal diffusivity $\alpha = \lambda/c_p\rho$ of oriented polyethylenes that the intrinsic anisotropy A_i of the orienting, partly crystalline lamellar clusters increases with the degree of crystallinity (A_i = 7 to 26, linear extrapolation yields A_i = 2 for the fully amorphous and A_i = 50 for a completely crystalline cluster); the average degree of orientation of the lamellae, $<\cos^2\vartheta>$, as determined from thermal measurements agrees very well with X-ray data; the observed increase of orientation with draw ratio is more rapid than it would be in affine deformation.

39

Chapter 3

Statistical, Continuum Mechanical, and Rate Process Theories of Fracture

I. Introduction

Test specimens loaded under laboratory conditions, but also regular engineering components fracturing in service, provide many different data which lend themselves to evaluation of the fracture process. These data are, for instance, the time to onset and completion of fracture, details of the fracture pattern (ductile or brittle fracture, appearance of the breaking specimen and of the fracture surface), crack dynamics, and change in physical or chemical properties. Naturally the most straightforward evaluation of a test or a set of data is the direct correlation of the property of interest (e.g., time under stress) to the environmental parameter(s) of interest (e.g., stress and temperature). In Figure 1.4 a set of data with just these variables has been plotted (PVC pipes under internal pressure). If one uses such a plot a number of questions arise:

– what is the statistical significance of an individual data point
– what are the failure conditions and how do they depend on external parameters
  and material properties
– what is the likely cause of failure
– what conclusions can be drawn with respect to extrapolating the set of curves
  into hitherto non-accessible regions of time, pressure, or temperature?

The first of these points will be commented upon subsequently in connection with an interpretation of the statistical variability of fracture events. To answer this and the second question the highly developed tools of statistical analysis (e.g., 1—2) are available; their application forms a backbone of industrial quality control and material design. Understandably, the complex input (mechanical, thermal, and environmental attack) acting upon a complex system (e. g., a highly structured polymer) has a complex, non-deterministic output. The determination and evaluation of only a small number of different data will necessarily reveal only a partial aspect of the fracture process. For this reason the questions as to cause and kinetics of fracture development can rarely be answered unambiguously. Any mathematical interrelations established between variables (e.g., time and stress) are valid for extrapolation only if the basis does not change.

In this chapter an overview of conceptually different fracture theories is presented which have in common that they do not make explicit reference to the characteristic properties of the molecular chains, their configurational and supermolecular order and their thermal and mechanical interaction. This will be seen to apply to the classical failure criteria and general continuum mechanical models. Rate process fracture theories take into consideration the viscoelastic behavior of polymeric materials but do not derive their fracture criteria from detailed morphological analysis. These basic theories are invaluable, however, to elucidate statistical, non-morphological, or continuum mechanical aspects of the fracture process.

II. Statistical Aspects

Experimentally determined quantities (observables) such as strength, time to fracture, or concentration of free radicals show a wide scatter of values; they are stochastic variables. As an extreme example of a stochastic variable in Figure 3.1 the service lives t of 500 HDPE-tubes under identical test conditions are plotted in the form of a histogram [3]. The distribution shown can be represented by a logarithmic normal distribution (Fig. 3.2) with a mean of lg t/h equal to 2.3937 and a variance of $s = 0.3043$. The expectation value of the time to fracture of a specimen subject to the test condition employed is the time which corresponds to the mean value of the (logarithmic) distribution, here 247.6 h. The actually determined values of t evidently scatter widely around this expectation value. Nevertheless even such a distribution can be characterized by testing just a few randomly selected specimens. For a normally distributed observable any three random values cover on the average a range of 1.69 s — which contains in 86% of all cases the mean value of the (infinite) distribution also [3]. This extreme example has been chosen to emphasize the meaning of stochastic variables with which one is dealing subsequently. It was also intended to show the significance of small samples — containing 3 to 10 members — which are generally used in practice to establish correlations between stochastic variables and environmental or material parameters. At this point we are not going to talk further about the statistical reduction of such a set of data. Here we are interested in the information which may be drawn from such a statistical analysis.

Fig. 3.1. Scatter of times to fracture of 500 PE tubes under identical loading conditions. 80 °C, $\sigma_v = 4$ MN/m^2 (after [3], from [78]).

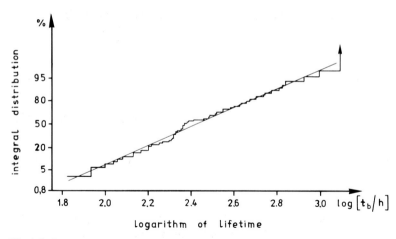

Fig. 3.2. Cumulative normal distribution of the logarithms of the lifetimes plotted in Fig. 3.1.

From a statistical standpoint, the variability of a material property, e.g., fracture strength, is explained by referring to one or more of the following three arguments:

1. The beginning of fracture is a statistical event, the occurrence of which is described by probability laws.

2. Apparently identical specimens of a set are inherently different, e.g., they contain a large number of flaws of different size or severeness and the most severe of these flaws determines the fracture strength.

3. The load to which a sample is subjected until it fractures activates a large number of molecular processes. The uncertainty of the cumulative event, e.g., mac-

roscopic failure, then results from the uncertainty of the molceular events and the mode of correlation of these events.

The first argument (fracture is a statistical event) would lead to the following general equation for the rate of change of the number N of members belonging to one ensemble:

$$d N/dt = -N K \tag{3.1}$$

where K may be a function of N, stress, stress history, temperature, or other environmental parameters. For a number N of statistically equal, independent and separately loaded units the rate function K in Eq. (3.1) is independent of N, but generally a function of time. Equation (3.1) is readily solved with respect to the number N(t) of surviving units:

$$N(t) = N_0 \exp \left[-\int_0^t K(\tau) \, d\tau \right] \tag{3.2}$$

where N_0 is their initial number. The fraction of units broken after some time t_b is obviously $1 - N(t_b)/N_0$ and the cumulative distribution function of lifetimes, $Q(t_b)$, therefore is:

$$Q(t_b) = 1 - \exp \left[-\int_0^{t_b} K(\tau) \, d\tau \right] . \tag{3.3}$$

The average lifetime, $<t_b>$, is solely determined by K and becomes for constant K:

$$<t_b> = \frac{1}{N_0} \int_0^\infty t N_0 K \exp(-Kt) \, dt = K^{-1} . \tag{3.4}$$

It may be noted that in this case the life expectancy of a surviving unit is at any time independent of N and t. This is equivalent to the assumption that the residual strength is equal to the initial strength, i.e., that no aging or deterioration has taken place in the stressed sample. The rate K of failure occurrence can be determined experimentally from the scatter of t_b [4]. Using this assumption Kawabata and Blatz [5] developed a simple stochastic theory of creep failure. They argued that the first moments of t_b for an ensemble of ruptured specimens should reveal whether or not K is independent of time. If K depends only on initial load and is independent of N the equation

$$K = (m!/<t_b^m>)^{1/m} \tag{3.5}$$

should hold for any m, that is for any order of moments of t_b. An investigation of the variability of fracture times in the special case of a rubber vulcanizate gave excellent agreement with the assumption of a time-independent K as used in Eq. (3.4).

43

The same conclusion was drawn by Narisawa et al. [82] from the exponential decrease of N(t) with t of uniaxially stressed PA 6 monofilaments.

In other experiments Kawabata et al. [6] investigated the failure statistics of carbon-filled styrene butadien rubber (SBR). They arrived at the conclusion that either the coefficient relating stress and rate of failure is increasing with time or that several local fracture events must be independently initiated before the rubber specimen breaks. The best fit of their theory with experiments was obtained for a critical number of 3 to 4 microscopic fracture events as nucleus for unstable crack formation. For the asymmetric distribution of times-to-fracture shown in Figure 3.2 the relation Eq. (3.5) is also not valid for higher orders of m (m equal to or larger than two). This means either that the probability density K for pipe failure is smaller for specimens having a longer service life or that K is a function of loading time. In the first case it must be assumed that from the beginning the specimens had not been statistically identical, in the second that they underwent structural changes affecting K. The first assumption seems to be valid for pipe failure through creep crack development. The investigations of Stockmayer [83] recently have shown that nature and location of material inhomogeneities largely determine the creep lives t_b of LDPE pipes. Morphological parameters thus account for more than half of $<t_b^2>$ [84]. The conclusion that structural differences exist and affect K must be drawn for other materials as well which exhibit a log-normal distribution of t_b values [3, 83–85]. These aspects will have to be discussed in later sections in terms of detailed structural models of the failing specimens.

Without reference to any such models and based purely on statistical considerations, Coleman and Marquardt developed their important theory of breaking kinetics of fibers (reviewed in ref. 7). They especially investigated the distribution of lifetimes of fibers under constant or periodic load and the effects of fiber length, rate of loading, and bundle size on the strength of filaments or bundles thereof (Figs. 3.3 and 3.4). Two statistical effects should be noted: the lower strength of a bundle as compared with that of a monofilament (due to the accelerating increase in failure probability K after rupture of the first filament within a bundle) and the increase of strength with loading rate here derived from the reduced time each filament spends at each load level. The authors [8] determined the average tensile strengths of 15-denier 66 nylon monofils and that of an infinite bundle; the strength values are plotted as a function of loading rate in Figure 3.3.

As mentioned above the large scatter of strengths and the dependency on sample geometry may be explained by introducing the *concept of flaws of different degrees of severity.* This means substituting one statistical variable (e.g., underlying strength) for another one (e.g., macroscopic strength or time to failure). Even though such a substitution leaves the nature of the microscopic defects out of consideration, it still allows one to obtain, from statistical analysis, some information on the size, number, location, and severity of these defects (or flaws). The term flaw needs some deliberation. First, a flaw can be considered as an *ellipsoidally shaped void* which may act as a stress concentrator and as the eventual cause of instability and failure. Secondly, it should also be understood as a *weak region* containing molecular irregularities. When being stressed the weak region may fail, leading to a void, a craze, and/or to a propagating crack and macroscopic failure. Following Epstein's statistical

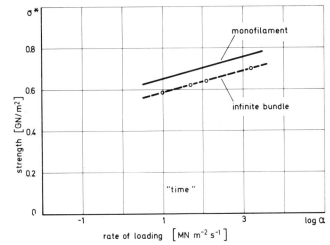

Fig. 3.3. Strength of 66 polyamide monofilaments and bundles of filaments as a function of loading rate (after Coleman et al., [8]).

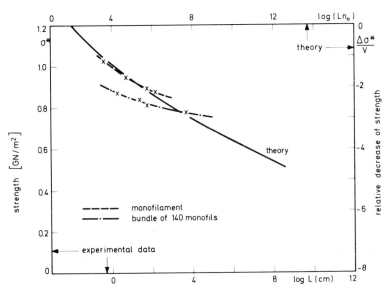

Fig. 3.4. Effect of specimen length L on 6-polyamide fiber strength (after [11]); theoretical curve from Eq. 3.6, experimental data from [10].

theory of extreme values [9] the strength of a material volume element depends on the "severeness" of the most critical flaw being present within that volume element. The macroscopic strength distribution g(L, x) of fibers of different length L but uniform cross section can then be expressed as

$$g(L, x) = Ln_0 f(x) [1 - \int_{-\infty}^{x} f(y) dy]^{Ln_0 - 1} \tag{3.6}$$

where n_0 is the number of flaws per unit length of fiber and $f(x)$ the underlying strength distribution. For $f(x)$ being a Gaussian distribution with mean μ and variance v, the most probable strength, σ^*, depends on Ln_0 as shown by the solid line (theory) in Figure 3.4. Also plotted are the experimental data of Levin and Savickij [10]. Even though it appears as if the experimental data would be fitted by the theoretical curve in the range of $10^3 < Ln_0 < 10^6$, some caution is required. If indeed the inherent distribution of flaws is a Gaussian distribution, then the decrease of $\sigma^*(Ln_0)$ with L and the variance of $\sigma(Ln_0)$ are both related to v and $n(= Ln_0)$ in a known manner and subject to proof. A proof is also possible for other underlying distributions. For the data of Levin and Savickij [10] neither a Gaussian distribution of x, nor a Laplacian, nor a Weibull distribution describes $\sigma^*(Ln_0)$ and $v(Ln_0)$ simultaneously [11]. As a result of this analysis one may conclude that either the underlying distribution $f(x)$ is truncated at the left-hand side (small values of x), or the variance of σ is much larger than to be expected from the number of flaws. The truncation at small strength values could be caused by a multi-step crack initiation process which would require a certain time. The effect of sample volume on strength was also not in agreement with extreme value statistics for viscose filaments, carbon fibers, and cotton yarn [12]. Fedors and Landel [13] obtained a doubly exponential distribution of strength values for eight SPR gum vulcanizates. Such a distribution derives, for example, from an underlying exponential distribution of flaw cross sections [14]. Here the volume effect was not studied.

When dealing with underlying distributions it should be recognized that different regions (core, surface) of a (seemingly) homogeneous specimen may have different underlying distributions. Evidence of this behavior was for instance found in polyamide fibers [15] and LDPE pipes [84]. Depending on processing conditions volume-related and surface-related breakages with quite different stress-lifetime characteristics could be distinguished. At this point the many attempts to determine underlying distributions are not discussed any further but reference is made to the publications on this subject [11–17]. If any conclusion may be attempted from statistical investigations of uniaxial failure of thermoplastics it is that the underlying distribution, $f(x)$, of natural flaws could be a Laplacian distribution possibly skewed at the left-hand side (small values of x). Moreover, it appears unlikely that a deterministic view may be adopted associating a particular flaw with a particular strength (or time to failure) but rather necessary to allow for the uncertainty of multistep flaw growth.

The third statistical argument (*fracture as the result of a large number of molecular processes*) is very old. Its application to polymers definitely was stimulated from many sides including fracture and fatigue of metals [4] or glasses [17] and the thermodynamics of reactions [19]. The interpretation of fracture as a sequence of individual steps will be mainly used in the following. It necessitates a consideration of the correlation of such steps and of the final failure criterion.

The discussion of the three different statistical arguments can be summarized by saying that in the first case (fracture as a statistical event) a property (probability of fracture) is attached to the material body as a whole. In the second case one flaw, i.e., one micro-heterogeneity (out of many), within the body is considered to dominate failure; in the third case many events interact, influence, and determine each other. *The very same lines of thought used here to explain the variability of strength, for example, are found in the continuum mechanics, fracture mechanics, and molecular approach to theories of strength, respectively.*

III. Continuum and Fracture Mechanics Approach

A. Classical Failure Criteria

In classical continuum mechanics homogeneous isotropic materials are taken into consideration. Failure criteria are established on the basis that there is indeed one strategic material property such as uniaxial tensile strength, shear strength, elastic or total extensibility, or energy storage capability that determines the failure of the stressed sample. If, in establishing such criteria, all environmental parameters (e.g., temperature T, strain rate, $\dot{\epsilon}$, or surrounding medium) are taken to be constant, then failure has to be expected whenever the components of multiaxial stress — usually the three principal stresses σ_1, σ_2, and σ_3 are considered — combine in such a way that the strategic quantity reaches a critical value C. For different T or $\dot{\epsilon}$ C may assume different values. The condition $f(\sigma_1, \sigma_2, \sigma_3) = C(T, \dot{\epsilon})$ represents a two-dimensional "failure surface" in three-dimensional stress space. The stable states of stress form a continuous body bounded by the failure surface against the unstable stress points.

It has already been indicated that failure in polymeric materials may occur in a number of phenomenologically different ways, e.g., as brittle fracture through propagating cracks, as ductile fracture through plastic deformation following shear yielding, or as quasi-brittle fracture after normal-stress yielding (crazing). It had correctly been anticipated that these different phenomena respond differently to the magnitude and the state of stress. This means that for different fracture phenomena separate failure surfaces exist which may overlap and penetrate each other. These facts are extensively investigated and discussed in the recent monograph by Ward [20] and in the general literature (e.g., 21–34).

The *classical criteria* were conceived to describe the failure of isotropic, elastic-plastic solids. The physical content of the six principal criteria is in brief the following:

1. The *maximum principle stress theory* (Rankine's theory) states that the largest principle stress component, σ_3, in the material determines failure regardless of the value of normal or shearing stresses. The stability criterion is formulated as

$$\sigma_3 < \sigma^*. \tag{3.7}$$

In this case σ^* is considered a basic material property which can be determined from, for example, a tension test. In stress space the surface of failure according to

47

the maximum stress theory is a cube. A modified form of this theory allows for a larger critical value σ^* if one of the stresses is compressive. Then the surface of failure is a cube again, but with the center displaced from origin.

2. The *maximum elastic strain theory* (St. Venant's theory) states that inception of failure is due if the largest local strain, ϵ_3, within the material exceeds somewhere a critical value ϵ^*. The failure criterion, therefore, is derived as

$$\epsilon_3 = \frac{1}{E} [\sigma_3 - \nu(\sigma_1 + \sigma_2)] < \epsilon^* \tag{3.8}$$

where E is the isotropic Young's modulus and ν Poisson's ratio. The term in square brackets thus has to be smaller than an equivalent stress, σ^*.

3. The *theory of total elastic energy of deformation* was first proposed by Beltrami. Since under high hydrostatic pressure large amounts of elastic energy may be stored without causing either fracture or permanent deformation, the elastic energy as such seems to have no significance as a limiting condition.

4. *Theory of constant elastic strain energy of distortion* (Huber, v. Mises, Hencky). The consideration of the poor significance of the elastically stored energy in hydrostatic compression or tension led to the idea of subtracting the "hydrostatic" part of the energy from the total amount of elastically stored energy. Thus, it is assumed that only the energy of distortion, W, determines the criticality of a state of stress. For small deformations one obtains:

$$\frac{6E}{1+\nu} W = [(\sigma_1 - \sigma_2)^2 + (\sigma_2 - \sigma_3)^2 + (\sigma_3 - \sigma_1)^2] < 2(\sigma^*)^2. \tag{3.9}$$

5. If a *limiting octahedral shearing stress, τ^*,* is postulated as a failure criterion the same mathematical expression as in Eq. (3.9) is obtained with

$$9(\tau^*)^2 = 2(\sigma^*)^2. \tag{3.10}$$

In this case the limiting condition of small deformations does not pertain.

6. The *Coulomb yield criterion* states that the critical stress, τ^*, for shear deformation to occur in any plane increases linearly with the pressure, σ_n, applied normally to that plane:

$$\sigma_3 - \sigma_1 < \tau^* = \tau_0 + \mu\sigma_n. \tag{3.11}$$

Here the material constant τ_0 is given by the cohesion of the material and μ is a "coefficient of friction". If friction is neglected the Tresca yield criterion is obtained.

The above failure (or stability) criteria do not explicitly contain time as a variable. One may apply them to rate sensitive materials, however, if one recognizes that σ^* and τ^* will depend on stress history. The classical approach of Eyring and other rate theories will be discussed in Section IV of this chapter. The limits of the applicability of the classical criteria and of their extension to anisotropic materials are analyzed by Ward [20]. Also in recent years numerous papers (e.g., 21–28) and review articles [29–34] have appeared which concern the geometry of failure surfaces of polymeric materials. Multiaxial failure experiments on elastomers which seem to corroborate the St. Venant criterion have been carried out [21] by Gent and Lindley on endbonded penny-shaped samples of a natural rubber gum vulcanizate, by Ko on polyurethan rubber, by Oberth and Bruenner on castable elastomers containing solid inclusions, and by Lim [22] on tubular samples of polyurethane and of a butadiene/acrylic acid copolymer. Figure 3.5 depicts Lim's data and a failure envelope for $\nu = 0.37$ resulting from the St. Venant criterion.

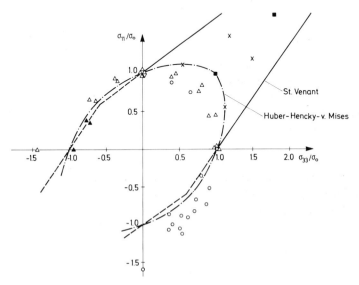

Fig. 3.5. Failure in multiaxial stress, ○ PMMA tubes (Broutman et al., [23]), △ 6 PA tubes, ▲ buckling (Ely, [24]), x PUR tubes (Lim, [22]), ■ SBR membranes (Dickie et al., [25]) – – – maximum strain failure criterion, – · – · – octahedral shear stress failure criterion.

Of the other theories, the Huber-v. Mises-Hencky criterion and the Coulomb yield criterion have received the most experimental support. The failure surface in the stress space described by Eq. (3.9) is a cylinder with its axis along the space diagonal. Traces of this cylinder with the (σ_1, σ_2)-plane are ellipses symmetric with respect to the point of origin. One such ellipse is shown in Figure 3.5. The failure surface corresponding to the Coulomb criterion is a cone. Its traces with the (σ_1, σ_2)-plane are also ellipses but displaced with regard to the origin. Elliptic traces seem to form the best available failure criterion in the biaxial yielding experiments with tubular specimens of PMMA [23, 33], PVC [26, 27], PC [27] and of PS, PE, PP, PA, ABS, and CAB [33]. In the case of multiaxial fracture without yielding (Trennbruch) elliptic and parabolic traces are obtained which pertain to a cone or a paraboloide in stress space [30, 33]. In any event there is no strict distinction between the form of failure surfaces describing failure through flow and the form of the surfaces referring to failure through material separation (brittle fracture). This indicates that in these isotropic polymers the *initiation* of failure was strongly influenced by shear stresses irrespective of the *ultimate* material behavior. No unique σ^* or ϵ^*

proved to be a critical quantity in the experiments of Dickie and Smith [25] on the fracture of SBR in equal biaxial tension and of Dong [28] on a highly filled fluor elastomer (Viton ®).

More recently there has been a strong interest in the deformation and fracture behavior of plastics under large hydrostatic pressures [31–32]. One should expect – and one observes – that the rigidity of a polymer increases with pressure. Sauer et al. [32] report that a pressure of 3.5 kbar raises the initial Young's moduli of amorphous thermoplastics (PC, PI, PSU, PVC, CA) by a factor of 1.2 to 1.9, that of crystalline polymers by 1.4 (POM) to 7.5 (PUR). Despite the increased rigidity, ductile fracture occurs. The effects are not yet understood in all generality. Following the two major review articles on this subject by Radcliffe [31] and Sauer and Pae [32] the Coulomb criterion corresponds best to most pressure-yield experiments. In the case of PTFE and PC with their nonlinear increase of yield strength with hydrostatic pressure ($\sigma_n = p$) even the consideration of a second-order term of p appeared to be necessary [32]. Whereas the application of pressure restricts the cold drawing of many ductile polymers, it enables yielding of normally brittle polymers, e.g., of PS at room temperature at a pressure of 2 to 3 kbar, and of PSU and polyimides at 3 to 7 kbar [32].

In this chapter no molecular interpretations of fracture phenomena are given. The above discussion has already shown, however, that a consideration of the three-dimensional state of stress is not sufficient to elucidate the possible role of chain scission or disentanglement in failure of polymers. This is particularly true if the phenomenon of crazing ("normal stress yielding") is included. Before investigating the molecular aspects, however, in the later chapters, the presentation of general, non-molecular fracture theories should be continued.

B. Fracture Mechanics

Real solids are inhomogeneous. Even if they are one-phase materials they will contain defects, voids, inclusions, flaws, cracks, and/or other inhomogeneities which are able to distort an otherwise homogeneous stress field. A continuum mechanical approach that analyzes these (singular) stress-strain fields in the vicinity of flaws or cracks and derives its criteria of crack stability from an energy balance consideration is the fracture mechanics approach. Griffith [35] was the first to find wide acclaim for his considerations of the energetic changes occurring as a consequence of the widening of a crack (of length $2a$). He equated the resulting decrease dU of elastic energy stored within a stressed specimen with the surface energy $\gamma_c dA$ necessary to enlarge the crack by an infinitesimal area dA:

$$- \frac{dU}{da} = \gamma_c \frac{dA}{da}.$$

(3.12)

For a plate containing an elliptical hole and stressed at right angles to the major axis a, he obtained the stability criterion

$$\sigma^* = (2\gamma_c E/\pi a)^{1/2}.$$

(3.13)

The size of the largest material defect together with the material constants modulus of elasticity (E) and surface work parameter (γ_c), thus, supposedly determine the strength of the material, i.e., that macroscopic value σ^* of uniaxial stress at which irreversible crack propagation begins. Equation (3.13) gives a mathematical form to the earlier used concept that flaw size (or severity) determines the inherent strength of a sample. It also explains that the actually observed macroscopic strength is much smaller than the theoretical strength of flawless specimens.

The application of fracture mechanics to viscoelastic media is caught between the limitations that are brough about by the deviation from infinitesimal strain condition, by molecular anisotropy, local concentration of strains, and the time dependency of stress and strain functions. It has been notably successful in investigations of crack propagation. The analytical extension of Griffith's work to linearly viscoelastic materials has been reviewed by Williams [36] and more recently by Knauss [37]. In Chapter 9 a more detailed account on the application of fracture mechanics to crack propagation is given. There the morphological aspects and the effect of time-dependent plastic deformation, craze initiation and growth, and chain scission on the energy of cohesive fracture of high polymers will be considered.

At this point attention will be drawn to an application of the Griffith approach to the effect of matrix orientation on strength. Such an interpretation has been forwarded by Mikitishin et al. [79] and Sternstein et al. [80]. These authors suggest that during the orientation process a rotation, elongation, and widening of possible defects takes place. The change of shape and orientation changes the effect of these flaws as stress concentrators. They become less dangerous in a direction parallel to the draw direction (z) and more critical in the perpendicular direction (x). If only the increase in length of (elliptical) voids which are oriented parallel to the draw direction is considered, the lateral strength σ_x is derived from Eq. (3.13) as:

$$\sigma_x = [2\,\gamma_c\,E/(1+\lambda)\pi a]^{1/2} \tag{3.14}$$

where λ is the draw ratio. Sternstein et al. [80] have made comprehensive calculations in two dimensions of the flaw-rotation and widening effect on axial and lateral strength for a statistical distribution of flaw orientations and various flaw geometries. They compare their predictions of the lateral strength of polystyrene with the experimental data of Retting [81]. For moderate molecular weights and draw ratios up to 1.5 they find a correspondence to within 20%. For higher molecular weights and larger draw ratios the discrepancies are above 70%.

If a crack travels at the interface of an *adhesive joint* one observes an adhesive debonding rather than cohesive failure. Except for minor differences in boundary condition the analgy to the fracture mechanical treatment of cohesive failure is complete [38–40] and the normal debonding stress σ_a^* is obtained as:

$$\sigma_a^* = (2\,\gamma_a\,E/\pi a)^{1/2}. \tag{3.13a}$$

The interpretation of γ_a and E is rather involved. These quantities not only depend on the static surface free energies and elastic moduli of the interacting (polymeric) materials but also on the work of deformation which the adhesive layer undergoes in debonding and on the surface activity of a possible third phase having access to the debonded area [40].

C. Continuous Viscoelastic Models

Under the heading of this section belongs the very fruitful model of a common "failure envelope" proposed by Smith [41] It is based on a consideration of the viscoelastic behavior of continuous high polymer bodies, i.e., on the concept that a reduction according to the time-temperature superposition principle of the environmental parameters stress, strain-rate, and temperature should lead to corresponding molecular states. If a fracture criterion refers indeed to unique limits of molecular load-carrying capability, for example, then the plotting of reduced stress at break versus strain at break for different experimental conditions should lead to one master curve, the failure envelope (Fig. 3.6). For a large number of natural and synthetic rubbers and vulcanizates under similar modes of mechanical excitation, this concept was found to hold. It thus generally allows the extrapolation of the ultimate behavior into unknown regimes of time, strain-rate, or temperature. This model, however, does not allow prediction of the shape of the fracture envelope or

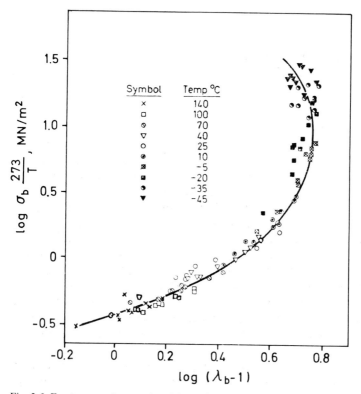

Fig. 3.6. Fracture envelope: reduced breaking stress vs. breaking elongation for an unfilled butyl rubber vulcanizate (after Smith, [41]).

the conversion of data obtained under different modes of excitation (uniaxial, equal biaxial, or other multiaxial loading).

An argument to resolve the discrepancy between the failure envelopes obtained for different modes of straining is indicated by the work of Blatz, Sharda, and Tschoegl [42]. These authors have proposed a generalized strain energy function as constitutive equation of multiaxial deformation. They incorporated more of the nonlinear behavior in the constitutive relation between the strain energy density and the strain. They were then able to describe simultaneously by four material constants the stress-strain curves of natural rubber and of styrene-butadiene rubber in simple tension, simple compression or equibiaxial tension, pure shear, and simple shear.

The continuous viscoelastic models discussed in this section make use of the theory of rubber elasticity and of the time-temperature superposition principle. They implicitly very well recognize the molecular origin of the viscoelastic behavior of a material — but explicitly they do not refer to noncontinuous quantities such as the discrete size, structure, and arrangement of molecules, the anisotropy of molecular properties, or the distribution of molecular stresses or stored strain energy. If individual molecular events or discontinuities of stress or strain are either not discernible or not of particular importance, a representation of a solid as a continuum is quite adequate.

IV. Rate Process Theories of Fracture

A. Overview

As opposed to the continuum mechanical theories, "molecular" rate process theories of fracture recognize the presence of discrete particles or elements forming the material body. Rate process theories intend to relate in a straightforward manner breakage, displacement, and reformation of these elements to deformation, defect development, and fracture of the structured material.

All theories of a first group to be described subsequently have in common the assumption that macroscopic failure is a rate process, that the basic fracture events are controlled by thermally activated breakages of secondary and/or primary bonds, and that the accumulation of these events leads to crack formation and/or to breakdown of the loaded sample. Here the basic events are defined in a general manner and not experimentally related to particular morphological changes. A second group of theories is based on the physical evidence of molecular damage obtained by spectroscopy and X-ray scattering methods it will be discussed in Chapters 7 and 8. A third group of theories concerns the role of cracks and craze initiation and is treated in Chapter 9.

One of the early applications of the theory given by Glasstone, Laidler, and Eyring [43] was the prediction of the flow and diffusion behavior of liquids. From the original concept of flow, the thermally activated jump of molecules across an energy barrier, various fracture theories of solids have emerged. Tobolsky and

Eyring [44] considered the general decrease of *secondary* bonds, Zhurkov [45—47] and Bueche [48—50] that of *primary* bonds to be the strength controlling factor. In their recent monographs on deformation and fracture of polymers Bartenev and Zuyev [51], Andrews [52], and Regel, Slutsker, and Tomashevskii [53] have reviewed a considerable amount of experimental data under this aspect; reference to their works should be made if information is sought on the effect of time and temperature on the fracture behavior of polymers of different composition and structure and under various environments.

Following the earlier works of Eyring, Zhurkov, and Bueche, these and other authors have continued to develop different aspects of the kinetic theory of fracture. Particularly in the USSR fracture data have been interpreted in terms of the failure of regular, nonmorphological model lattices (e.g., 54—58). Gubanov et al. [54], Bartenev [55], and Perepelkin [56] focussed on the potential energy of interaction between adjacent members of a polymer chain, Mikitishin et al. [57] on the defects introduced into the lattice by chain ends, and Dobrodumov [58] on the increase of the load and — consequently — of the rate of scission of bonds adjacent to a broken bond.

Hsiao and Kausch [59—61] studied the effect of orientation-dependent local strain on the total rate of chain scission. Holzmüller [62], Bartenev et al. [63], and Salganik [64] analyzed the amount of thermal energy and the direction of relative motion imparted by statistically fluctuating phonons to adjacent segments. Different statistical aspects of the accumulation of molecular defects were treated by Orlov et al. [65], Goikhman [66], and Gotlib [67] who simulated the isolated production, growth, interaction, and coalescence of defects. An energy probability theory was forwarded by Valanis [68] who combined the role of strain energy density, the stochastic nature of fracture, and the fracture hypothesis of Zhurkov.

Whereas all of the above theories [54—68] refer in one way or another to the thermally activated breakage of one kind of molecular bonds as primary fracture event, the Bueche-Halpin theory for the tensile strength of gum elastomers [69] employs the idea that the viscoelastic straining of rubber filaments together with the degree of their ultimate elongation determine the kinetics of crack propagation in an elastomeric solid.

In the following the characteristic aspects of these generalized, nonmorphological rate process fracture theories are to be discussed, i.e. the nature of the basic fracture events, their distribution in space and time, the law of the accumulation of these events, and resulting fracture criteria.

B. Eyring' Theory of Flow

Flow of a liquid, melt, or solid is the result of a stress-biased thermodynamically irreversible displacement of molecules past each other. A molecule in thermal equilibrium with its surroundings is in thermal motion, which in the case of a liquid or solid is predominantly a vibration around a temporary equilibrium position. The amplitude of vibration is constantly changing. Eyring [43] postulated that a displacement (or jump) of a molecule from an initial to a neighboring equilibrium position may occur if the thermal energy of the molecule is sufficient to reach the "activated" stage, i.e., the top of the energy barrier separating the initial and the final equilibrium positions. The rate of decay of activated states towards the final position was obtained by him as

$$k_0 = \kappa \, \frac{kT}{h} \, \exp\left(-u_0/kT\right) \exp \Delta s/k. \tag{3.15}$$

Here κ is the transmission constant which indicates how many of the activated complexes actually disintegrate, k is Boltzmann's, h Planck's, and R the gas constant, T is absolute temperature, u_0 the height of the potential barrier, and Δs the difference in entropy between ground state and activated state (the quantities u and s in small letters refer to one particle, the capital quantities to a mole of them). In the absence of external forces initial and final equilibrium states are assumed to have the same potential energy. Then the rates of flow of particles across the dividing potential barrier are the same in the forward and backward directions. Under the influence of external forces, however, local stress fields are generated. A particle thermally activated to move in the direction of, say, stress Ψ by a distance $\lambda/2$ then gains an amount of energy w:

$$w = \Psi \lambda \, q/2 \tag{3.16}$$

where q is the average cross section occupied by the particle normal to the direction of motion. A particle moving in the $-\Psi$ direction will loose the same amount of energy. The first particle, therefore, needs less thermal energy to reach the activated state than the second particle. Consequently the rates of flow of particles across an energy barrier in opposite directions are biased by the local stress:

$$k_1 = k_0 \exp\left(w/kT\right) \tag{3.17}$$

$$k_2 = k_0 \exp\left(-w/kT\right). \tag{3.18}$$

The net flow rate, therefore, is:

$$K = k_1 - k_2 = 2 \, k_0 \sinh\left(w/kT\right). \tag{3.19}$$

It should be noted very carefully that w is equal to the product of local stress times an activation volume; this is the volume drawn up by particle cross section and step width in the process of activation. Activation volume and activation energy necessarily must refer to one and the same activation process.

C. Tobolsky-Eyring

The concept of slipping of secondary bonds was applied by Tobolsky and Eyring in 1943 to the breaking of polymeric threads under — uniaxial — load [44]. The rate of decrease of the number N of such bonds per unit area under constant uniaxial stress Ψ_0 is obtained from Eqs. (3.1) and (3.19):

$$-\frac{1}{N}\frac{dN}{dt} = 2\,k_0\,\sinh\,(\Psi_0\lambda/2\,NkT). \qquad (3.20)$$

For large values of stress the flow of bonds takes place almost exclusively as breakage and not as reformation. In that case one may use Eq. (3.17) instead of (3.19). If one substitutes there $\Psi_0\lambda/2\,NkT$ by y one obtains a solution in terms of the exponential integral $-Ei(-y)$:

$$-Ei\,(-y) = \int_{y_0}^{\infty} dy'\exp(-y')/y' = k_0\,t_b. \qquad (3.21)$$

The lower limit of integration is $y_0 = \Psi_0\lambda/2\,N_0kT = w/kT$. Equation (3.21) offers an implicit relation between the expectation value of the time to fracture of the thread, t_b, the rate of flow of unstressed secondary bonds, k_0, and the stress Ψ_0 acting on the N_0 initial bonds. With $w \gg kT$ the exponential integral may be approximated by:

$$-Ei\,(-y) = [(\exp - y)/y](1 - 1/y). \qquad (3.22)$$

Within the range of validity of this approximation the logarithm of lifetime, $\log t_b$, is almost linearly related to w, i.e., to the applied stress. Exactly this behavior is exhibited by a large number of stressed metals, ceramics, or polymers. Together with Eyring's original interpretation this has laid the foundations of the kinetic theory of fracture, to which — as indicated above — subsequently a large number of researchers have contributed.

D. Zhurkov, Bueche

Independently from each other Zhurkov et al. [45–47] in the USSR and Bueche [48–50] in the USA expressed the idea that in fracture of polymers — and also of metals for that matter — the breakage of primary (chemical) bonds plays an important role. They found that the time to fracture under uniaxial stress σ_0 of Zn, Al, and for example, PMMA and PS below their glass transition temperatures could be expressed by an exponential relationship involving three kinetic parameters:

$$t_b = t_0\exp\,(U_0 - \gamma\sigma_0)/RT. \qquad (3.23)$$

The three parameters were — based on the available data — interpreted as energy for activation of bond scission (U_0), as the inverse of molecular oscillation frequency (t_0), and as a structure sensitive parameter γ. In Figure 3.7 experimental data and the theoretical curves according to Eq. (3.23) are reproduced.

Equation (3.23) can be considered as a general expression of the kinetic nature of material disintegration. A closer discussion of the meaning of the molecular oscillation frequency, of the structure-sensitive parameter γ, and also of the activation

56

Fig. 3.7. Times to fracture under constant uniaxial load: cellulose nitrate (CN) at T = 70 (1), 30 (2), −10 (3), and −50 °C (4); polyamide 6 (PA 6) at T = 80 (1), 30 (2), 18 (3), −60 (4), and −110 °C (5) (after Savitskii et al. [86].

energy for bond scission in view of the molecular structure and of the sequence of molecular processes will be given in Chapters 7—9. At this point one would like to check further arguments that the existence (and volume concentration) of the strong primary bonds determines the rigidity and strength of (polymeric) solids.

Holliday et al. [70] have looked under this aspect on the degree of connectivity CN (average number of main network bonds per chain atom, e.g., 2 in hydrocarbon chains, 3 in graphite, and 4 in diamond) and on the relative volume density of main bonds NCR (which here is density times number of network atoms in the repeat unit divided by molecular weight of a repeat unit). The term NCR varies between less than 0.01 for bulky polymeric chains and 0.58 for diamond. The bulk moduli, Young's moduli, hardness values, and cubical coefficients of thermal expansion of the more than 50 organic and inorganic polymers investigated at room temperature exhibit a strong correlation with NCR. In Figure 3.8 the best fit curves drawn by Holliday for Young's modulus and Brinell hardness values BHN are reproduced (the actual NCR values — except for those of the graphitelike structures — fall into a narrow band along the curves having a width of ± 0.05). The nearly identical shapes of the upper portions of both curves suggest that hardness (compressive yield strength) and elastic modulus depend in the same way on the concentration of primary bonds. In fact for the points on the curve with NCR > 0.08 (mostly organic thermosetting resins or inorganic polymers) the ratio of BNH value and E has a constant value of 0.0625.

For values of NCR < 0.08 (thermoplastic materials) the correlation between mechanical properties and bond density is much less pronounced although the steep slopes of the shown curves tend to make the scatter of the actual data points less apparent. The main reason for the lack of good correlation certainly is the fact that *in high polymer solids the stiffness and strength can be utilized only to that extent permitted by intermolecular attraction and segment orientation* (cf. Table 1.1 for the differences in elastic moduli of chains, highly oriented fibers, and isotropic solids; in section E below and in Chapter 5 the role of orientation and intermolecular attraction on chain strain is discussed in detail).

In order to investigate the effect of the density of main chain bonds on the mechanical properties it is necessary to select experimental conditions such that

57

Fig. 3.8. Young's modulus E and Brinell hardness values BHN of various solids vs. volume concentration of primary bonds (after Holliday [70]).

comparable mechanical excitation of said bonds does indeed take place. According to Vincent [71] commensurable reference temperatures to measure a critical tensile strength are the brittle point temperatures, where brittle failure changes to a ductile failure. As indicated in Figure 3.9 he has related the number of backbone bonds per unit area (i.e., sample density times length of repeat unit over weight of repeat unit) to the critical tensile strength of (isotropic) polymers. A linear relationship exists between critical sample strength and bond packing density which does not seem to depend on the chemical nature of the bond. From the slope of the straight line the critical strength (in MN/m^2) of the isotropic polymers is obtained as 36.8 times the number of backbone bonds per nm^2. In highly oriented textile fibers this figure is much larger and amounts to between 100 and 250; in aromatic polyamide fibers (Kevlar 49) it reaches about 650. This should be compared with the theoretical value for fully oriented chains of infinite length which is about 3000, since the axial load to break a molecular chain of any one of the species listed in Figure 3.9 is of the order of 3 nN.

In other words, in isotropic thermoplastics at the point of brittle failure only a minute fraction – less than 1% – of all main chain bonds is fully strained. Under these conditions, onset of unstable crack propagation is determined by the magnitude of intermolecular attraction. Reinterpreting Vincent's data one must say that it is not the number of backbone bonds per unit area and their accumulated strength

Fig. 3.9. Critical tensile strength (strength at ductile-brittle transition) vs. chain cross-section (after Vincent [71]).

which determine macroscopic strength but the level of intermolecular forces; the latter benefit from a closer packing.

The effect of *molecular weight* on fiber strength has been widely investigated (cf. 20, 51, 52, 54, 72–74). The general observations have been briefly summarized and illustrated by Figure 1.13 for PA 6 fibers. In low molecular weight material chain slip occurs readily so that the sample strength depends solely on the strength of intermolecular attraction. Noticeable macroscopic strength is only obtained if the molecular weight is sufficient to permit physical cross-links through entanglements or chain folds between several chains.

In the following section it will be discussed in more detail to what extent the orientation distribution of chain segments, the ensuing distribution of axial chain stresses, and the gradual accumulation of defects through successive breakage of chains can be held responsible for the discrepancy between actual and theoretical strength.

E. Hsiao-Kausch

In the fracture theory of Hsiao-Kausch [60, 61] the state of orientation of a polymeric solid is explicitly recognized. The theory combines the kinetic concept of Zhurkov or Bueche and the theory of deformation of anisotropic solids developed by Hsiao [59]

59

for a model made up of rodlike elastic elements. The fracture theory is based on the simplifying assumption that the mechanical properties of the anisotropic solid are predominantly determined by the state of orientation and the properties of the (one-dimensional) elastic elements. The intermolecular interaction is neglected. In Chapter 2 the mathematical model (Fig. 2.12) has been described and the constitutive stress-strain relations (Eq. (2.5)) have been derived. The kinetic aspects are introduced into this model through the assumption that the rodlike elastic elements can undergo breakage, that the probability of failure of an element is determined by the axial stress acting upon that element, and that an immediate redistribution of the load carried up to then by the breaking element occurs. Again the models is characterized by the fact that it *distinguishes between the macroscopic state of stress (σ) and the microscopic (axial) stress Ψ of an element*. The axial stress of a one-dimensional elastic element deformed in a homogeneous strain field ϵ_{mn} is determined by its axial elongation:

$$\Psi = E \, \epsilon_{mn} s_m s_n.$$ (3.24)

The summation convention for repeated indices is used. E is the modulus of elasticity of the elastic elements, and s_m, s_n are components of the unit vector in the direction of orientation identified by ϑ and ϕ ($s_1 = \sin \vartheta \cos \phi$, $s_2 = \sin \vartheta \sin \phi$, $s_3 = \cos \vartheta$). For uniaxial stress, $\sigma_{33} = \sigma_0$, and transverse symmetry about the 33-axis Eq. (3.24) becomes:

$$\Psi(\vartheta, t) = E \, (\cos^2 \vartheta - \nu \sin^2 \vartheta) \epsilon_{33}(t)$$ (3.25)

where ν is Poisson ratio $\epsilon_{11}/\epsilon_{33}$.

In a partially or nonoriented system, the microscopic axial stresses are larger than σ_0 by a factor of up to E/E_λ where E_λ is the longitudinal modulus of elasticity of the subvolume containing the elastic element. If all subvolumes of a sample behave identically, then E_λ is the longitudinal modulus of the (partially) oriented sample. Naturally the largest axial stresses are experienced by those elements oriented in the direction of uniaxial stress. In a completely oriented system E and E_λ are equal by definition, and macroscopic and microscopic stress (in the direction of orientation) are equal also.

As stated above it is then first assumed in accordance with Zhurkov's kinetic considerations that elements break and secondly that the orientation-dependent rate of breakage of elements is given by Eq. (3.17) as

$$-\frac{d \, f(\vartheta, t)}{f \, d \, t} = k(\vartheta) = k_0 \exp(\beta \Psi / RT)$$ (3.26)

where β is the activation volume for breakage of the elastic elements in question and $f = N(\vartheta, t)/N(\vartheta, 0)$ the fraction of unbroken elements.

Thirdly it is assumed that after breakage of an element mechanical equilibrium is established within the affected subvolume and that strain ϵ_{33} increases. This means that the elastic elements of all orientations take over an appropriate share of the load carried previously by the broken element to balance the load acting on the boundaries of the subvolume. It has been shown in Chapter 2 that the stress tensor acting on a small volume element of the elastic network can be calculated once the orientation distribution function, ρ, is known

$$\sigma_{ij} = \int \rho(\vartheta)\, f(\vartheta, t)\, s_i\, s_j\, \Psi(\vartheta, t)\, d\Omega \tag{3.27}$$

where $d\Omega$ is the infinitesimal solid angle.

In uniaxial stressing, mechanical equilibrium within the elastic network is established if

$$\sigma_{11} = \sigma_{22} = 0 = \int_0^{\pi/2} E\rho(\vartheta)\, f(\vartheta, t)\, [\epsilon_{11} \sin^2 \vartheta + \epsilon_{33} \cos^2 \vartheta]\sin^3 \vartheta\, d\vartheta \tag{3.28}$$

and

$$\sigma_{33} = 2\pi \int_0^{\pi/2} E\rho(\vartheta)\, f(\vartheta, t)\, [\epsilon_{11} \sin^2 \vartheta + \epsilon_{33} \cos^2 \vartheta]\cos^2 \vartheta \sin \vartheta\, d\vartheta. \tag{3.29}$$

The system of Eqs. (3.26), (3.28), and (3.29) has been evaluated for a completely oriented and an unoriented network under constant uniaxial stress σ_0. The case of a completely oriented network is that treated by Tobolsky and Eyring. All elements are thought to be subjected to the same stress Ψ which increases inversely proportional to the decreasing number of unbroken elements. Breakdown of the subvolume occurs with breakage of the last element within the subvolume, i.e., after f has reached zero. The time to breakdown of the subvolume, t_b, is taken from Eqs. (3.20), (3.21), and (3.26) as

$$k_0 t_0 = - \text{Ei}\,(-\beta\Psi_0 /RT) \tag{3.30}$$

where Ψ_0 is the initial stress carried by elements of orientation $\vartheta = 0$.

In the case of random network orientation the system of Eqs. (3.26), (3.28), and (3.29) cannot be solved analytically. An approximation method has been proposed, however [60]. Without repeating details of the method it may be said that the changes of f for a time interval Δt as calculated from Eq. (3.26) are being fed back into Eqs. (3.28) and (3.29) giving the corresponding changes of $\epsilon_{11}(t)$ and $\epsilon_{33}(t)$, which in turn determine the changes of $\Psi(t)$. The iterative integration of Eqs. (3.26), (3.28), and (3.29) can be carried out numerically.

Two results of these calculations seem to be of interest with respect to the effect of network orientation on defect accumulation and strength: that the range of orientation angles within which elements break preferentially is narrow and that the increase in strength resulting from improved uniaxial orientation is limited. The first effect is illustrated in Figure 3.10 for a randomly oriented network where the initial distribution $\rho = 1/2\,\pi$, of elements is independent of ϑ. Elements oriented in the direction of uniaxial stress (cos $\vartheta = 1$) experience the largest initial stress (Ψ_0), and they break preferentially. The angular distribution of f (cos ϑ, TIME) is shown for three different stages of defect development, namely after f (1, TIME) has dropped to 0.5 and zero, respectively, and for the point of impending network failure — where the numerical calculations had been terminated because breakage

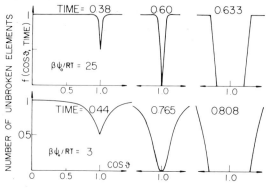

Distribution of Elements

Fig. 3.10. Decay within a population of randomly oriented elements; at high stresses ($\beta \Psi_0 > 10$ RT) essentially those elements oriented in the direction of uniaxial stress break (upper graph). TIME = linear time scale according to Eq. (3.31) (Kausch, Hsiao [60]).

of the remaining elements occurred at too high a rate. For ease of comparison the time scale is given in terms of TIME where

$$\text{TIME} = t/t_z = k_0 t \exp(\beta \Psi_0/RT) \tag{3.31}$$

and t_z the time to failure according to Zhurkov's Eq. (3.23). It should be noted that more than one-half of the loading period has to pass before one-half of the most heavily stressed elements is broken. More than 90% of the loading period has to elapse before all elements within a small angular section are broken. At this second stage the majority of all elements is still unbroken. The step from the second to the third stage of defect accumulation is rather short ($< 6\%$ of loading time) and from there to network failure practically negligible. It should also be noted that TIME (failure) of a randomly oriented network composed of successively breaking interacting elements is only somewhat smaller than the time given by the empirical Eq. (3.23) where TIME (failure) for a stress Ψ_0 is unity; on the other hand it is larger than $RT/\beta \Psi$, the value predicted by the exponential integral (Eq. (3.22)) for a completely oriented network loaded to such an extent that again the elements experience the same initial stress Ψ_0.

The numerical calculations according to the above model resulted in an estimate of the time to failure of arbitrarily oriented networks under constant load:

$$k_0 t_b = -[\text{Ei}(-\beta \Psi_0/RT)]C \tag{3.32}$$

where C is a slowly varying correction term the limits of which are expressed by:

$$1 \leq C \leq 1 + 0.63 \, \beta \Psi_0/RT. \tag{3.33}$$

The lower limit refers to completely oriented, the upper to randomly oriented networks. With C being determined it has become possible to compare Eqs. (3.23) and (3.32) and to express the structure sensitive kinetic parameter γ in terms of the element properties. In the range of validity of Eq. (3.22) one obtains for a fully oriented network:

$$\gamma = \beta \frac{\Psi_0}{\sigma_0} \left(1 - \frac{RT}{\beta \Psi_0} \ln \frac{RT}{\beta \Psi_0} \right) \tag{3.34}$$

and for a randomly oriented one

$$\gamma = \beta \frac{\Psi_0}{\sigma_0} \left(1 - \frac{RT}{\beta \Psi_0} \ln 0.63 \right). \tag{3.35}$$

The term in parentheses is an orientation-dependent and minor kinetic contribution derived from the assumption of successive failure of many elements (in the technically relevant range of $\beta \Psi_0$ values from 40 to 200 kJ/mole the term in parentheses assumes values between 1.17 and 1.0). If the nature of the breaking elements is not changed during sample treatment or fracture development, then β can be considered to be constant. The major effect on γ is derived from the local stress concentration Ψ_0/σ_0 which here is equal to the stiffness ratio E/E_λ. This theory essentially predicts, therefore, that the increase in strength is equal to the increase in stiffness; it is based on the presumption that the elements break indeed at a critical local strain (kinetic version of St. Venant's criterion of maximum strain). A different interpretation of the strength of oriented samples which is based on the rotation of flaws has been presented in an earlier section of this chapter.

In Figure 3.11 the experimental data obtained by Regel and Leksovskii [75] for the lifetimes of partially oriented PAN fibers are compared with theoretical curves derived from Eq. (3.32). It should be emphasized that the gain in strength

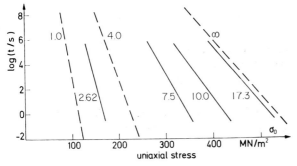

Fig. 3.11. Time to fracture of differently oriented PAN fibers. ——— experimental curves (Regel et al. [75]), – – – – theoretical curves (Eq. 3.32), parameter: draw-ratio λ.

through improved orientation of the PAN fibers (or of their model representation) amounts to a factor of $\Psi_0/\sigma_0 = 5$. Similar values of between 2 and 5 for the increase in strength with orientation have been reported for PE, PP, PS, PVC, PMMA, or PA [51, 54]. It should be pointed out that the rather limited increase in stiffness in these experiments indicates that the orienting "elements" are not simply highly extended chain segments but larger molecular domains of limited anisotropy. This does not preclude the possibility that breakage of an element is essentially identical to breakage of the most highly stressed chain molecule(s). This will be the case if the breaking chain(s) carry practically the total load of an element or initiate its breakage.

A comparison of different reaction rate models was made by Henderson et al. [76] who found that the bond rupture theories (Tobolsky-Eyring, Zhurkov-Bueche) gave better agreement with fracture data on filled polybutadiene and filled and unfilled PVC than the bond slippage hypothesis. Primary uncertainties exist, however, with respect to a repair reaction and to the temperature dependence of β and γ. The question of the true meaning of the activation volumes β and γ, of the stress ratio Ψ_0/σ_0, and of the role of chain stretching and chain scission in macroscopic fracture will be resumed in Chapters 7 to 9.

F. Gotlib, Dobrodumov et al.

In none of the previously treated molecular fracture theories (A − E) allowance has been made for possible stress concentration in the neighborhood of a breaking element. The first ruptures in a large ensemble of initially equally stressed molecules certainly will occur at random within the stressed volume. The breakage and retraction of a particular element must lead, however, preferentially to an increase in axial stresses of those elements to which it is coupled (through secondary forces). The probability density for rupture of those elements will, therefore, be slightly higher than that for others. But the higher probability density applies only to the small number of elements in the vicinity of the existing fracture points. So the (integrated) probability for rupture of one of these elements is initially smaller than the probability for rupture of one of the large number of chains hitherto not affected by a ruptured element. The rupture events, therefore, will continue for some time to occur at random [67]. With their growing number, chances are increasing that rupture events occur in an immediate vicinity of each other thereby forming the nucleus of a crack and enhancing the fracture probability of neighboring elements.

Whenever the random accumulation of defects in one site has grown to such a size that subsequent fractures will preferentially occur at that site, one has reached the end of the phase of fracture initiation and the beginning of (thermal) crack growth.

In Figure 3.12 a computer simulation by Dobrodumov et al. of the fracture development within a regular model lattice is shown [58]. The authors have taken the ratio of the elastic constants in and perpendicular to the direction of orientation as 10:1 and an element fracture probability as given by Eq. (3.23). The distribution of local stresses was determined by a numerical iteration procedure. It is readily observed that for small loads ($\beta\Psi_0/RT = 5$) a large number of isolated defects appear before (from the accidental accumulation of a few defect points) a "crack" starts growing. At higher loads ($\beta\Psi_0/RT = 20$) a smaller number of defect points initiate

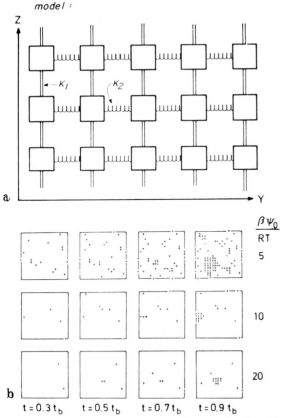

model :

Fig. 3.12 a and b. Simulation of brittle fracture of a three-dimensional network model (after [58]). Upper part: model representation, elastic constants $K_1:K_2 = 10$. Bottom part: successive fracture events in x-y plane as a function of t/τ (τ = time to complete failure); note the increase in cooperativeness with $\beta \Psi_0/RT$ similar to that in Figure 3.10.

a crack and very little damage is done to the remainder of the plane during fracture initiation. These results are qualitatively similar to those of Hsiao and Kausch shown in Figure 3.10. From these studies the authors [57, 67] conclude that in polymers only weak stress concentration on a molecular scale can exist (axial stresses adjacent to a site comprising j defects should be larger by a factor of $\sqrt[8]{j}$ than the stress further away). It should be noted that this model — as the previous one — strictly operates with elastic elements and with their rupture as the only way to relieve local stress. The application of these models, therefore, seems to be restricted to the brittle fracture of organic glasses or to other systems not involving large anelastic or plastic deformation in fracture initiation.

G. Bueche — Halpin

The limited viscoelastic extensibility of rubber strands determines according to the theory of Bueche and Halpin [69] the fracture of elastomers. The authors assume

65

that a large number of filaments at the tip of a growing crack have to be strained successively to their critical elongation λ_c. The sample will fracture at a gross strain λ_b after q filaments have been broken in a time $t_b = qt'$. The values λ_b and λ_c are connected through the creep function of the material and the stress concentration factor. The theory allows one to calculate the elongation at break, λ_b, from the knowledge of the creep function. It does not take into consideration any change of stress concentration with growing crack length or the decrease of t', the time necessary to fracture one filament, with continuing creep of the sample. It is assumed that all filaments will have to be stretched from practically zero elongation to λ_c. That will, in the first place, affect any numerical values of q which may be calculated from fitting experimental failure envelopes. A group of q filaments subject to the statistical condition that fracture of one filament may start once the fracture of the preceding one is completed has an average lifetime, $<t_b>$, of qt' and a Poisson distribution of t_b:

$$p(t_b)\, dt_b = \frac{dt_b}{t'(q-1)!} \left(\frac{t_b}{t'}\right)^{q-1} \exp(-t_b/t'). \qquad (3.36)$$

The same concept of filament deformation was independently employed by Bartenev et al. [77] to the opening and growth of crazes (silver cracks) in PMMA.

The molecular theories treated in this chapter have considered the thermally activated breakage of elements or filaments as a source of crack initiation and macroscopic failure. It is the purpose of the following chapters to investigate the mechanical strength of primary bonds and of molecular chains and to study the occurrence of chain breakages. With that information available it will be possible to resume the discussion on the nature of the "elements" used in molecular theories of fracture.

References for Chapter 3

1. R. M. Bethea, B. S. Duran, T. L. Boullion: Statistical Methods for Engineers and Scientists, New York: Marcel Dekker 1975.
2. A. Hald: Statistical Theory with Engineering Applications, New York: Wiley 1952.
3. H. H. Kausch: Materialprüf. 6, 246 (1964).
4. T. Yokobori: Kolloid-Z. 166, 20 (1959).
5. S. Kawabata, P. J. Blatz: Rubber Chem. Technol. 39, 923 (1966).
6. S. Kawabata, S. Tatsuta, H. Kawai: Proc. Fifth Internat. Congr. Rheology, Kyoto, Japan 3, 111 (1968).
7. B. D. Coleman: J. Appl. Phy. 29, 968 (1958).
8. B. D. Coleman, D. W. Marquardt: J. Appl. Phys. 28, 1065 (1975).
9. B. Epstein: J. Appl. Phys. 19, 140 (1948).
10. B. Ya. Levin, A. V. Savickij: Chim. Volokna 5, 48 (1968).
11. H. H. Kausch: Proc. Internat. Conf. Mech. Behavior of Materials, Kyoto, Japan, Aug. 1971, Vol. III, pg. 518.

12. L. J. Knox, J. C. Whitwell: Text. Res. J. *41*, 510 (1971).
13. R. F. Fedors, R. F. Landel: Trans. Soc. Rheol. *9*, 195 (1965).
14. S. Kase: J. Polym. Sci., *11*, 424 (1953).
15. S. Kaufmann: Faserforsch. u. Textiltechn. *22*, 406 (1971) and ibid. 463 (1971).
16. K. E. Hansen, W. C. Forsman: J. Polym. Sci., A-2, *7*, 1863 (1969).
17. D. A. Stuart, O. L. Anderson: J. Am. Ceramic Soc. *36*, 416 (1953).
18. H. H. Kausch: J. Polym. Sci., C. (Polymer Symposia) *32*, 1 (1971).
19. A. Tobolsky, H. Eyring: J. Chem. Phys. *11*, 125 (1943).
20. I. M. Ward: Mechanical Properties of Solid Polymers, London: Wiley-Interscience 1971, p. 280.
21. A. N. Gent, P. B. Lindley: Proc. Roy. Soc. (London) *A249*, 195 (1968).
    W. K. Ko, Ph. D. Dissertation, California Institute of Technology, 1963.
    A. E. Oberth, R. S. Bruenner: Trans. Soc. Rheology *9*, 165 (1965).
22. C. K. Lim, Ph. D. Thesis: The Pennsylvania State University, University Park, PA, 1967.
23. L. J. Broutman, S. M. Kirshnakumar, P. D. Mallick: J. Appl. Polym. Sci. *14*, 1477 (1970).
24. R. E. Ely: Polymer Engng. and Sci. *7*, 40 (1967).
25. R. A. Dickie, T. L. Smith: J. Polym. Sci. A-2, *7*, 687 (1969).
26. J. C. Bauwens: J. Polymer Sci. A-2 *8*, 893–901 (1970).
27. R. Raghava, M. Caddell, G. S. Y. Yeh: J. Materials Sci. *8*, 225–232 (1973).
28. R. G. Dong: Trans. Soc. Rheol. *18*, 45 (1974).
29. N. W. Tschoegl: J. Polym. Sci. C *32*, 239–267 (1971).
30. T. L. Smith: ibid., 269–282.
31. S. V. Radcliffe: Deformation and Fracture of High Polymers, H. H. Kausch, J. A. Hassell, R. I. Jaffee, eds. New York: Plenum Press 1973, p. 191–208.
32. J. A. Sauer, K. D. Pae: Colloid a. Polymer Sci. *252*, 680–695 (1974).
33. W. Schneider, R. Bardenheier: Z. Werkstofftechnik *6*, 269–280 and 339–348 (1975).
34. W. Retting: Angew. Makromol. Chem. *58/59*, 133 (1977).
35. A. A. Griffith: Phil. Trans. Roy. Soc. *221*, 163 (1921).
36. M. L. Williams: Fracture of Solids, D. C. Drucker and J. J. Gilman, eds. New York: Interscience, 1963, p. 157.
37. W. G. Knauss: Appl. Mechanics Rev., January 1973, p. 1.
38. M. L. Williams: J. Adhesion *4*, 307 (1972).
39. G. P. Anderson, K. L. DeVries, M. L. Williams: Int. J. Fracture *9*, 421 (1973).
40. D. H. Kaelble: J. Appl. Polymer Sci. *18*, 1869–1889 (1974).
41. T. L. Smith: Rheology, Vol. 5, F. R. Erich, ed., New York: Academic Press 1969, p. 127.
    T. L. Smith, IV, V, and VI Microsymposia on Macromolecules, Main Lectures, London: Butterworths, 1970, p. 235–253.
42. P. J. Blatz, S. C. Sharda, N. W. Tschoegl: Trans. Soc. Rheol. *18*, 145 (1974).
43. S. Glasstone, K. J. Laidler, H. Eyring: The Theory of Rate Processes, New York: McGraw Hill 1941.
44. A. Tobolsky, H. Eyring: J. Chem. Phys. *11*, 125 (1943).
45. S. N. Zhurkov, B. N. Narzullayev: Zhur. Techn. Phys. *23*, 1677 (1953).
46. S. N. Zhurkov, E. Ye. Tomashevskii: Zhur. Techn. Phys. *25*, 66 (1955).
47. S. N. Zhurkov, T. P. Sanfirova: Dokl. Akad. Nauk SSSR *101*, 237 (1955).
48. F. Bueche: J. Appl. Phys. *26*, 1133 (191955).
49. F. Bueche: J. Appl. Phys. *28*, 784 (1957).
50. F. Bueche: J. Appl. Phys. *29*, 1231 (1958).
51. G. M. Bartenev, Yu. S. Zuyev: Strength and Failure of Viscoelastic Materials, Oxford: Pergamon Press 1968, p. 143ff.
52. E. H. Andrews: Fracture in Polymers, Edinburgh: Oliver and Boyd 1968.
53. V. R. Regel, A. I. Slutsker, E. E. Tomashevskii: Kinetic Nature of Durability of Solids (in Russian), Isdat. Nauka, Moscow 1974.
54. A. I. Gubanov, A. D. Chevychelov: Fiz. Tverd. Tela *4*, 928–933 (1962), Sov. Phys.-Solid State *4*, 681–684 (1962).

55. G. M. Bartenev: Mekh. Polim.(USSR) 6, 458–464 (1970), Polym. Mech. (USA) 6, 393–398 (1970).
56. K. E. Perepelkin: Fiz. Khim. Mekh. Materialov 6, 78–80 (1970).
57. S. I. Mikitishin, Yu. N. Khonitskii, A. N. Tynnyi: Fiz. Khim. Mekh. Materialov 5, 69–74 (1969).
58. A. V. Dobrodumov, A. M. El'yashevich: Fiz. Tverd, Tela 15, 1891–1893 (1973). Sov. Phys. – Solid State, 15, 1259–1260 (1973).
59. C. C. Hsiao: J. Appl. Phys. 30, 1492 (1959).
60. H. H. Kausch, C. C. Hsiao: J. Appl. Phys. 39, 4915–4919 (1968).
61. H. H. Kausch: Kolloid-Z. Z. Polymere 236, 48–58 (1970).
62. W. Holzmüller: Kolloid-Z. Z. Polymere 203, 7–19 (1965).
63. G. M. Bartenev, I. V. Razumovskaya: Fiz. Khim. Mekh. Materialov 5, 60–68 (1969).
64. R. L. Salganik: Int. J. Fracture Mech. 6, 1–5 (1970).
65. A. N. Orlov, V. A. Petrov, V. I. Vladimirov: phys. stat. sol. (b) 47, 293–303 (1971).
66. B. D. Goikhman: Strength Mater. (USA) 4, 918–922 (1972).
67. Yu. Ya. Gotlib, A. V. Dobrodumov, A. M. El'yashevich, Yu. E. Svetlov: Fiz. Tverd. Tela 15, 801 (1973), Sov. Phys. – Solid State, 15, 555–559 (1973).
68. K. C. Valanis: Report G 378 – ChME – 001 (1974), University of Iowa, Iowa City, USA.
69. F. Bueche, J. C. Halpin: J. Appl. Phys. 35, 36–41 (1964)
70. L. Holliday, W. A. Holmes-Walker, J. Appl. Polym. Sci. 16, 139–155 (1972).
71. P. I. Vincent: Polymer 13, 557–560 (1972).
72. D. C. Prevorsek: J. Polym. Sci. C 32, 343–375 (1971).
73. V. M. Poludennaja, N. S. Volkova, D. N. Archangel'skij, A. G. Zigockij, A. A. Konkin: Chim. Volokna 5, 6–7 (1969).
74. H. Krässig: Textilveredelung 4, 26–37 (1969).
75. V. R. Regel, A. M. Leksovsky: Int. J. Fracture Mechanics 3, 99 (1967).
76. C. B. Henderson, P. H. Graham, C. N. Robinson: Int. J. Fracture Mechanics 6, 33–40 (1970).
77. G. M. Bartenev, I. V. Razumovskaya: Fiz. Tverd. Tela 6, 657–661 (1964), Sov. Phys. – Solid State 6, 513–516 (1964).
78. H. H. Kausch: J. Polym. sci. C 32, 1–44 (1971).
79. S. I. Mikitishin, A. N. Tynnyi: Soviet materials sci. 4/3, 187–188 (1968).
80. S. S. Sternstein, J. Rosenthal: Toughness & Brittleness of Plastics, Adv. Chemistry Series, A.C.S., Washington, D. C. (1974).
81. W. Retting: Orientation in Polymers Programme, Polystyrene, Rep. Results of BASF, 2nd part, March 1972, Macromol. Division, I.U.P.A.C.
82. I. Narisawa, T. Kondo: J. Polymer Sci. – Polymer Phys. Ed. – 11, 1235–1245 (1973).
83. P. Stockmayer: Stuttgarter Kunststoff-Kolloquium 1977, Kunststoffe 67, 470 (1977).
84. H. H. Kausch: Materialprüfung.
85. J. W. S. Hearle: J. Textile Inst. 68/4, 155–157 (1977).
86. A. V. Savitskii, V. A. Mal'chevski, T. P. Sanfirova, L. P. Zoshin: Polymer Sci. USSR 16, 2470–2477 (1974), Vysokomol. Soyed. A 16, 2130–2135 (1974).

Chapter 4

Strength of Primary Bonds

I. Covalent Bonds

In the last chapter, in the section on molecular theories of fracture, almost throughout an Arrhenius equation has been used to describe the activation of element breakage. The energy of activation, U_0, frequently turned out to be equal to (or was
assumed to be equal to) the dissociation energy of the weakest main chain bond.
Before further analyzing the kinetics of element — and possibly chain — breakage
a definition of the mechanical strength of a bond and of a chain have to be given.
In order to do this basic results of quantum chemistry [1, 2] are recalled in this
chapter which concern the "strengths" of intramolecular bonds and factors in-
fluencing the binding potential such as electronic excitation or ionization.

A. Atomic Orbitals

Covalent chemical bonds between atoms of the same or a different species rely on
the interaction of the outermost — or valence — electrons. Even though one speaks
of electrons one should rather think of electron clouds, i.e. of electronic density
distributions. The radial and angular distribution of the electron density is described
by "one electron" wave functions Ψ — also called atomic orbitals — which are
derived as a solution of the quantum mechanical Schrödinger equation:

$$\left(-\frac{h^2}{8\pi^2\, m_r}\, \nabla^2 + V\right)\Psi = E\,\Psi. \tag{4.1}$$

Here h is Planck's constant, m_r the reduced mass of an electron with charge e, ∇ the Nabla operator, and E the energy eigenvalue of an electron moving within a potential V. The term in brackets constitutes the Hamiltonian operator **H**.

The Schrödinger equation has a solution only for certain discrete energy values E. Different values of E mean different forms of the electronic density distributions which are characterized by the principal quantum number n, the angular quantum number ℓ and the magnetic quantum number m.

For the simplest atomic system, the hydrogen atom, which consists of one proton and one electron the potential V is $-e^2/r$. In this case energy levels are obtained, which are degenerate, i.e. the energy levels for different l and m coincide. The corresponding wave functions, however, do depend on the three quantum numbers n, l and m. It should be noted that the radial distribution, $R(r)$, and the angular distribution $\Theta(\theta,\phi)$ may be separated:

$$\Psi_{n\ell m} = \Theta_\ell\,(\theta,\phi)\,R_{n\ell}(r). \tag{4.2}$$

The quantum numbers are subject to the condition that

$\ell = 0, 1, 2 \ldots n-1$
$m = 0, \pm 1, \pm 2 \ldots \pm\ell.$

It is customary to denote the atomic orbitals by the spectroscopical terms s, p, d ... if ℓ is 0, 1, 2 ... respectively.

The first few wave functions for a hydrogen-like atom with a charge of $+Ze$ of the nucleus then have the following form:

1s $\quad R_{10}(r) = (Z/a_0)^{3/2}\, 2\exp(-Zr/a_0)$ (4.3)
2s $\quad R_{20}(r) = (Z/a_0)^{3/2}\, (2 - Zr/a_0)\,(\exp - Zr/2a_0)\,(1/2\sqrt{2})$ (4.4)
2p $\quad R_{21}(r) = (Z/a_0)^{3/2}\, (Zr/2a_0\,\sqrt{6})\exp(-Zr/2a_0)$ (4.5)

and

s $\quad \Theta_{00}\,(\theta,\phi) = 1/2\sqrt{\pi}$ (4.6)
p_x $\quad \Theta_{lx}\,(\theta,\phi) = (\sqrt{3}/2\,\sqrt{\pi})\cos\phi\,\sin\theta$ (4.7a)
p_y $\quad \Theta_{ly}\,(\theta,\phi) = (\sqrt{3}/2\,\sqrt{\pi})\sin\phi\,\sin\theta$ (4.7b)
p_z $\quad \Theta_{lz}\,(\theta,\phi) = (\sqrt{3}/2\,\sqrt{\pi})\cos\theta.$ (4.7c)

Here a_0 is the radius of the first Bohr orbit in hydrogen ($0.529 \cdot 10^{-10}$ m). According to the Pauli exclusion principle every atomic orbital may be occupied by one or two electrons which in the latter case have to have their spins aligned antiparallel to each other (paired electron spins).

For many-electron atoms an exact analytical solution of the Schrödinger equation (Eq. 4.1) is not possible. Nevertheless atomic orbitals may be obtained by an iterative procedure, which starts by placing the appropriate number of electrons on hydrogen-like one-electron orbitals. With the potential obtained from such a charge distribution a better approximation of the first orbital can be calculated which in turn is considered in the recalculation of the second orbital and so on until no further corrections become necessary. Again these orbitals are denoted by 1s, 2s, 2p, 3d if the quantum numbers n and ℓ are 1 and 0, 2 and 0, 2 and 1, 3 and 2 and so on.

In a many-electron atom – as for instance Li – one finds a particular electron in any of the accessible orbitals (1s, 2s) and in any of the two possible states of spin (m_s = + 1/2 or –1/2). Since electrons are indistinguishable the energy of the atom must not be affected by interchanging the coordinates of any pair of electrons. The wave function belonging to one eigenvalue E will have to encompass, therefore, all terms resulting from an interchange of electron numbering.

In the case of Li the ground state has the wave function:

$$\Psi = \begin{vmatrix} 1s\,\alpha\,(1) & 1s\beta\,(1) & 2s\,\alpha\,(1) \\ .\ . & .\ . & .\ . \\ .\ \ . & .\ \ . & 2s\,\alpha\,(3) \end{vmatrix} \qquad (4.8a)$$

or

$$\Psi = 1s\,\alpha\,(1)\,1s\beta\,(2)\,2s\,\alpha\,(3) - 1s\,\alpha\,(2)\,1s\beta\,(1)\,2s\,\alpha\,(3) + \ldots . \qquad (4.8b)$$

Each of the six terms of the fully written determinant (4.8b) contains an orbital and a spin function for each of the three electrons. The spin function $\alpha\,(1)$ may denote that the electron 1 has a spin quantum number m_s = +1/2, $\beta\,(1)$ corresponds to m = –1/2.

B. Hybridization

It can easily be shown that an atom with one electron in each of the 2s-, $2p_x$-, $2p_y$-, and $2p_z$-orbitals has a total charge density distribution which is completely spherical, i.e. independent of Θ and ϕ. The s- and p-atomic orbitals are not the only functions that lead to a spherical distribution of the total charge. A large number of suitable linear combinations of the s- and p-orbitals do have the same property. These linear combinations are called hybrid orbitals or hybrids. Depending on the number and kind of atomic orbitals used in hybridization the hybrids are denoted as sp, sp^2, sp^3, dsp etc. The four normalized and orthogonal sp^3-hybrids (of a C-atom) have the form:

$$\Psi_1 = \frac{1}{\sqrt{4}}\,(2s + 2p_x + 2p_y + 2p_z) \qquad (4.9a)$$

$$\Psi_2 = \frac{1}{\sqrt{4}}\,(2s + 2p_x - 2p_y - 2p_z) \qquad (4.9b)$$

$$\Psi_3 = \frac{1}{\sqrt{4}}\,(2s - 2p_x - 2p_y + 2p_z) \qquad (4.9c)$$

$$\Psi_4 = \frac{1}{\sqrt{4}} (2s - 2p_x + 2p_y - 2p_z) \ . \tag{4.9d}$$

The individual hybrid is, of course, non-spherical, i.e. the electron density distribution has a characteristic shape. In this particular case the directions of maximum electron density point into the corners of a regular tetrahedron.

The electronic energy of a C-atom with electrons in hybrid orbitals is higher than that of the ground state. In carbon the ground state will be formed by two electrons in each of the lowest orbitals: $1s^2\ 2s^2\ 2p^2$. Hybridization involves the change from pyramidal orbitals (p^3) to tetragonal ones (sp^3); it requires, that three p-orbitals be occupied. The first step is, therefore, that one electron be promoted from 2s to 2p which necessitates an energy of promotion. The effect of promotion and hybridization is, however, that four rather than two electrons are available for bonding. The gain in binding energy of the two additional C—H bonds (90 to 100 kcal/mol per C—H bond) more than balances the energy expended for promotion and hybridization.

C. Molecular Orbitals

Within the limited scope of this book a mere terminology of molecular orbital theory can be supplied — intended only for the discussion of bond energies. Rigorous quantum mechanical calculations for any molecule being more complex than the hydrogen molecule are not feasible. Therefore, approximate methods for calculation of the orbitals of the bonding electrons have been developed [1, 2]. The two most important ones are the *molecular orbital* (MO) method relating to the work of Hund, Mulliken, and Hückel [1a] and the *bond orbital* method, the Heitler-London-Slater-Pauling (HSLP) method [1b]. In the MO approach electrons are placed in *one-electron*-wave functions which belong to the molecule as a whole with each nucleus in its proper configuration. In its crudest form these wave functions are obtained as a *linear combination of the atomic orbitals* (LCAO) of all atoms forming the molecule [2a].

In bond orbital theory the wave functions of the molecule are derived from the wave functions associated with the different bonds of a molecule, i.e. from the bond orbitals. In the LCBO model the molecular orbitals are a *linear combination of bond orbitals* which in turn are a combination of the atomic orbitals or hybrids forming the bond in question [2b]. In the HLSP-method the *many-electron* wave functions are a sum of product functions which contain a Heitler and London type of factor (space function and spin function) for each bond of the molecule.

Bond orbital and molecular orbital approach lead to the common conclusion that bonding will occur only at directions where the atomic orbitals have large values and only if the atoms are close enough to each other so that the atomic orbitals overlap, i.e. if the "electron clouds" partly penetrate each other.

A covalent bond is associated with an electron pair and hence subject to the quantum mechanical phenomenon of electron exchange: in a molecule A—B the electron from A may also be found near the nucleus of atom B and vice versa. It is this additional exchange energy term which significantly deepens the interatomic potential and leads to a distinct potential minimum – prerequisite for a stable binding. Thus a bond can be formed if a bonding molecular orbital exists and if it is occupied by an electron or an electron pair. In the ground energy state of a molecule paired electrons occupy the molecular orbitals with the lowest energy.

D. Multiple Bonds

So far multiple bonds between atoms of the same or a different species such as carbon, nitrogen, and oxygen have not been explicitly discussed). At this point attention should be drawn to the fact that the two (or three) molecular orbitals involved in forming a double (or triple) bond are usually not of the same kind but of different character. In ethylene $(H_2C=CH_2)$ all six atoms lie in one plane. Therefore, one may take three sp^2-hybrid orbitals of each C-atom to form the four C–H bonds and the one axially symmetric C–C bond, the σ bond.

The two remaining p_z orbitals are combined into the so called (molecular) π bond orbital. Since the p_z orbitals are directed perpendicular to the π bond which they are forming their overlap is smaller than that of orbitals forming a σ bond. Also the regions of maximum overlap are not in the xy plane between the C-atoms but rather above and below this plane. Because of this orientation of the p_z orbitals the interaction between conjugated π bonds is so strong that in fact it leads to a delocalization of the π electrons. In view of the geometry of the π bond it is easily realized that rotation of a group around multiple bonds is forbidden [1c].

II. Bond Energies

A. Electronic Energy and Heat of Formation

The term bond energy requires a definition. The potential energy-scheme of an arbitrary bond A–B in a polyatomic molecule given in Figure 4.1 serves to illustrate this point. As pointed out previously one has to take into account that in polyelectronic atoms the valency state of the atom may lie above the respective ground state. If this is the case no binding can occur between two atoms in their ground states; if they encounter each other their potential energy will rise. At some interatomal distance the potential energy of the system will approach the energy of the atoms in their valency states (Fig. 4.1, dashed line) and a transition to the binding state can occur. The intrinsic binding energy (wahre Bindungsenergie) E is, therefore, the difference between the energy of the molecular ground state and that of the valency states at infinite interatomal distance. The energy of dissociation D is smaller than E, namely by the zero point energy (Nullpunktsenergie) $h\nu/2$ of vibration and by the sum P of energies of promotion, hybridization, and polar and steric rearrangements necessary to attain the valency state. The difference between zero point energy and the maximum of the potential energy curve is the activation

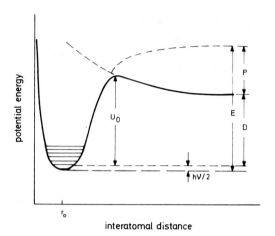

Fig. 4.1. Scheme of the potential energy of an arbitrary bond in a polyatomic molecule.
D energy of dissociation, E intrinsic binding energy, P energies of promotion, hybridization
and of polar and steric rearrangements, U_0 activation energy of bond dissociation, r_0 equilibrium
bond length, $h\nu/2$ zero point energy of vibration.

energy U_0 of bond dissociation. In diatomic molecules U_0 and D generally have the
same value and represent the bond energy.

In molecules with three or more atoms the assignment of a specific increment
of energy to a particular bond is complicated by the presence of the above mentioned
energy term P for promotion and electron reorganization, by the occurrence of delo-
calized electrons, and by the fact that the dissociation energies of successively broken
identical bonds are different. The latter statement may be illustrated by the disintegration of CH_4. It must be assumed that the four hydrogen atoms in CH_4 are equivalent, all C–H bonds are formed by hydrogen 1s orbitals and carbon $2sp^3$ orbitals.
Abstraction of the first hydrogen atom requires an energy of 425.8 kJ/mol
(101.7 kcal/mol) and changes the methane atom into the methyl radical CH_3. From
symmetry considerations confirmed by ESR investigations it has been concluded
[4] that the methyl radical is planar, with the unpaired electron undoubtedly in a
2p orbital of the C-atom. The bond character of the C–H bonds remaining in what
now is the methyl radical has changed, therefore, from the tetrahedral configuration
of sp^3 hybrid orbitals into the planar one of sp^2 hybrids. The dissociation energy
of one of these C–H bonds is 477.3 kJ/mol (114 kcal/mol). The third hydrogen
has a dissociation energy of 418.7 (100 kcal/mol), the fourth of 341.2 kJ/mol
(81.5 kcal/mol). The total energy gained in forming CH_4 from its constituent atoms
is, therefore, 1663 kJ/mol (397.2 kcal/mol), the average C–H bond energy
415.7 kJ/mol (99.3 kcal/mol) [5]. It is this average derived from the energy of formation of the molecule which is usually termed bond energy (mittlere Bindungsenergie).

The bond energies of saturated and unsaturated homologues of *hydrocarbons*
may be treated in some detail since they will give us some understanding of the correlation of bond energies in *polyolefins*. In saturated hydrocarbons with localized
electrons it is reasonable to equate the energy of a bond with the energy of the

electrons forming that particular bond. Theoretically the electronic energy of a molecule may be calculated using any of the mentioned approximations, for instance the LCBO method [2b, 6]. In that case the *molecular* orbitals for the 2N σ-electrons of a hydrocarbon with N bonds may be written as a linear combination of N bond orbitals ϕ_i:

$$\Psi_j = \sum_{i=1}^{N} c_{ij} \, \phi_i. \tag{4.10}$$

One assumes that the wave functions Ψ_j are a solution of the equation

$$\mathbf{H} \, \Psi_j = E \, \Psi_j \tag{4.11}$$

or

$$(\mathbf{H} - E) \, \Psi_j = 0. \tag{4.12}$$

If one defines the notations

$$H_{ij} = \int \phi_i \, \mathbf{H} \, \phi_j \, dv \qquad \text{(resonance integral)} \tag{4.13a}$$
$$S_{ij} = \int \phi_i \, \phi_j \, dv \qquad \text{(overlap integral),} \tag{4.13b}$$

introduces Eq. (4.10) into Eq. (4.12), and multiplies that equation from the left hand side by ϕ_j one obtains

$$\sum_{i=1}^{N} c_{ji} \, (H_{ji} - ES_{ji}) = 0 \quad j = 1, 2, \ldots N. \tag{4.14}$$

The system (4.14) of equations has a nontrivial solution only if the determinant of the quantities in brackets is zero. Since in principle the coefficients H_{ij} and S_{ij} are derivable from the electron wave functions the *secular determinant* of the system (4.14) may be solved to yield the energy eigenvalues E_1 to E_N. The total electronic energy E of the molecule then is

$$E = \sum_{j=1}^{N} 2 \, E_j. \tag{4.15}$$

Following the treatment of Brown [6] it is assumed that the *Coulomb integral,* H_{mm}, of an orbital ϕ_m depends only on the type of bond in which it resides, thus all C–H bond orbitals will be assigned the same value α of the Coulomb integral. The term is one half of the binding energy of an isolated C–H bond. The resonance integral, H_{mn}, will be assumed to be zero unless m and n denote adjacent bonds. In that case H_{mn} still depends on the type of neighboring bonds. In Table 4.1 the assignments of Brown for the bond interaction parameters in saturated hydrocarbons are listed.

Table 4.1. Bond Interaction Parameters

Coulomb parameter		Resonance parameter		Overlap integrals	
H_{mm}		H_{mn}		S_{mn}	
Bonds	Parameter	Bonds	Parameter	Bonds	Parameter
C–H	α	C–H, C–H	β	C–H, C–H	S
C–C	$\alpha + h\gamma$	C–H, C–C	$\Theta\beta$	C–H, C–C	ΘS
		C–C, C–C	$\eta\beta$	C–C, C–C	ηS

Table 4.2. Electronic Energies of Paraffins

Paraffin	Electronic Energy E	E_{theor} $(S_\gamma = 6.5)$ kJ/mol	$E_{exp} + W$ kJ/mol
Methane	$8\alpha - 24S\gamma$	-1765	-1665
Ethane	$14\alpha + 2h\gamma - (24 + 24\theta^2)S\gamma$	-2923	-2828
Propane	$20\alpha + 4h\gamma - (28 + 40\theta^2 + 4\eta^2)S\gamma$	-4088	-4001
Butane	$26\alpha + 6h\gamma - (32 + 56\theta^2 + 8\eta^2)S\gamma$	-5253	-5175
isoButane	$26\alpha + 6h\gamma - (36 + 48\theta^2 + 12\eta^2)S\gamma$	-5260	-5182
Pentane	$32\alpha + 8h\gamma - (36 + 72\theta^2 + 12\eta^2)S\gamma$	-6418	-6350
isoPentane	$32\alpha + 8h\gamma - (40 + 64\theta^2 + 16\eta^2)S\gamma$	-6425	-6358
neoPentane	$32\alpha + 8h\gamma - (48 + 48\theta^2 + 24\eta^2)S\gamma$	-6438	-6370
Hexane	$38\alpha + 10h\gamma - (40 + 88\theta^2 + 16\eta^2)S\gamma$	-7583	-7525
2-Methylpentane	$38\alpha + 10h\gamma - (44 + 80\theta^2 + 20\eta^2)S\gamma$	-7590	-7532
3-Methylpentane	$38\alpha + 10h\gamma - (44 + 80\theta^2 + 20\eta^2)S\gamma$	-7590	-7529
2,3-Dimethylbutane	$38\alpha + 10h\gamma - (48 + 72\theta^2 + 24\eta^2)S\gamma$	-7597	-7535
2,2-Dimethylbutane	$38\alpha + 10h\gamma - (52 + 64\theta^2 + 28\eta^2)S\gamma$	-7603	-7543

In Table 4.2 and 4.3 the electronic energies of a few hydrocarbons and of their radicals are given in terms of Brown's parameters. It will be noted that to a zeroth approximation (S = 0) the total electronic energy E of a molecule or a radical is given by the sum of the energies of the corresponding number of isolated C–C and C–H-bonds. In the first approximation also the energetic interactions of neighboring bonds are considered (terms to the first power of S as listed in Tables 4.2 and 4.3) and higher powers of S are neglected. Such terms would be necessary, however, to account for the effect of third neighbors within the molecule. That effect is clearly revealed by analysis of thermochemical data [7].

If it were possible to determine the numerical values of the bond interaction parameters by simply using molecular geometry and atomic constants one would have a direct and independent way to calculate bond energies. So far, however, it has proven very cumbersome to derive the explicit values of the bond energy parameters α, h, γ, Θ^2, η^2, and S directly and sufficiently correctly from molecular wave functions. Usually thermochemical data are employed to obtain the electronic energy of a molecule. For example the electronic energy E of a hydrocarbon can be

Table 4.3. Energy* of Hydrocarbon Radicals R

Radical R	Electronic Energy	$E(R-H)-E(\dot{R})$	$D(R-H)$* kJ/mol	$\Delta H_f(298\ K)$* kJ/mol
Methyl $\dot{C}H_3$	$6\alpha - 12S\gamma$	$F_1 + 3F_2$	436	134
Ethyl $H_3C-\dot{C}H_2$	$12\alpha + 2h\gamma - (16 + 20\theta^2)S\gamma$	$F_1 + 2F_2$	414	110
n-Propyl $H_3C-CH_2-\dot{C}H_2$	$18\alpha + 4h\gamma - (20 + 36\theta^2 + 4\eta^2)S\gamma$	$F_1 + 2F_2$	411	86
iso-Propyl $H_3C-\dot{C}H-CH_3$	$18\alpha + 4h\gamma - (24 + 32\theta^2 + 4\eta^2)S\gamma$	$F_1 + F_2$	394	72
n-Butyl $H_3C-(CH_2)_2-\dot{C}H_2$	$24\alpha + 6h\gamma - (24 + 52\theta^2 + 8\eta^2)S\gamma$	$F_1 + 2F_2$	411	67
iso-Butyl $H_3C-CH(CH_3)-\dot{C}H_2$	$24\alpha + 6h\gamma - (28 + 44\theta^2 + 12\eta^2)S\gamma$	$F_1 + 2F_2$	411	57
sec.-Butyl $H_3C-CH_2-\dot{C}H-CH_3$	$24\alpha + 6h\gamma - (28 + 48\theta^2 + 8\eta^2)S\gamma$	$F_1 + F_2$	394	50
tert.-Butyl $(CH_3)_3\dot{C}$	$24\alpha + 6h\gamma - (36 + 36\theta^2 + 12\eta^2)S\gamma$	F_1	381	28

* Energy values from Refs. [8a] and [9].

calculated from the enthalpies H_a of atomization of hydrogen, L of sublimation of graphite, and ΔH_f of formation of the hydrocarbon from H_2 and C_n:

$$E(C_nH_{2n+2}) = -n\,L(C) - (n+1)H_a + \Delta H_f - W. \tag{4.16}$$

The term W accounts for the fact that the minimum of a binding potential is by the zero point vibrational energy $(1/2\,h\nu)$ lower than the corresponding dissociation energy (Fig. 4.1). Since the relevant frequencies are of the order of $2.5 \cdot 10^{13}\ s^{-1}$ (C–C skeleton) and $9 \cdot 10^{13}\ s^{-1}$ (C–H) the zero point vibrational energies range between 5 kJ/mol and 18 kJ/mol.

A quantitative analysis of the quantum mechanical calculations through thermochemical data will be attempted, although considerable uncertainties exist. These are caused by the limitations of the used first order approximation, the lack of consideration of steric interaction of atoms, and of the difficulties to determine W. The analysis is firstly based on the fact that the differences between the electronic energies of the listed hydrocarbons and their radicals can be expressed by only four quantities F_i:

$$F_1 = -2\alpha + 12\,\Theta^2\,S\gamma \tag{4.17}$$
$$F_2 = 4(1-\Theta^2)\,S\gamma \tag{4.18}$$
$$F_3 = -4(1 - 2\,\Theta^2 + \eta^2)\,S\gamma \tag{4.19}$$
$$F_4 = -2\alpha - 2\,h\gamma + 24\,\eta^2\,S\gamma. \tag{4.20}$$

Brown [6] determined F_2 to be 15.5 kJ/mol (from the radical dissociation energies listed in Table 4.3) and F_3 to be -6.7 kJ/mol (from the energy differences found for 5 pairs of isomers listed in Table 4.3).

Since these difference values are not so much affected by the systematic uncertainties mentioned above they will be adopted here along with the value of 300 kJ/mol for F_4; F_4 is the electronic energy of a C–C bond interacting solely with C–C bonds as in $(CH_3)_3C–C(CH_3)_3$. Further the standard values (at 298.15 K and 1 bar) L = 717.2 kJ/mol and H_a = 436.2 kJ/mol [7] and a ΔH_f (CH_4) = $= -75$ kJ/mol have been used [6]. The zero point vibrational energy of the 9 vibrational modes of the –C–H and H–C–H bonds of methane is estimated at 100 kJ/mol. The total electronic energy of methane then is derived from Eq. (4.16) as E (CH_4) = -1765 kJ/mol. From Eqs. (4.17) and (4.18) it follows that $6.2 < S\gamma < 8.4$ kJ/mol. Employing E (CH_4), F_2, F_3, and F_4 as given above and a value of

$S\gamma$ =	6.5 kJ/mol	one obtains
α =	-200 kJ/mol	
$h\gamma$ =	55 kJ/mol	
Θ^2 =	0.4 kJ/mol	
η^2 =	0.06 kJ/mol .	

On the basis of these values the electronic energies of the hydrocarbons have been determined and listed as E_{theor} in Table 4.2. For comparison the thermochemically determined enthalpies of formation are referenced as $E_{exp} + W$. The latter values are consistently larger than E_{theor} although the difference decreases rather than to increase with the number of vibrational modes available to a molecule.

Table 4.4. Dissociation Energies of Primary Bonds in Organic Molecules and Polymer Chains*

The table is a two-axis matrix of bond dissociation energies. The upper-left triangle (after van Krevelen) is indexed by the top column headers and the left row labels; the lower-right triangle (after Stepan'yan et al.) is indexed by the bottom column headers and the right-hand row labels. A staircase line separates the two regions.

**Column headers (top axis):**

	$-CH_3$	$-CH_2CH_3$	$-CH(CH_3)_2$	$-C(CH_3)_3$	$-CH{=}CH_2$	$-C{\equiv}CH$	$-O{-}CH_3$	$-C(CH_3){=}O$	$-O{-}C(CH_3){=}O$	$-CH{=}O$	$-CH_2{-}CN$	C_6H_5-	$C_6H_5{-}C({=}O)-$	$-CH_2{-}C_6H_5$	$-CH(CH_3){-}C_6H_5$	$R{-}CH_2{-}CH{-}R'$
H–	431	406	394	373	436	507	427	360	469	365		427		347		300
F–	453	444	440	427		245						524		377		448
Cl–	352	339	339	331	360	205	344					419		285		314
OH–	381	381	385	381				427	419			469				356
O=CH–	314	297														
NH₂–	331	327	323	323				411				419				
N≡C–	461											545				451

Lower region (left row labels, after Stepan'yan et al.):

	$-CH_3$	$-CH_2CH_3$	$-CH(CH_3)_2$	$-C(CH_3)_3$	$-CH{=}CH_2$	$-C{\equiv}CH$	$-O{-}CH_3$	$-C(CH_3){=}O$	$-O{-}C(CH_3){=}O$	$-CH{=}O$	$-CH_2{-}CN$	C_6H_5-	C_6H_5CO-	$-CH_2C_6H_5$	$-CH(CH_3)C_6H_5$	$R{-}CH_2{-}CH{-}R'$
$C_6H_5{-}CH(CH_3)-$	276															355
$C_6H_5{-}CH_2-$	302	260	230		293			264		210		323		199		324
$C_6H_5{-}C({=}O)-$										377			364*			
C_6H_5-	427	381	348	327	423	499	423			377	355	432	316		335	345
$CN{-}CH_2-$	306													290		311
$\underset{O}{H}{>}C-$	314	297						251		251					336	
$CH_3{-}C({=}O){-}O-$	379*								126							311
$CH_3{-}C({=}O)-$	344	323								251						328
$CH_3{-}O-$	335	335	339	327			142									326
$CH{\equiv}C-$	419	457	432			461										342
$CH_2{=}CH-$	377	377	356	339	423											298
$(CH_3)_3C-$	335	314	306	293				433						377		301
$(CH_3)_2CH-$	348	327	318													270
$CH_3{-}CH_2-$	356	337						224								
CH_3-	369									328						254

Additional lower-region values: 285, 272, 206.

Right-hand row labels (after Stepan'yan et al.), top to bottom:
$-CH_2-R$; $-C(CH_3)_2-R$; $R-CH_2-CH-R'$; $-O-R$; $-CH_2O-R$; $-(CH_2)_2O-$; $-CH(OH)-R$; $-CH(CH_3)-R$; $-CHCl-R$; $-CHF-R$; $-CH(CN)-R$; $-C_6H_4-R$; $-CH(C_6H_5)-R$; $RNH-C_6H_4-NH-$; $-C(CH_3)(C_6H_5)-R$; $RC({=}O)-C_6H_4-C({=}O)-O-CH_2-$; $-(CH_2)_m-CO-R$; $-CH_2-C_6H_4-R$

Bottom column labels (after Stepan'yan et al.), left to right:
$R-C({=}O)-C_6H_4-C({=}O)-$; $R-C_6H_4-$; $RNH(CH_2)_nNHCOCH_2-$; $ROCH_2-$; $R-CH_2-CH-R'$; $CH_3-CH(R-CH_2)-$; $R-(CH_2)_4-$; $R-(CH_2)_3-$; $R-(CH_2)_2-$; $R-CH_2-$

---

* Upper and left hand side: after van Krevelen [9]
  Lower and right hand side: after Stepan'yan et al. [10]

The treatment may be extended to deal with the energy of alkyl radicals if one assumes that the single unpaired elect on on the "radicalized" carbon atom occupies of $2p\pi$ molecular orbital and does not change the interaction of other electrons on other orbitals. As mentioned earlier this assumption is in accord with ESR-investigations [4]. In electronic terms the dissociation energy D of a bond $R_1 - R_2$ can then be formulated as

$$D(R_1 - R_2) = E(R_1 - R_2) - E(\dot{R}_1) - E(\dot{R}_2) - 1/2\ h\nu(R_1 - R_2). \qquad (4.21)$$

In Table 4.3 the electronic energy of hydrocarbon radicals and R–H dissociation energies are listed together with thermochemically determined values. The difference values of electronic and thermochemical data are in correspondence with each other. The absolute value of $F_1 = D(R{-}H) + 1/2\ h\nu(R{-}H)$ can be calculated from Eq. (4.17) to be 431 kJ/mol; since D(R–H) for tertiary carbon atoms is 381 kJ/mol a contribution of 50 kJ/mol must be assigned to the zero point vibrations. This contribution is certainly too high by $\simeq$30 kJ/mol. The discrepancies between calculated and measured dissociation energies of C–C bonds, however, lie between 0 and 6 kJ/mol for the bonds covered by Tables 4.2 and 4.3. The experimentally determined dissociation energies of small molecular groups are listed in the upper and left hand side of Table 4.4 [after van Krevelen, 9]. For these molecular fragments the activation energy of bond rupture and the dissociation energy are assumed to correspond. In the lower right hand side we have listed the dissociation energies obtained for sp^2 and sp^3 bonds by Stepan'yan et al. [10]. In their semi-empirical approach the authors calculate the bond dissociation energies $R_1{-}R_2$ by extrapolation from those of the series of compounds R–X, X representing F, H, Cl, Br, and I. Judging the available experimental and theoretical determinations of bond energies of chain molecules it must be stated that still considerable differences with respect to the depth of the binding potentials for main chain bonds exist.

Before leaving the subject of bond energies one should briefly discuss the meaning of the term bond strength. In quantum chemistry the term bond strength, or more explicitly the bonding strength of atomic orbitals, refers to the excentricity of a (hybrid or nonhybrid atomic) orbital and is measured by the maximum angular intensity of the wave function as compared with the spherically symmetrical s-orbital [1]. So the *strength* of an sp^3 hybrid is $\Psi_1(\Theta = 54°\,44'; \phi = 45°)/\left(\dfrac{1}{4\pi}\right)^{1/2} = 2.$

Other hybrids of interest have the following bond strengths: the (pyramidal) p^3 hybrid 1.732, the (plane triangular) sp^2 hybrid 1.991, and the (linear) sp hybrid 1.932.

Throughout our further discussions, however, the term bond strength will have a more technological meaning, i.e. that of the force required to break the bond (cf. Chapter 5).

B. Binding Energy of Excited States and Radicalized Chains

In as much as the covalent binding force between any two atoms rests on the electron density distribution between (and around) these atoms any change of the electron distribution necessarily affects the bond strength. Since the states of different electronic energy of a molecule are states of different electron distribution they are also states of different interatomal forces.

This is also true, of course, for molecular ions, i.e. for molecules where one or more electrons have been removed completely. As already seen for methane the character of a bond and its energy are also affected through the radicalization of the atom group to which it belongs.

1. Electronic Excitation

The electronic excitation of a polymeric network chain can be caused by electromagnetic (light, UV, γ) or particle irradiation. The transfer of energy from the impinging particle or quantum of radiation to the electron requires that sufficient energy is available to reach an excited state and that an interaction mechanism exists. In the case of visible light a photon of, say, a wavelength of 330 nm has enough energy to break a C–C bond. However, the photon will not be absorbed by alkanes and there are no electronic states of that or a smaller energy which could be excited. To be effective a photon has to be absorbed and to interact with a binding electron. This interaction occurs either directly or indirectly through the mechanisms of energy transfer by exciton diffusion, single step transfer or absorption of fluorescence light emitted by the same or another (impurity) molecule [11]. The nature and sequence of these important processes which determine the photochemical stability – or instability – of polymers will not be treated here in detail. One is interested, however, to know the levels of energy where the excitation of electrons – or the ionization of molecules – begins and to know the changes of binding energy caused by excitation – or ionization – respectively.

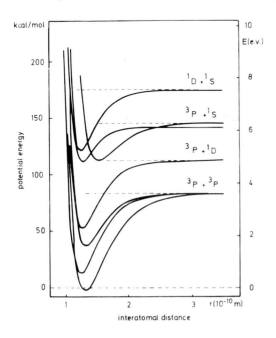

Fig. 4.2. Interatomal energy of excited states of a C_2 molecule [12].

The potential curves of different excited states of the C_2-molecule illustrate the effect of electronic excitation on bond energies (Fig. 4.2). We would like to point out that the *dissociation energies* of the different molecular states shown range from 350 to 119 kJ/mol (84 to 28 kcal/mol). These dissociation energies are generally smaller than the energy differences caused by the electronic excitation. The excited states of the (many-electron) C-atoms are denoted in the usual manner as S, P, D. Since two excited atoms may combine in a number of ways there are more than one molecular state corresponding to any two atomic states.

Transitions from the ground state of a molecule to an excited state usually occur from a bonding orbital (σ or π electron) or a non-binding orbital (n electron) to an antibonding σ^* or π^* orbital. In Table 4.5 are listed the electronic transitions for some common high polymers as given by [13]. If an electronic excitation of a network chain occurs, it leads to decreased bond energies. This will reduce the mechanical stability of a loaded polymeric network and thus contribute to chain scission, fracture initiation or facilitated crack propagation. It is readily seen from Table 4.5,

Table 4.5. Electronic Transitions for some Polymers[a]

Polymer	λ_{max}, nm	$h\nu_{max}$ kJ/mol	kcal/mol	Transition
Polyethylene	<150	>802	>191.5	$\sigma \rightarrow \sigma^*$
Polybutadiene	180	670	160	$\pi \rightarrow \pi^*$
Polystyrene	187–260	461–645	110–154	b
Poly(methyl methacrylate)	214	561	134	$n \rightarrow \pi^*$
Poly(ethylene terephthalate)	290	415	99	$n \rightarrow \pi^*$
Poly(ethylene terephthalate)	240	502	120	b

[a] As solid films (after Brenner and Kapfer, [12]).
[b] π-electron system involved.

however, that the energy values of the lowest lying excited states are larger than that of any C–C bond. One does not expect, therefore, that the recombination of any chain radicals will lead to energy quanta which are large enough to cause electronic excitation of the reacting radicals or of neighboring molecules.

2. Ionization

The complete removal of an electron from a molecule, i.e. the ionization, requires more energy than electron excitation and has an even stronger effect on the dissociation energy. In Table 4.6 the dissociation and ionization potentials of the first few hydrocarbons have been collected from various sources. It is assumed that ionization is first accomplished through removal of the least bound electron from the highest occupied orbital and also that this orbital embraces the whole molecule.

Table 4.6. Dissociation and Ionization Energies of Hydrocarbons

Molecule (R_1–R_2)	$D(R_1–R_2)^a$ kJ/mol	$I(R_1–R_2)^b$ eV	kJ/mol	Radical Ion (R_2^+)	$I(R_2)^c$ eV	kJ/mol	$I(R_1–R_2) – I(R_2)$ eV	kJ/mol	$D(R_1–R_2^+)$ kJ/mol
Hydrogen H_2	436	15.43	1490	H^+	13.6	1314	1.83	176	260
Ethylene C_2H_6	352	11.76	1134	CH_3^+	9.82	950	1.94	184	167
	414	11.76	1134	$C_2H_5^+$	≤ 8.4	≤ 812	≥3.36	326	88
Propylene C_3H_8	347	11.21	1084	CH_3^+	9.82	950	1.39	134	213
	350	11.21	1084	$C_2H_5^+$	≤ 8.4	≤ 812	≥2.81	≥272	≤ 75
	406	11.21	1084	$nC_3H_7^+$	< 8.1	< 783	>3.11	>301	<109
	393	11.21	1084	$isoC_3H_7^+$	< 7.5	< 724	>3.71	>360	< 33
n Butane C_4H_{10}	347	10.8	1042	CH_3^+	9.82	950	0.98	92	255
	347	10.8	1042	$C_2H_5^+$	≤ 8.4	≤ 812	≥2.4	≥230	≤117
	347	10.8	1042	$C_3H_7^+$	< 8.1	< 783	>2.7	>260	< 88

a From Table 3.4.
b From Landolt Börnstein I/3, 6. Auflage, Springer Verlag, Berlin 1951, p. 363.
c From F. A. Elder, C. Giese, B. Steiner, M. Inghram, J. Chem. Phys. 36, 3292 (1962).

The dissociation energies $D(R_1 - R_2^+)$ in the last column of Table 4.6 can be calculated from the ionization energies according to the relation

$$D(R_1 - R_2^+) = D(R_1 - R_2) - I(R_1 - R_2) + I(R_2). \tag{4.22}$$

The values derived in that (indirect) way are given in parenthesis because of the ambiguities that are still involved in obtaining the ionization potentials from electron impact or photo ionization efficiency measurements. For the molecules listed here it is clearly apparent that the dissociation energies of the positive molecular ions are much smaller than those of the non-ionized molecule. (It should be noted that the dissociation energies of excited states are not necessarily *always smaller* than those of the corresponding ground states. But in the organic compounds here of interest dissociation energies are reduced by electron removal.)

It has been emphasized here that the ionization should lead to a *positive* molecular or radical ion because the electronic energy of negative ions may be quite different. These differences are not expected and not explained in molecular orbital theory. If no electron repulsion term is explicitly introduced then the same amount of energy will be predicted for the removal of an electron and for placing an electron onto such an orbital. These differences can be quite considerable. For H_2 the state $H^+ + H^-$ lies above the state $H + H$ by an amount of 12.8 eV (1230 kJ/mol). The energy difference between $R_1^+ + R_2^-$ and $R_1 + R_2$ is given by $I(R_1) - E'(R_2)$ where $E'(R_2)$ is the *electron affinity* of R_2 (i.e. the energy which is released when the atom or radical R_2 combines with an electron to form a negative ion). As is illustrated by Table 4.7 the electron affinities of saturated hydrocarbons are of the order of 1 eV (96 kJ/mol) or less, the differences between $I(R)$ and $E(R)$ are of the order of 7 eV (672 kJ/mol) or more. The dissociation of a hydrocarbon molecule $R_1 - R_2$ into the two radical ions R_1^- and R_2^+ generally requires an amount of 672 kJ/mol or more *in addition* to the energy required for homolytic scission into the radicals R_1 and R_2. Homolytic chain scission, therefore, is a much more probable mode of molecular disintegration (of saturated hydrocarbons) than ionic dissociation.

3. Radicalization

The removal of a hydrogen atom from a hydrocarbon leads to the formation of a radical. Its electron distribution differs from the original one near the radical

Table 4.7. Electron Affinity and Ionization Potential

Radical R	Electron Affinity* $E'(R)$ [eV]	Ionization potential $I(R)$ [eV]	$I(R) - E(R)$ eV	kJ/mol
H	0.80	13.60	12.8	1230
CH_3	1.08	9.82	8.74	840
C_2H_5	0.89	8.4	7.5	720
$n-C_3H_7$	0.69	8.1	7.4	710

* Values of E(R) taken from Handbook of Chemistry and Physics, 53 rd Edition, CRC Press, Cleveland 1972/73, p E 55.

C atom. In saturated hydrocarbons the 4 sp^3 hybrid orbitals change to 3 sp^2 hybrids. The radical free electron is considered to be located in a completely separated π eigenfunction which has no interaction with the σ-system. The strength of a bond connecting a C atom with the radicalized C atom, $\dot{C}$, is now mainly determined by the interaction of an sp^2 with an sp^3 hybrid. The energy of a bond formed by an overlap of an sp^3 hybrid and an sp^2 hybrid is not smaller than that formed by two sp^3 hybrids. This is concluded from the dissociation energies of the C–C bonds in RCH_2–CH_2R (337 kJ/mol) and RCH_2–C_6H_5 (381 kJ/mol). The assumption that replacing sp^3 orbitals by sp^2 orbitals in bonds will not reduce the bond energy per se is further supported by the observation that the bond energy of the $C_{(sp^2)}$–H bond in $H_2C=CH_2$ is the same as the value of the $C_{(sp^3)}$–H bond in H_3C–CH_3 (436 kJ/mol). The increased energy of dissociation of the $C_{(sp^2)}$–H bond in the methyl radical (477 kJ/mol) leads to a similar conclusion.

The C–C bonds in α position to the carbon carrying a free electron are, therefore, slightly stronger than the same bonds in the original molecule. The C–C bonds in β position, however, are adversely affected by the described electronic reorganization. In thermal or mechanical degradation they are found to break preferentially. Reactions involving the scission of a radicalized carbon backbone chain lead to the formation of a double bond:

$$R_1-CHX-\underset{\beta}{CH_2}-\underset{\alpha}{\dot{C}X}-CH_2-R_2 \;\rightarrow\; R_1-\dot{C}HX + CH_2{=}CX-CH_2-R_2.$$

The considerable amount of energy gained in transforming a $-CH_2-\dot{C}X-$ bond into a double bond (250–280 kJ/mol) should decrease the activation energy of such a reaction correspondingly. Kagiya [14] has derived a number of activation energies, D, mostly for vinyl polymers. He uses the assumption that D is equal to the heat of polymerization, Q, and the activation energy E_p of the propagation reaction:

$$D = Q + E_p. \tag{4.23}$$

His values are reported in Table 4.8. Again it must be mentioned that a stress-induced scission reaction may require an energy of activation which is considerably higher than the listed values of D.

Table 4.8. Activation Energy of Main Chain Scission of Radicalized Chains*

Polymer	Activation Energy kJ/mol	Polymer	Activation Energy kJ/mol
PE	118	PVF	116
PP	102	PVDF	131
PIB	76	PTFE	147
PS	85	PVAC	118
Pα-MS	95	PMA	98
PVC	87	PMMA	85
		PA 6	117**

* Ref. 14.
** T. Kagiya, personal communication.

III. Form of Binding Potential

In Figure 4.1 it was *qualitatively* indicated how the potential energy of two bonded atoms changes with their interatomal distance r. Knowledge of the *exact* form of the interatomal potential could — in principle — be obtained from a calculation of the total electronic energy E of a molecule as a function of r. For polyatomic molecules, however, $E(r)$ cannot be calculated by any of the mentioned approximation methods with reasonable accuracy. So the potential functions are described by *empirical, phenomenological* functions. These functions will, of course, necessarily utilize those molecular constants which are otherwise obtainable, e.g. bond length (r_0), bond stretching force constant, and dissociation energy D. One of these functions is the Morse potential function which has the virtue to lead to a solvable Schrödinger Equation:

$$V = D \exp\left[-2a(r - r_0)\right] - 2D \exp\left[-a(r - r_0)\right]. \tag{4.24}$$

The Morse constant a can be expressed through force constant and anharmonicity constant x of the oscillator formed by the bond atoms. The Morse function yields the energy eigenvalues of an anharmonic oscillator which are correct within a sizeable range of amplitudes (in the case of H_2 for $0.4 < r/r_0 < 1.6$). For large r, however, V becomes progressively too small [8b].

References for Chapter 4

1. K. S. Pitzer: Quantum Chemistry. New York: Prentice Hall, 1953 a) 153, b) 157, c) 163.
2. R. Daudel, R. Lefebvre, C. Moser: Quantum Chemistry, Methods and Applications. New York: Interscience, 1959 a) 48, b) 91, c) 3–24.
3. A. G. Shenstone: Phys. Rev. *72*, 411 (1947).
4. T. Cole, H. O. Pritchard, N. R. Davidson, H. M. McConnell: Mol. Phys. *1*, 406 (1958).
5. H. A. Stuart, Ed.: Molekülstruktur. Berlin — Heidelberg — New York: Springer 1967, 35.
6. R. D. Brown: J. Chem. Soc. 2615 (1953).
7. R. M. Joshi: J. Macromol. Sci.-Chem. *A4* (8), 1819 (1970).
8. J. G. Calvert, J. N. Pitts, Jr.: Photochemistry. New York: Wiley, Inc. 1967, a) 815ff, b) 136f.
9. D. W. van Krevelen: Properties of Polymers, 2nd Edition. Amsterdam: Elsevier 1976.
10. A. Ye. Stepan'yan, Yu. G. Papulov, Ye, P. Krasnov, G. A. Kurakov: Vysokomol. Soyed. *A14*, 2033–2040 (1972), Polymer Sci. USSR *14*, 2388–2397 (1972).
11. M. Heskins: Encyclopedia of Polymer Science and Technology, Supplement Vol. 1. H. F. Mark, N. G. Gaylord, Eds. New York: Interscience 1969, 763.
12. G. Herzberg: Spectra of Diatomic Molecules, 2nd Edit., New York: D. van Nostrand 1953.
13. W. Brenner, W. Kapfer: Encyclopedia of Polymer Science and Technology, Vol. *11*, H. F. Mark, N. G. Gaylord, Eds. New York: Interscience 1969, 763.
14. V. T. Kagiya, K. Takemoto, M. Hagiwara: IUPAC Macromol. Symposium "Long-Term Properties of Polymers and Polymeric Materials". Stockholm, Sweden, Aug. 30 – Sept. 1, 1976.

Chapter 5

Mechanical Excitation and Scission of a Chain

I. Stress-Strain Curve of a Single Chain

A chain molecule as part of a thermoplastic body is in thermal contact with other chains and — at room temperature — constantly in motion. The atoms vibrate and take part in the more or less hindered rotations of groups and even of chain segments. With no external forces acting all molecular entities try to approach — and fluctuate around — the most stable positions attainable to them. The action of external forces causes — or maintains — displacements of the chain from those positions and evokes retractive forces. Let us consider a chain or a bundle of chains in thermal contact with the surrounding and at constant volume. The condition of thermodynamic stability of such a system is that the free energy

$$F = U - TS \tag{5.1}$$

assume a minimum. The differential changes of the free energy of the system are given by the changes of its internal energy U and its entropy S.

$$dF = dU - T\,d\,S. \tag{5.2}$$

An increase of U can be effected by heat dQ flowing into the system from the outside or by mechanical work dW done onto the system. The change of entropy may be split up into a portion $-d_e S$ which is equal to dQ/T and into the internal entropy production $d_i S$ caused by irreversible processes as for instance by stress relaxation or generation of heat through friction:

$$dF = dQ + dW - Td_eS - Td_iS = dW - Td_iS. \tag{5.3}$$

From Eq. (5.3) one derives the force f acting upon a chain with end-to-end distance r as:

$$f = \left(\frac{\partial W}{\partial r}\right)_{V,T} = \left(\frac{\partial F}{\partial r}\right)_{V,T} + T \left(\frac{\partial_i S}{\partial r}\right)_{V,T} \tag{5.4}$$

The uniaxial force on a relaxing chain molecule depends, therefore, on the changes of internal energy and entropy associated with a change of its end-to-end distance and on the entropy production term (which is always positive). These three different force contributions will be examined in the following.

A. Entropy Elastic Deformation

The kinetic theory of rubber elasticity was essentially developed between 1930 and 1943. A detailed description of the development of this theory and its present state is given by e.g. the classical treatises of Flory [1] or Treloar [2]. The necessary conditions for the occurrence of rubberlike elastic behavior are the presence of a long chain molecule possessing internal flexibility (freely rotating links) and the absence of strong secondary bonds acting between a segment and surrounding segments of the same or other chains. The stress-strain relation for a single finite chain then is derived from the distribution of chain conformations. Following the treatment of Treloar it will briefly be recalled which chain properties enter the stress-strain relations and what levels of force are attainded.

First the conformations of certain idealized chains are considered. For the *chain of n equal links of length ℓ joined entirely at random* (i.e. neglecting valence angles and hindrances to internal rotation) the probability distribution $p(x, y, z)$ to find a chain end at a point (x, y, z) – if the other end is fixed at the origin – is obtained as a Gaussian distribution:

$$p(x, y, z) = (b^2/\pi^{3/2}) \exp\left[-b^2(x^2 + y^2 + z^2)\right]. \tag{5.5}$$

The only parameter, b, is the inverse of the most probable chain end separation and is under the above conditions $\sqrt{3/n\ell^2}$. The extended length of the chain is, of course, equal to $n\ell$, whereas the average chain end separation (the "root-mean-square length" $\sqrt{\overline{r^2}}$ equals $\ell\sqrt{n}$.

More realistic is the model of the *statistically kinked chain with valence angles and hindrances to internal rotation*. Here the mean square length is derived as:

$$\overline{r^2} = n\ell^2 \frac{1 + \cos\alpha}{1 - \cos\alpha} \cdot \frac{1 + a}{1 - a} \tag{5.6}$$

where α is the semi-angle of the cone of bond rotation and a is the mean value of $\cos\phi$, ϕ being the angle of bond rotation counted with respect to the trans-conformation. The ratio $\overline{r^2}/n\ell^2$ is called the characteristic ratio. If the rotation of a

88

bond is considered to be independent of the position of other bonds and free within its cone of rotation then a is zero.

A given (hydrocarbon) chain can be treated as a statistically kinked chain with valence angles and also as freely jointed chain. The root-mean-square lengths and extended lengths of both models correspond to each other then and only then when a random link of appropriate length has been used. This "equivalent random link" of length ℓ_r is rather short and contains only a few chain atoms. For paraffin-type chains Treloar obtained a length ℓ_r of 0.377 nm, for cis-polyisoprene 0.352 nm. These equivalent random links are about two times smaller than the (shortest) segment lengths necessary to form a fold of $180°$.

The entropy S of a single chain is according to Eq. (5.5) and the Boltzmann relation

$$S = k \ln p = C_1 - kb^2 r^2 \tag{5.7}$$

where C_1 is a constant. An important aspect of the entropy of a chain is — in the words of Treloar [2a] —: "A chain does not possess an entropy by the mere fact of having a specified conformation, but only on the virtue of the large number of conformations accessible to it under specified conditions of restraint". If these conditions consist only in the location of the end-points of the chain at specified positions, then the appropriate expression for the entropy is Eq. (5.7). In the presence of external hindrances to bond rotation the entropy of a chain is not given by Eq. (5.7) since $p(x, y, z)$ must then be corrected for the number of inaccessible states.

The retractive force f acting between the chain ends — and in the direction of the line joining them — is derived from Eqs. (5.2), (5.4) and (5.7). If one considers a reversible deformation and postulates that all conformational states posses the same internal energy U so that U does not depend on conformation then

$$f = 2 kT b^2 r. \tag{5.8}$$

The random coil in thermal equilibrium behaves as a spring (of length zero) with a spring constant $2 kT b^2$. The level of force exerted by such a spring with its ends separated by the most probable distance $r = 1/b$ is fairly small. For a hydrocarbon chain of molecular weight 14'000 with an extended length of 125.5 nm and a most probable chain end distance of 7 nm, this force is $17 \cdot 10^{-13}$ N which corresponds to a force per cross-section of the chain, f/q, of 10 MN/m^2 (10^8 dyn/cm^2). As one should expect this "chain stress" is very close to the modulus $G = NkT$ of a rubber network having chains of exactly this length between crosslink points and for which G is 17.5 MN/m^2.

The Gaussian theory considers the number of possible conformations of a chain having a specified end-to-end distance. A more accurate non-Gaussian statistical treatment of the random chain is based on the distribution of $\sin \Theta_1$, i.e. of the angle between the direction of a random link and of the end-to-end vector. From the probability of finding n_1 links in the range $\Delta\Theta_1$, n_2 in $\Delta\Theta_2$ and so on, the entropy of a single chain is derived [2b] as

$$S = C_2 - kn \left[\frac{r}{n\ell} \beta + \ell n \frac{\beta}{\sinh \beta} \right] \tag{5.9}$$

where C_2 is a constant and β the so-called inverse Langevin function $\mathscr{L}^{-1}$ $(r/n\ell)$. Using (5.2) and (5.4) one obtains

$$f = \left(\frac{kT}{\ell} \right) \mathscr{L}^{-1} \ (r/n\ell) \tag{5.10}$$

which can be expanded into

$$f = \frac{kT}{\ell} \left\{ 3 \left(\frac{r}{n\ell} \right) + \frac{9}{5} \left(\frac{r}{n\ell} \right)^3 + \frac{297}{175} \left(\frac{r}{n\ell} \right)^5 + \ldots \right\}. \tag{5.11}$$

It is of special interest to note that this stress-strain relation can also be derived from a consideration of the potential energy of the statistical links in a unidirectional force field. James and Guth [3] and Flory [1] have used the fact that an amount of work of $f \cdot \ell (1 - \cos \Theta_i)$ must be performed to rotate a link from a position of $\Theta = 0$ to $\Theta = \Theta_i$; if the orientational energy $-f \ell \cos \Theta_i$ is distributed according to Boltzmann statistics then the mean extension of a link in the direction of force is

$$\bar{z}_i = \ell \mathscr{L}(f\ell/kT) \tag{5.12}$$

and the mean total chain end-to-end distance is

$$r = n\bar{z}_i = n\ell \mathscr{L} \ (f\ell/kT) \tag{5.13}$$

which by inversion directly yields Eq. (5.10). It should be pointed out that so far this result has been derived without introducing — or permitting — any internal or external hindrances to rotation of a link. The absence of external hindrances also means that in longitudinal extension or contraction of the chain or of parts thereof no frictional forces are encountered. The forces exerted by the chain ends are solely the consequence of the imbalance of exchange of momentum between the chain and its surrounding.

If a hydrocarbon chain behaves as described by Eq. (5.13) and if such a chain were to reach a breaking force (of, say, $17 \cdot 10^{-10}$ N) at room temperature then $f\ell/kT$ must exceed 150. For an argument that large the Langevin function is equal to $1 - kT/f\ell$ and f is:

$$f = nkT(n\ell - r). \tag{5.14}$$

In the chosen example of a hydrocarbon chain of 125.5 nm extended length, breaking stresses will only be reached if r is larger than 124.7 nm. In other words only 2 of the 333 random links of 0.377 nm length may point in a direction perpendicular to the chain end-to-end vector while all the remaining ones then have to be fully aligned. Even at this extreme elongation the Langevin function gives a good approximation of the stress-strain behavior of the randomly kinked chain. This

90

becomes evident by comparison with the so called exact solution of Treloar [2c] which is purely based on geometrical (and combinatorial) considerations and from which one derives for extreme elongations:

$$f = -kT \, d\ln p/dr = -kT/r + (n-2) \, kT/(n\ell - r). \tag{5.15}$$

The force-elongation relations (5.8), (5.10), and (5.15) have been derived under the condition that all conformational states have the same internal energy, that all states are accessible, and that transitions between different states are easily accomplished. No chain molecule — and especially not within a solid — does fully comply with the above conditions. For one thing there are the *intramolecular* interactions (between nearest and second nearest atoms of one chain) which cause the potential energy of the chain to be a function of the angle of rotation with distinct relative minima in more than one position. All this is well known and documented (e.g. 1—5) and will not be discussed here. The direct or indirect consequences with respect to the retractive forces between (PE or PA) chain ends are the following:

— the internal energy of (in-plane) trans-conformations is lower than that of (out-of plane) gauche-conformations
— the probability to find a conformation with internal energy ΔU is weighted by the Boltzmann factor $\exp(-\Delta U/kT)$
— the average end-to-end distance increases with a according to Eq. (5.6), the length ℓ of the equivalent random link also grows in proportion to $(1+a)/(1-a)$, while the number n of random links in a chain of finite length decreases
— the force between chain ends at their most probable distance $(r = 1/b)$ is inversely proportional to r and thus also to the square root of the characteristic ratio, the force is the smaller the larger the characteristic ratio is.

The characteristic ratio of PE at room temperature is — according to Abe et al. [4] — 8.0 as compared to a value of 2 calculated under neglect of hindrances to bond rotation. The force between PE chain ends, therefore, is by a factor of 2 smaller if the effect of the potential energy on chain conformation is taken into consideration. For polyamide chains the detailed calculations of Flory and Williams [5] led to characteristic ratios of 6.08 (PA 6) and 6.10 (PA 66) in excellent agreement with the experimental value of 5.95 for PA 66 found by Saunders [6]. The calculated force between PA chain ends, then, will be smaller by a factor of about 1.7.

A second cause to invalidate Eq. (5.7) are strong *intermolecular* interactions between chain segments which do have the effect that only a restricted number of conformational states are accessible to a chain segment, that transitions between different states may not be independent from each other, and that energy is required to activate these (cooperative) transitions. In recent years considerable progress has been made to apply the methods of statistical thermodynamics also to systems subject to the above restrictions and to explain e.g. their dynamic stress-strain behavior [7] or their melting and crystallisation [8]. With respect to chain loading one is interested in the entropy changes associated with the elongation of a highly extended chain. In the following ΔS will be calculated for PE and PA chain segments of 3 to 6 nm length.

It is convenient to start with Pechhold's [7] treatment of the polyethylene molecule. He derived the partition function of the rotational isomers of a chain with n independent C–C bonds. The chain is allowed to rotate around its C–C bonds, it will, therefore, encounter three minima of potential energy of which the absolute minimum belongs to the extended trans-conformation (t). Two relative minima at an angle $\pm 120°$ out of plane allow two gauche-conformations $(g, \bar{g})$ with a conformational energy of $+ \Delta U$. The kink-isomers are the subclass of the rotational isomers where only $n/2$ of the bonds — and no adjacent ones — may have a g or $\bar{g}$ position. For kink isomers entropy, internal energy, and free energy as a function of the average concentration $\bar{x}$ of gauche-conformations are derived [7] as:

$$\bar{x} = \exp(-\Delta U/kT) / [1 + 2 \exp(-\Delta U/kT)] \tag{5.16}$$

$$S = -0.5 \, nk \, \ln(1 - 2x) + n\bar{x} \, \Delta U/T \tag{5.17}$$

$$U = n\bar{x} \, \Delta U \tag{5.18}$$

$$F = 0.5 \, nkT \, \ln(1 - 2\bar{x}). \tag{5.19}$$

The group of kink isomers still comprises chains with quite different deviations from a planar zig-zag conformation, the simplest of those kinks being ttgt$\bar{g}$ttt (2g1 kink), tgtt$\bar{g}$t (2g2 kink), tgtgtgt (3g2 kink). Each kink causes a contraction in longitudinal direction which is a multiple Δz of $\sqrt{2/3} \, a$ with a the length of a C–C bond (Pechhold accordingly denotes a kink containing q gauche positions and causing a contraction Δz as a qg Δz-kink). A chain with n C–C bonds and m_i qgi-kinks has a projected length in z-direction of $(n - im_i) \sqrt{2/3} \, a$. Vice versa a projected length corresponding to that of n C–C bonds can be realized as well by an extended chain with n C–C bonds, by a chain with n + 1 bonds containing a 2g1 kink, by chains with n + 2 bonds containing either two 2g1 kinks or a 2g2 kink or by other appropriate combinations. The statistical weight of each conformation is given by the number of ways *that* conformation can be realized and by the Boltzmann factor of its conformational energy. The by far leading term in the partition function for a given r is always that listing the number of possible conformations of chains containing only 2g1 kinks. The number of ways of arranging $n_g/2$ pairs of g and $\bar{g}$ bonds on $n/2$ available positions clearly is

$$Z(n, n_g) = 2^n g/2 \, \prod_{i=1}^{n_g/2} \left(\frac{n}{2} - 2i + 1 \right) \exp(-2i\Delta U/kT). \tag{5.20}$$

In Table 5.1 the statistical weights of PE chains with a gauche-concentration of up to 20% are given according to Eq. (5.20). Also listed are the statistical weights of 6 PA chains subject to the condition that the two gauche positions of any one gt$\bar{g}$t kink are confined to the (six) C–C bonds of one monomeric unit and that the carbonamide group adheres to a trans-planar conformation [5]. As potential energy ΔU of a gauche position the "preferred value of 0.5 kcal/mol" [5] has been used.

The change of $\ln Z(n, i)$ with r is essentially the same as the change of the complete partition function (which accounts for mgi-kinks of all types). One can cal-

Table 5.1. Statistical Weight of (n, n_g) Conformations

Polyethylene chains, $\Delta U = 0.5$ kcal/mol

n	$n_g = 0$	2	4	6	8	10
38	1	6.80	41.1	217	985	3722
40	1	7.18	46.1	261	1282	5329
42	1	7.55	51.4	311	1642	7444
44	1	7.93	56.9	366	2072	10174
46	1	8.31	62.8	427	2580	13646
48	1	8.69	68.9	495	3177	18000

6 Polyamide chains, ΔU of gauche positions of C–C bonds 0.5 kcal/mol

n	$n_g = 0$	2	4	6	8	10
35	1	5.67	25.7	87.3	198	224
42	1	6.80	38.5	174.6	594	1346
49	1	7.93	53.9	305.6	1385	4710
56	1	9.07	71.9	489.0	2771	12560

culate, therefore, from the data of Table 5.1 the entropy elastic contribution to the retractive force as $- kT \Delta \ln (Z)/\Delta r$. The energy elastic contribution, f_e, derives here from the change of conformational energy and is $\Delta n_g \Delta U/\Delta r$. For the chosen examples of a PE chain with 40 C–C bonds, a 6 PA chain of 6 monomer units and gauche energies ΔU of 0.5 and 0.75 kcal/mol one obtains the force contributions listed in Table 5.2. For comparison the forces derived for the statistically kinked chain according to the Langevin and the "exact" approximation are also given. It is immediately apparent that in kink isomers the energy-elastic contribution f_e is of the same order of magnitude — and opposite sign — as the entropy-elastic force. The molecular stresses, Ψ, are obtained by dividing the axial forces f through the molecular cross-section q, whereas the elastic moduli are understood as:

$$E = r \Delta f/\Delta rq. \tag{5.21}$$

This investigation may be summarized by saying that the retractive forces experienced by highly extended chain segments at room temperature due to conformational changes are small in comparison to the ultimate forces that can be carried by a chain. External hindrances to segment rotation have not been treated here. At present is suffices to state that they will only further diminish the —already — small entropy elastic forces.

B. Energy Elastic Chain Deformation

So far the energy-elastic deformation of the (assumedly rigid) chain skeleton has not been considered. In response to axial forces the chain skeleton will deform,

Table 5.2. Entropy-Elastic Forces and Moduli

Approach		Eq.	L Å	r Å	f_e 10^{-10}N	f 10^{-10}N	ψ MN/m²	E GN/m²
Langevin	(ℓ = 0.377 nm)	(5.14)	49	45.2	0	1.43	843	5.5
Exact	(ℓ = 0.377 nm)	(5.15)	49	45.2	0	1.20	712	4.6
PE-Kinkisomer	($\Delta U = 0.50$)	(5.20)	50.3	49	-0.55	0.10	55	1
PE-Kinkisomer	($\Delta U = 0.75$)	(5.20)	50.3	49	-0.83	-0.45	-247	(0.9)
PA-Kinkisomer	($\Delta U = 0.50$)	(5.20)	52.1	50.8	-0.55	0.08	49	1.9
PA-Kinkisomer	($\Delta U = 0.75$)	(5.20)	52.1	50.8	-0.83	-0.47	-278	(1.6)
PE bond stretching and bending:						23.8	14000	(235)

Table 5.3. Stiffnesses of Molecular Deformation Modes

Mode of Deformation	Polymer	Calculated Modulus $[GN/m^2]$	Derived from	Ref.
Hindered rotation (helix)	Polyesters	4–7	X-ray diffraction on various polyesters	12
Hindered rotation of kinked chain	PE	$2.6 \, n/n_g$	Eq. 5.21, sinusoidal rotation potential	–
Pure bending of C—C bonds	PE	80	–	12
Pure stretching of C—C bonds	PE	740	–	12
Stretching of zig-zag chain	PE	182–340	Force constants	9
	PE	182, 299	Urey Bradley force field	13
	PE	297	Minimum energy electron wave function	11
	PE	272	Deformation of rigid valence tetrapods	47
	PA 6	180–200	Spacing of (070) planes	46
	PA 6	157–196	Force constants	9
	PA 6	244	Urey Bradley force field	13
	PA 66	181	Urey Bradley force field	13
Crystal lattice deformation	PE	235	Spacing of (002) planes	9
	PA 6 (α form)	20, 24.5	Spacing of (020) planes	46, 9

however, through hindered rotation, bending, and stretching of main chain bonds (9). In Table 5.3 the stiffnesses characterizing these individual modes of deformation and their combination in a loaded chain are listed. The C–C bond bending and stretching moduli have been calculated from the corresponding force constants. The modulus for the hindered rotation of a kinked chain is derived from the rotational potential of CH_2 groups and the associated axial chain elongation as formulated by Pechhold [7] who used an intramolecular rotational barrier of 2 kcal/mol above gauche energy. The resulting modulus is a function of the angle of rotation and ranges between 4 and 7 GN/m^2 in good agreement with a value of Bowden [10] of 5.7 GN/m^2. In axial stretching the *hindered* rotation of gauche bonds contributes to axial elongation, the chain stiffness thus decreases with the concentration of gauche bonds.

Longitudinal chain moduli, defined as in Eq. (5.21), have been determined theoretically and experimentally in a number of ways:

— through calculation from spectroscopically determined bond bending and stretching constants
— from the Raman shifts of extended chain hydrocarbons
— from x-ray observations of load-induced changes of the unit cell dimensions in highly oriented polymers.

These methods, their results, and the arguments to explain a certain disagreement between the numerical results have been reviewed by e.g. Sauer and Woodward [9]. More recently the direct calculation of the total (electronic) energy (of a PE chain) with changing atomic geometry has been attempted by Boudreaux [11]. In his self-consistent field molecular orbital (SCF − MO) calculations he searched for "that set of numbers which, when used as coefficients of the assigned atomic orbitals, produces an approximation to the wave function which minimizes the total energy". Hahn et al. [14] presently investigate a model based on the elastic interaction of essentially rigid sp^3 valence tetrapods (shell-model).

From Table 5.3 it is apparent that the weakest mode of skeleton deformation (4.7 GN/m^2) is that derived from the hindered rotation of (non-extended) bonds whereas a fully extended chain segment has a 50 times larger stiffness. The elastic modulus of a partially extended kinked chain, then, is:

$$1/\overline{E} = \sum_i x_i/E_i \tag{5.22}$$

where the E_i are the stiffnesses of the individual (trans-planar, kinked, or helical) segments of a chain and the x_i are the corresponding (volume) fractions.

One is now in a position to determine the change of the free energy F of a partly extended chain as a function of the end-to-end distance r. As an example the quasi-static deformation of PE segments will be treated within the frame of the bond-bending and stretching model. The minimum free energy of a segment containing n C–C bonds and n_k 2g1 kinks is attained at an end-to-end distance $r = (n - n_k)\sqrt{2/3}\,a$. This minimum free energy is equal to $n_k \Delta U - RT \ln Z$. From the statistical weight of the (n, n_k) conformations of PE segments with n = 40 (Table 5.1) the minimum free energy values have been calculated. They are indicated in Figure 5.1 at the proper chain-end distances. If the chain ends are axially dis-

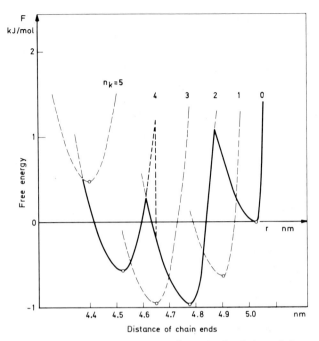

Fig. 5.1. Free energy of uniaxially deformed polyethylene chain segment. Number of C–C-bonds n = 40, number of kinks n_k = 2 n_g, extended length L = 5,03 nm.

placed from these equilibrium positions energy elastic forces arise which have an unsymmetric potential. In *stretching* completely extended sections the trans bonds are deformed in the plane of the zig-zag chain according to the chain modulus E_{str}. The gauche bonds are performing a hindered rotation out of the zig-zag plane (E_{rot}). The stretching modulus $\bar{E}$ of the kinked segment then is derived from Eq. (5.22). The chain is the stiffer the fewer gauche bonds it contains. Employing the rotation potential mentioned above [7] and an extended chain modulus of 200 GN/m^2 the *tensile* sections of the free energy curves have been calculated. The presence of only 5 kinks softens the chain segment quite noticeably ($\bar{E}$ decreases to 9.5 GN/m^2).

The axial *compression* of an extended chain differs in so far as the trans bonds now aquire the freedom to perform a hindered rotation out of the zig-zag plane. This freedom is not exercised in tension because it would be energetically less favorable than bond bending and stretching. The axial compression modulus of a zig-zag chain is essentially determined by the hindered rotation mode of deformation, it corresponds to the extension modules of a highly kinked chain (n_k = n/4) and has been taken to be 5 GN/m^2.

For each conformational state an asymmetric parabola is obtained. In quasi-static straining an individual, kinked chain will tend to assume that state of minimum free energy accessible to it. Transitions between different conformational states will occur (solid line in Fig. 5.1). The transition to a more extended conformation will cause an immediate release of axial stress for the chain segment in question — and

97

thus constitute a thermodynamically irreversible process. Macroscopically the step-wise and uncorrelated release of stress in a large number of chains is felt as stress relaxation. The random thermal motion of chain atoms and the existence of ex-ternal hindrances to bond rotation are the reason that the deformation potentials of individual chain segments deviate from each other and from the solid line – as arbitrarily indicated by the heavy dashed line. The deformation potential shown in Figure 5.1 essentially represents also that of other (planar) zig-zag chains such as PA 6 or 66 which have the same kind and concentration of gauche bonds and about the same length.

C. Non-Equilibrium Response

The stress-induced conformational transitions of a kinked chain as represented by the solid line in Figure 5.1 are thermodynamically irreversible, internal energy is converted into heat. At this point one is interested in the time-constant of the tran-sition process: if it is small compared to the time within which chain stretching is accomplished then the stress-strain curve will not deviate very much from the solid line, if the time constant is extremely large then transitions may be considered to be impossible, the chains deform energy-elastically. In the intermediate region of time constants, however, the largest stresses of not fully extended chains will depend on the rate at which stress releasing conformational transitions take place. A detailed dis-cussion of this effect would require a consideration of the geometry and interaction of molecular chains, of the principles of irreversible thermodynamics [15], and analy-sis of the potential of the secondary or van der Waal's bonds between segments [16]. It would then lead to a description of the anelastic *deformation* of polymers which is not intended in this book. Nevertheless one is still interested in some in-formation with respect to the rate of stress relieving transitions between different kink-isomers. Since any transitions leading to the motion of just a single kink gene-rally do not cause any axial chain elongation one has at least to consider the simul-taneous activation and annihilation of two kinks. Such process involves the rotation of four gauche bonds and the rotational translation of the segment between the kinks, the energy of activation can be estimated to be 8 kcal/mol for the rotation of gauche bonds [7] and 0.5 to 1.2 kcal per mol of rotating CH_2 groups to overcome the intermolecular barriers [17, 18]. A $tgt\bar{g}\,tt\bar{g}\,tgt$-kink requires, therefore, an acti-vation energy of 11 to 15.2 kcal/mol to transform into the all-trans conformation. One may assume that in stress-free PE chains such transformations do occur with a rate of at least 1 s^{-1} at a temperature somewhat below the melting point, i.e. at 400 K. One may then calculate the rate of this process at 300 K from Eq. (3.22) as 0.0018 s^{-1}. If the chain is strained the activation energy for segment rotation only decreases and the rate of stress relieving transitions increases [19]. With such a time-scale one may consider stress-induced rotational transitions as being easily accomplished before a chain segment reaches elastic energies sufficient for chain scission at room temperature, i.e. more than 25 kcal/mol.

It remains the entropy production term in Eq. (5.4) to be discussed. Under the conditions of this section – quasi-static axial loading of the ends of a segment and determination of the ensuing retractive forces – this term is essentially zero.

II. Axial Mechanical Excitation of Chains

A. Secondary or van der Waal's Bonds

In Section A the stress-strain behavior of a single chain segment loaded pointwise at
its ends has been discussed. In (uncross-linked) thermoplastic polymers, however,
large axial forces cannot be applied pointwise to a main chain bond but only be
built up gradually along the chain through — the much weaker — intermolecular
forces. The forces between molecules are a superposition of the close range (nucle-
onic) repulsion and the (electronic) van der Waal's attraction, which includes elec-
trostatic forces between ions, dipoles or quadrupoles, induction forces due to polari-
zation of atoms and molecules, and the generally most important quantum-chemical
dispersion forces. The van der Waal's attraction is the cause of the solidification and
crystallization of high polymers, it is theoretically quite well understood and has
been treated in detail by Langbein [16]. With reference to this work and the general
references in Chapter 1 it can be stated that the secondary forces are not saturated
and are nondirectional, i.e. not confined to precise locations of the neighbor atoms
at e.g. tetrahedral bond angles. To the extent that these assumptions are valid the
potential of the intermolecular forces acting on a chain or segment can be replaced
by the sum of the interaction potentials of all appropriate pairs of atoms. The pair
potentials contain an attractive term — which is theoretically derivable and decreases
with the 6th power of the interatomal distance [16] and a repulsive term for which
only semi-empirical expressions can be given. The total energy of intermolecular
interaction, i.e. the cohesive energy of a solid then is to be understood as the sum of
the pair interactions. In thermodynamic equilibrium a chain segment attains a position
such that the free energy of interaction is a minimum. If a chain segment is displaced
with respect to this position the free energy is raised and retractive forces arise. Such
axial displacements are realized in quasi-static pulling a chain segment which is part
of a crystal lamella against the crystal lattice or of extended chains against a glassy
matrix. Dynamic chain stretching occurs in the shear deformation of molecules in
solution or the melt. In the following it will be discussed which level of axial chain
stress can be attained in these cases, that is in the loading of primary bonds through
secondary or van der Waal's forces.

B. Static Displacements of Chains Against Crystal Lattices

The first investigations of the loading of chains displaced with respect to their
equilibrium positions within a crystal lattice have been carried out by Chevychelov
[20]. He used a continuum approach, i.e. replaced the string of discrete chain atoms
by a continuous mass distribution. Each mass point experiences a periodic repulsive
crystal potential if displaced from its equilibrium position. Later Kausch and Langbein [21] and Kausch and Becht [22] extended these calculations to treat the static
and dynamic interaction of chains of discrete atoms with arbitrary periodic potentials.

The model used in these investigations is represented by Figure 5.2. It consists of a tie molecule (t) which forms part of a crystal lamella (c), leaves it perpendicularly to the crystal fold surface, (I), extends straightly through an amorphous region (a) and enters the adjacent crystallite of which it forms a part. The crystal boundaries are assumed to be ideally sharp. No interaction between the tie chain and the amorphous material outside of the crystallite is taken into account.

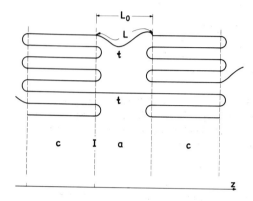

Fig. 5.2. Model of tie molecules (t) connecting folded chain crystal lamellae (c). a = amorphous region, L_0 = length of amorphous region, L = contour length of tie molecule, I = crystal boundary.

A tie molecule thus embedded in two different crystalline layers is streched if the crystal lamellae are separated by macroscopic stresses. Because of the large elastic modulus of the chain molecule very high axial stresses are exerted onto the chain if it experiences the same strain as its amorphous vicinity. This high axial stress tends to pull the tie molecule against the intracrystalline potential and out of the crystal lamella.

In the following calculations the assumedly rigid solid is represented by an interaction potential the depth of which corresponds to two times the average (lateral) cohesive energy U with which the chain segments are bound within lamellar crystals or amorphous regions. Upon axial displacement in z-direction (chain axis) a chain will rotate and translate along a temperature-dependent minimum energy path. For polyethylene (PE) Lindenmeyer [17] has given the maximum potential energy to be overcome by a CH_2 group in twisted translation as $\Delta U_{tt} = 500$ cal/mol at room temperature. The potential energy of a CH_2 group within the amorphous phase can be determined from the difference of the enthalpies of crystalline and amorphous PE at absolute zero temperature, $H_{am, o} - H_{cr, o} = 661$ cal/mol [23]. It should also be taken into account that the internal energy of a partly oriented PE may be smaller than that of an amorphous one. The energy difference will be comparable to the enthalpy difference $(H_{am} - H_{def})$, which was estimated by Fischer and Hinrichsen [24] to be 140 cal/mol. As function of its z-position within an otherwise undisturbed crystal a CH_2 group would encounter a potential energy as shown in Figure 5.3.

Each point along a tie chain is characterized by its distance z from the crystal boundary, where z refers to the unstressed situation. The displacement of the molecular point z under stress is denoted by $u = u(z)$ and the crystal potential per unit

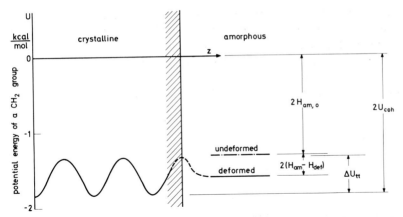

Fig. 5.3. Lattice potential in displacement of a CH_2 group in crystalline and amorphous poly-ethylene (after [22]). U_{coh} = cohesive energy per CH_2 in PE crystal, ΔU_{tt} = potential maximum in twisted translation, $H_{am, 0}$ = enthalpy of PE crystal at 0 K, $H_{am} - H_{def}$ = difference of enthalpies of undeformed and deformed amorphous phase of PE.

length by V(u). The axial stress, s, acting along the chain decreases with increasing |z| due to the forces exerted on the chain through the crystal potential V(u). Equilibrium of forces at each section of the chain is established if

$$\frac{ds}{dz} = \frac{dV(u)}{du} \,. \tag{5.23}$$

This relation implies no assumptions with respect to the form of V(u) or the response s(u) of the chain.

One does assume, on the other hand, the tie-chain segments within the crystallites to resemble an elastic spring, i.e. one equates the axial stress s to the local chain deformation

$$s = \kappa \, du/dz \tag{5.24}$$

yielding

$$\kappa \frac{d^2 u}{dz^2} = \frac{dV(u)}{du} \,. \tag{5.25}$$

Here κ is the constant of chain elasticity (not normalized with respect to the chain cross-section).

For a semi-infinite crystal and a sinusoidal potential V(u) = v(1 − cos K u(z)) Chevychelov integrated Eq. (5.25) to obtain

$$\frac{1}{2} \left(\frac{du}{dz} \right)^2 = A - \frac{v}{\kappa} \cos K u \tag{5.26}$$

and, after having determined $A = v/\kappa$ from $du/dz = 0$ for zero displacement,

$$\frac{du}{dz} = 2\sqrt{v/\kappa}\ \sin K\ u/2 \tag{5.27}$$

and

$$u(z) = \frac{4}{K}\ \arctan\ \{C\ \exp\ [(K\ \sqrt{v/\kappa})\ z]\}. \tag{5.28}$$

The integration constant C is tan $Ku(0)/4$ and K is 2π divided by the intra-chain repeat unit length d. For a polyethylene chain with a spring constant κ of 34 nN Eqs. (5.28) and (5.24) have been evaluated [22]. The crystal potential depth v has been derived as 0.0138 nN from the cohesive energy density δ^2 through

$$v = 2\delta^2\ M/\rho\ d\ N_L \tag{5.29}$$

where N_L is Avogadros number, M the molecular weight of a repeat unit, and ρ the density. In static mechanical equilibrium the stresses and displacements of tie chain atoms within the crystallite are a unique function of the axial tension s_a existing at the crystal boundary $(z = 0)$. In Figure 5.4 tension and displacement of a PE chain within a PE crystallite are shown for maximum axial chain tension s_0 equal to 1.372 nN. For any force larger than s_0 no static equilibrium can be established within a defect-free crystallite. It should be noted that this force which pulls out of the crystal a chain of full crystal length is 15.8 times larger than the force κ v d which is necessary to pull out a single monomer unit. If the tension s_0 is divided

Fig. 5.4. Decrease of tension and displacement of a highly stressed tie chain within a polyethylene folded chain crystal lamella (after [22]).

102

by the chain cross-section (0.1824 nm^2) one arrives at the maximum axial stress which a perfect PE crystallite is able to exert on a tie molecule: 7.5 GN/m^2. As can be seen from Figure 5.4, the maximum mechanical excitation of a chain penetrates about 5 nm into the crystallite. At a distance of 6 nm from the crystal boundary it is smaller than the average thermal amplitude at room temperature, which is about 0.008 nm.

Kausch and Langbein [21] extended the above calculations to various boundary conditions at the lower stress end, $-Z$, of the chain. If the chain terminates at $-Z$ or continues into the adjacent amorphous region, it experiences no or only very weak axial forces. This case is covered by putting $A < v/\kappa$. If, on the other hand, a stressed tie-chain folds back at $-Z$, any axial displacement there implies a deformation of the fold. Such a deformation entails a considerable distortion of the crystal lattice and requires strong axial forces. In this case the integration constant A generally becomes larger than v/κ. After having derived the equations necessary for describing the elastic interaction between a chain and a crystal – represented by one unique potential – for arbitrary boundary conditions the authors [21] took into account also the fact that in a crystal containing hydrogen bonds – like that of 6-polyamide – the intermolecular interaction differs in different crystal regions. It is much stronger at the sites of C=O and N–H chain groups than it is at the sites of CH$_2$ groups. The problem of calculating the elastic displacement of a continuous chain passing through alternating regions of strong and weak attraction can be solved through alternate application of appropriate boundary conditions. In this case the boundary condition is that chain displacement and chain tension are continuous at all points z_i where the potential changes from weak attraction ($v = v_1$) to strong attraction ($v = v_2$) or vice versa (refer to Fig. 5.5 for notation). Here it was assumed that the difference

Fig. 5.5. Geometry and notations used in calculating stresses and displacements of 6 polyamide chains.

between the cohesive energies U_{coh} of PA (18.4 kcal/mol) and PE (6.3 kcal/mol) is caused by the presence of hydrogen bonds between the carbonamide groups of neighboring molecules within the grid planes of the (monoclinic) PA crystallites. This excess energy contributes to the intermolecular potential v_2 acting between carbonamide groups. The potential v_2 thus is ($12.1 + 0.50$) kcal/mol, which is equivalent to 0.369 nN.

103

Chain tension and displacement of a 6-PA tie molecule within a 6-PA crystal are shown in Figure 5.6. The strong attractive effect of the hydrogen bonds results in a rapid drop of chain tension at the position of the carbonamide groups. The decrease of both tension and displacement is much more rapid than in the case of PE. At a distance of 2.1 nm from the crystal boundary the displacement has already dropped to the average level of thermal vibration at room temperature. With the crystal boundary positioned – as shown in Figure 5.6 – at the end of a $(-CH_2-)_5$ segment the maximum chain tension is $s_0 = 3.94$ nN. This tension is only 1.7 times the force necessary to remove one carbonamide group from the crystal, i.e. 59% of the maximum chain tension is expended to break the hydrogen bonds of the

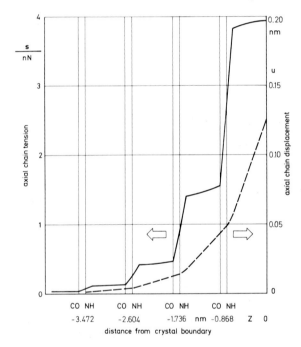

Fig. 5.6. Decrease of tension and displacement of highly stressed PA 6 tie molecules within folded chain crystal lamellae (after [22]).

first CONH group. It should be noted that the value of s_0 depends upon the geometric arrangement of the atoms at the crystal boundary. With a carbonamide group at the crystal boundary the maximum tension s_0 is about 20% smaller than in the case where the crystal boundary runs through the middle of the $(-CH_2-)_5$ segment [21].

If s_0 is divided by the molecular cross section (0.176 nm^2) a stress of 22.4 GN/m^2 is obtained, which may be termed the static strength of a perfect (monoclinic) 6-PA crystal.

The above strength values have been derived for the interaction between a periodic potential and a continuous elastic chain. Kausch and Langbein [21] have shown, however, that there are only insignificant changes if the model using a continuous chain is replaced by the more accurate one using a string of discrete atoms.

C. Thermally Activated Displacements of Chains Against Crystal Lattices

As stated repeatedly crystal and chain are in thermal motion; the chain atoms, therefore, vibrate around the new equilibrium positions $z + u$ given e.g. by Eq. (5.28).
The effect of the correlated or uncorrelated motion of the chain atoms on the intermolecular potential energy can be calculated if the dispersion of vibrational amplitudes along the chain is known [22]. A chain subjected to maximum tension s_0
is not resistant to translation. Even the smallest thermal vibration will cause further
displacement of the chain and a decrease in axial tension at the boundary. The point
of maximum tension will rapidly propagate into the crystallite (Fig. 5.7). The "dis-

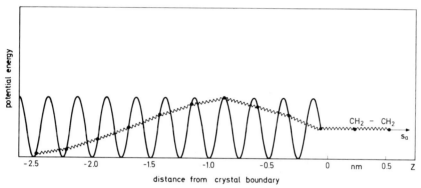

Fig. 5.7. Schematic representation of a stress-induced defect. The point of maximum chain
stress travels *into* the crystal as the leading chain group is pulled *out* of the crystal (after [22]).

location" is fully developed once u_a has reached the value of d. The energy W_d of
the defect introduced into the crystal by displacement of the atoms of one chain
is easily calculated as the sum of the potential energies of the chain atoms and of
the elastic chain energy. Both are a function of external tension. In Figure 5.8 the
dislocation energy has been plotted as a function of displacement u_a and tension
s_a at the crystal boundary. The energy of a completed defect (displacement of the
chain by d) in PE amounts to 32.85 kcal/mol. At tensions smaller than s_0 the displacements u_a are smaller than d/2, the defect energy is smaller than $W_d(s_0)$, and
the defect has a stable position. At displacements larger than d/2 the system becomes
unstable since the chain tension decreases with increasing displacement u_a.

As shown in Figure 5.8, transitions between the stable and the unstable state
can be caused by thermal activation. The required activation energy U_d is given by

$$U_d(s_a) = 2 \left[W_d(s_0) - W_d(s_a) \right]. \tag{5.30}$$

In Table 5.4 the activation energies required for translation of CH_2-CH_2 segments are listed for various displacements u_a and tensions s_a. One observes that
at chain tensions which are as little as 4% less than the maximum tension still an
activation energy for chain translation of 9.8 kcal/mol is required. One may safely

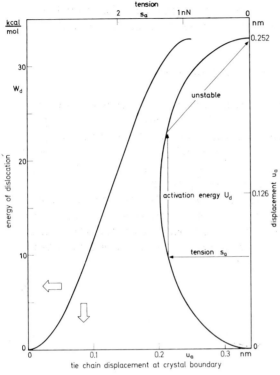

Fig. 5.8. Energy W_d of a chain displacement defect in a polyethylene crystal as function of tie-chain displacement u_a at crystal boundary (after [22]).

Table 5.4. Activation Energies for Chain Translation in PE

u_a Å	S_a 10^{-10} N	U_d 10^{-20} J	U_d kcal/mol
0	0	23.08	33.0
0.104	1.79	22.89	32.7
0.195	3.38	22.38	32.0
0.303	5.18	21.40	30.6
0.412	6.91	20.00	28.6
0.494	8.12	18.69	26.6
0.589	9.41	16.90	24.2
0.699	10.71	14.60	20.8
0.824	11.93	11.68	16.6
0.914	12.63	9.39	13.40
1.009	13.18	6.88	9.82
1.108	13.56	4.19	5.98
1.158	13.67	2.80	4.12
1.209	13.72	1.40	2.06
1.260	13.72	0	0

state, therefore, that even under consideration of longitudinal chain vibrations chain translation will occur at chain tensions close to the specified values.

D. Chain Displacements Against Randomly Distributed Forces

The rupture of chains in amorphous matrices or in solution clearly indicates that large axial forces can also be reached in the absence of periodic potentials. Again forces derive from the displacement of chains or of parts thereof with respect to the surrounding matrix. If one divides a non-extended chain into small sections of identical projected lengths one will obtain sections of quite different conformation. As well the sectional chain moduli E_i as — most probably — the interaction potentials $V_i = -w_i \cdot u_i^2/2$ will vary from section to section. To obtain the axial displacement of an arbitrary sequence of chain sections an interpolation given by Kausch and Langbein [21] may be used who have found that the changes of chain stress σ and displacement u within a heterogeneous matrix are interpolated by an exponential law:

$$u = u_0 \exp \gamma \, z/L \tag{5.31}$$

$$\text{with } \gamma = \sqrt{n \, \sum_{i=1}^{n} (w_i L_i/E_i \, q)}. \tag{5.32}$$

The chain stress is derived from Eq. (5.24) replacing the chain elasticity constant κ by the elastic constant of a series of elastic elements

$$\sigma = u_0 \, \gamma/\Sigma(L_i/E_i). \tag{5.33}$$

The above equation shows quantitatively that large chain stresses will only be obtained if the interaction force constants w_i are of about the same order they are in a crystal and if the elastic moduli of the chain sections are large throughout. One "weak" section will greatly enhance average displacement and reduce the stress. It should also be mentioned that it is not the chain length L and the absolute number n of interaction sites which raise the chain stress but the intensity of the interaction forces per unit chain length.

As an example of the static excitation of a chain under these conditions a nearly extended chain of end-to-end distance L embedded in a glassy matrix may be considered. If the matrix (index m) and the chain (no index) were homogeneously strained in chain axis direction they would experience the following stresses:

$$\sigma_m = \epsilon_m E_m \tag{5.34}$$

and

$$\sigma = \epsilon_m L/\Sigma L_i/E_i. \tag{5.35}$$

Due to the different elastic response the free chain would have to act against the lattice potential until chain stress and lattice interaction were in equilibrium at any point of the chain. This interaction can be described by identically the same mathematical formalism developed above. Since the chain ends are assumed to be free they cannot bear any axial stresses. This condition will exactly be met if one superimposes the hypothetical stress σ (Eq. 5.35) by a compressive stress according to Eq. (5.33) where

$$u_0 = \epsilon_m L/\gamma. \tag{5.36}$$

The largest stresses will be found in the center section of the strained chain and here the superposition yields:

$$\sigma(L/2) = \frac{\epsilon_m L}{\Sigma L_i/E_i} [1 - \exp(-\gamma/2)]. \tag{5.37}$$

The distribution of axial stresses along the chain is characterized by the two stress-free chain ends, two terminal sections of length L/γ where the stress increases with a rate determined by γ/L, and a central chain segment subjected to the maximum stress given by Eq. (5.37). The *continuous* straining of a chain segment through interaction with (harmonic) lattice potentials — as required for the terminal sections — is limited to a section of finite length because the largest displacement u may not exceed the range of validity of the intermolecular potential which is of the order of 0.1 nm. For breaking stresses of 20 GN/m^2 to be reached under such a condition the constant γ/L must be larger than 0.01 nm *and* the average chain modulus at these loads must be that of a fully extended chain; since the displacements of PE and 6 PA chains within their rigid crystal lamellae led to γ/L-values of 0.004 and 0.015 nm respectively [22] one is forced to the conclusion that the displacements of extended chains within the less tightly packed random aggregates of misaligned chains can only lead to smaller than the above values of γ/L. If chain scission were to occur under the above static conditions it would require a sufficiently stable physical crosslinking of the (then non-extended) terminal chain sections to the matrix through folds or kinks.

E. Dynamic loading of a chain

So far the quasi-static interaction between a chain and a surrounding matrix has been investigated. It has been indicated that the level of axial forces to be obtained in that manner is limited by the occurrence of chain slip. In dynamic loading these force levels can be exceeded if frictional forces or forces of inertia become effective.

If the chain displacement does not occur in static equilibrium and not through a single thermally excited step then the translation will not occur reversibly along a path of least energy but will require more energy than in the previous cases. The rate-dependent energy expended per unit distance of forced translation of a chain segment i is equivalent to a frictional shearing force τ_i. The response of chains to shearing forces in solution has extensively been investigated and discussed [25]. A number of different molecular theories of the viscoelastic behavior of polymer chains in solution have been developed. These predict a relation between e.g. molecular weight M (or degree of polymerisation P), solvent viscosity η_s, intrinsic viscosity $[\eta] = \lim (\eta - \eta_s)/c\eta_s$, molecular friction coefficient ζ_0, and mean square end-to-end distance $\overline{r_0^2}$. Following the bead spring model of Rouse [25] one obtains

$$[\eta] \, \eta_s = N_L \, \overline{r_0^2} \, \zeta_0/36 \, M_0 .$$
(5.38)

The quantity ζ_0 represents the frictional force encountered by each monomeric unit within the chain which is moved with unit velocity through its surrounding. The sectional force f_i, therefore, is:

$$f_i = \zeta_0 \cdot v_i = \zeta_0 \, (\dot{u}_{im} - \dot{u}_{ic}) .$$
(5.39)

In the case of additivity of all sectional forces and in a homogeneous shear field one obtains the following distribution of axial stress ψ within a chain of length L:

$$\psi = E d u_c/dz = \frac{\zeta_0 \dot{\epsilon} \rho N_A}{M_0} \left(\frac{L^2}{8} - \frac{z^2}{2} \right) .$$
(5.40)

The largest stresses are encountered in the middle $(z = 0)$ and amount to

$$\psi_{max} = \zeta_0 \dot{\epsilon} M^2 L_{mon}/8 M_0^2 .$$
(5.41)

Equation (5.41) predicts that for a given system (ζ_0, L_{mon}, M_0) axial stresses are proportional to shear strain and square of molecular weight. Chain rupture occurs if ψ_{max} reaches or exceeds the chain strength ψ_b resulting in a gradual decrease of M. The decrease of the average molecular weight diminishes if M approaches the limiting value $M(\infty)$ which is derived from Eq. (5.41) for the chain strength ψ_b. The Eqs. (5.41), (5.38), and (5.6) together predict that $[\eta] \dot{\epsilon} M(\infty)$ is constant for a given polymer/solvent system. Kadim [26] obtained for a 0.5 to 1% solution of PIB in decalin $[\eta] \tau M(\infty) = 1.34$ kJ/mol. He employed shear rates between 25 and 700 s^{-1}. Abdel-Alim et al. [27] report on a relation of $\tau^{0.41} M(\infty) = $ = const. for a 2% solution of polyacrylamide in water subjected to shear rates of between 3000 and 100000 s^{-1}.

Within more concentrated solutions the physical-chemical behavior of long chains (size and interpenetration of molecular coils, kinetics of entanglement formation) seems to be much more important than the rheological behavior characterized by ζ_0 and $\dot{\epsilon}$. A striking demonstration of this fact was given by Breitenbach, Rigler and Wolf [28], who prepared solutions of 3.6 to 14.2% by weight of polystyrene in cyclohexane. These systems showed a concentration-dependent phase separation at a temperature T_{tr} between 26.4 and 29.4 $^\circ$C. Shearing these solutions at a rate of 600 s^{-1} and at temperatures slightly above T_{tr} they observed a dramatic increase in the rate of degradation when approaching T_{tr}. At $T_{tr} + 11.6$ K no noticeable degradation occurred within 20 h. At $T_r + 0.6$ K a 13% decrease of the limiting viscosity $[\eta]$ was measured after one hour already. Within 20 h a reduction of the molecular weight from $\simeq 7 \cdot 10^5$ g/mol to $1.6 \cdot 10^5$ could be achieved.

A related subject of growing interest is the ultrasonic degradation of polymers in solution. In their recent review Basedow and Ebert [29] summarize that ultrasonic chain scission is the result of cavitation, i.e. of the nucleation, growth and collapse of (gas-filled) bubbles. Chain scission occurs in the convergent flow fields in the vicinity of collapsing bubbles and through the ensuing shock waves. The properties of the solvent do not seem to be of particular importance, but the presence of nuclei to initiate cavitation and a minimum intensity of the ultrasonic field (environ $4 \, \text{W/cm}^2$) are necessary conditions. Owing to the nature of the degradation process, i.e. to the *shear loading* of chains, the rates of degradation decrease with decreasing molecular weight (degree of polymerization) and become infinitely small at a so-called *limiting molecular weight*. For polystyrene in tetrahydrofuran degraded 88 h at 20 kHz Basedow and Ebert cite one limiting molecular weight of 24000 g/mol and a newer value of 15000 g/mol. In a quite recent publication Sheth et al. [39] report for the same system the appearance of appreciable amounts of polymer material at molecular weight of 1000 g/mol or lower in the gel permeation curve of the degraded sample. They indicate that the limiting degree of polymerization depends on the initial molecular weight distribution and is either much lower than previously reported or perhaps non-existent.

The stresses encountered by a chain displaced with respect to a *solid* matrix can also be described utilizing the concept of the monomeric friction coefficient ζ_0 [25]. The implications of such an assumption are discussed in detail by Ferry [25] who also lists the numerical values of the monomeric friction coefficients for many high polymers. Naturally the coefficients strongly depend on temperature. But even if compared at corresponding temperature, e.g. at the individual glass transition temperatures, the monomeric friction coefficients vary with the physical and chemical chain structure by 10 orders of magnitude. In the upper region one finds the 1740 Ns/m for Polymethylacrylate, the 19.5 Ns/m for Polyvinyl acetate, and the 11.2 Ns/m for Polyvinyl chloride, each at the respective glass transition temperature [25]. This means that a PVC segment pulled at 80 °C through a PVC matrix at a rate of 0.005 nm/s encounters a shearing force of 0.056 nN per monomeric unit. At lower temperatures the molecular friction coefficient increases essentially in proportion to the intensity of the relaxation time spectrum $H(\tau)$, the increase being about one to two orders of magnitude with every 10 K temperature difference [25]. If a PVC segment has a length of 2 L = 10 nm and is part of a PVC matrix strained at a rate of 0.001 /s at the glass transition temperature it experiences at the center an axial force of magnitude 1.1 nN. This force is about as large as that statically transferred upon extended PE chains by PE crystallites at room temperature. It may be repeated that the above estimate on the level of axial forces is based on the assumption that in sample straining an extended chain segment is subjected to shearing forces determined by the monomeric friction coefficients ζ_0 and that the segment does not relieve axial stresses through conformational changes. Using the same assumptions one expects forces up to 206 nN for PMMA and up to 0.01 nN for PS each at the respective glass transition temperature and a strain rate of 0.001/s.

This estimate shows that in dynamic straining of glassy polymers very large axial forces can be transferred onto extended chain segments if the straining occurs at a temperature not higher than the glass transition temperature (PMMA) or some 10 K (PVC) to 20 K (PS) lower than that. These forces are sufficient to cause chain breakage.

A loading mechanism not to be investigated in detail in the context of this monograph is the straining of chain molecules through forces of inertia, e.g. through propagating stress waves. Brittle thermoplastic materials (PS, SAN, PMMA) behave "classically" at velocities of uniaxial straining below 3 m/s or strain rates below 50 s^{-1} [30]. In this region increase the strength properties and decrease the elongation with increasing strain rate. At strain rates between 50 and 66 s^{-1} a transition to stress wave-initiated fracture was observed which was accompanied by a tenfold reduction of the apparent load-carrying capability [30]. Skelton et al. [40] studied PA 6, PETP, and aromatic polyamide (Nomex). At ambient temperatures and in the range of loading rates between 0.01 and 140 s^{-1} these fibers also behaved classically. At $-67\,^{\circ}$C and $-196\,^{\circ}$C a decrease in strength with loading rate was observed at a loading rate of 30 s^{-1}.

III. Deexcitation of Chains

The mechanisms of deexcitation of stressed chains have been discussed en passant while investigating bond strength and chain loading. These mechanisms are chain slip with respect to a surrounding matrix (enthalpy relaxation), change of chain conformation (entropy relaxation), or chain rupture.

The retraction of a stressed chain through slippage leads to an exponential stress relaxation

$$\psi = \psi_0 \exp\left(-t/\tau\right). \tag{5.42}$$

If the slippage is adequately described by the monomeric friction coefficient ζ_0 the relaxation time τ will be given by:

$$\tau = \zeta_0\, L_{mon}/E\, q \tag{5.43}$$

where q is the chain cross-section. For a PVC chain at the glass transition temperature one obtains a relaxation time of 0.05 s.

The effect of a change of conformation on the elastic chain energy had been demonstrated in Figure 5.1. The annihilation of 4 gauche conformations within a PE segment of 5 nm length corresponds to an increase in length of the segment of 0.25 nm. This increase of 5% will reduce the axial elastic forces acting on the chain by 0.05 E, i.e. by $\simeq 10$ GN/m^2. In static straining this will correspond to complete unloading since the largest elastically transmitted forces yielded axial stresses of only up to 7.5 GN/m^2. The stress-biased change of chain conformation will lead, therefore, to a considerable reduction of axial stresses. The rate of conformational

111

changes — even of high T_g polymers and in a stress-free state — is sufficiently large and it will be further increased under stress [19, 31].

Conformational changes under stress have been observed repeatedly (e.g. 41–45). Gafurov and Nowak [41] studied the IR absorption of PE fibers in the range of 1200 to 1400 cm^{-1}. The extinctions of four bands (at 1370, 1350, 1305, and 1270 cm^{-1}) observed there and related to the (wagging) vibrations of methylene groups in gauche conformations decreased with strain. The authors estimated that at 4% sample strain (which corresponded to much larger amorphous strains) the number of gauche isomers in the amorphous regions was decreased by 20%.

Zhurkov et al. [42] derived the degree of chain orientation from the dichroism of "deformed-IR bands" of PETP (cf. Chapter 8 I B and Fig. 8.5); they observed that for highly stressed segments an initial orientation with $\cos^2\theta = 0.75$ had turned into practically complete orientation ($\overline{\cos^2\theta} \simeq 1.0$). The problems posed by an application of this method are discussed in detail by Read et al. [43]. Bouriot [44] deduced from IR measurements a reversible cis-trans transformation of the ethylene glycol segment in PETP fibers under the effect of tensile stresses. In polyamide 66 he observed a reversible increase of free (non-associated) NH groups and an increase in the average chain orientation with chain tension.

Conformational changes due to straining of highly uniaxially oriented PETP films have been reported by Sikka et al. [45]. These changes have been inferred from a comparison of the IR spectrum of unloaded and loaded PETP films. Employing a highly sensitive Fourier Transform IR spectrometer Sikka noticed frequency shifts and splittings of the absorption bands in the subtraction spectra (stressed PETP IR spectrum minus unstressed PETP IR spectrum). At a stress of the order of 20% of the breaking stress those IR bands were affected which are assigned to the gauche conformations of the C–O bond next to the C–C bond. These conformations are most likely to exist in the amorphous regions of the highly uniaxially oriented semi-crystalline polymer. The stress-induced conformational changes seemed to include all angles of bond rotation from minute ones to the completed gauche-trans transition.

The third mechanism of releasing axial stresses is the breakage of a chain through homolytic bond scission, ionic decomposition or degrading chemical reactions.

It is assumed that chain scission is a thermo-mechanically activated process. A chain segment containing n_c weakest bonds of energy U_0, and carrying at its ends a constant stress ψ_0 will, on the average, break after time τ_c. One refers to ψ_c as the *time-dependent strength* of a chain *segment:*

$$\psi_c (U_0, n_c) = (U_0 - RT \, \ell n \omega_0 n_c \tau_c)/\beta. \tag{5.44}$$

The quantities ω_0 and β have been defined as bond vibration frequency and activation volume for bond scission respectively [32]. A 5 nm PA 6 segment, for instance, at room temperature, has an instantaneous strength of 20 GN/m^2 (taking $U_0 = 188$ kJ/mol, $\omega_0 = 10^{12}$ s^{-1}, $n_c = 12$, $\tau_c = 4$ s, $\beta = 5.53 \cdot 10^{-6}$ m^3/mol). If in straining a segment, a stress of 20 GN/m^2 is maintained for a period of 4 s the segment has a probability of failure of $1 - 1/e = 63\%$. In view of the different interpretations found in the literature concerning the significance of ω_0, β, and even U_0 the assumptions made in deriving Eq. (5.44) will briefly be discussed.

A chain in thermal contact with its surrounding can be represented by a system of coupled oscillators. The degree of excitation of individual oscillators (modes of vibration) is statistically changing. In the absence of external mechanical forces chain scission at a C–C bond will occur whenever an oscillator representing a C–C stretch vibration becomes excited beyond a critical level, U_0, the "strength" of the C–C bond. In Figure 4.1 the potential energy scheme was given illustrating different states of vibrational energy, the bond strength U_0, and the dissociation energy D. One is interested in the rate of dissociation events of oscillators activated beyond U_0.

To evaluate this problem it will be assumed that Boltzmann statistics can be applied to the system of, say, s oscillators representing the C–C stretch vibrations within a volume element of the polymer. The fraction of energy states in which this system s will have a total energy level $E_r = r\epsilon$ is proportional to the degeneracy g_r of the energy level E_r and the Boltzmann factor $\exp(-E_r/RT)$. (ϵ is the – constant – spacing between oscillator energy levels). The partition function, therefore, will be given by:

$$Q_t = \sum_r g_r \exp(-E_r/RT). \tag{5.45}$$

The number of states in which at least one oscillator has an energy $E_i \geqslant U_0$ is given by Q_a:

$$Q_a = \sum_r g'_r \exp[-(E_r + U_0)/RT]. \tag{5.46}$$

For thermodynamic equilibrium within the volume element comprising s oscillators the fraction K^+ of activated states, i.e. states with $E_i \geqslant U_0$, is given by:

$$K^+ = Q_a/Q_t. \tag{5.47}$$

This can be approximated [33] if $r \gg 1$ by:

$$\frac{Q_a}{Q_t} = \frac{g'_m}{g_m} \exp(-U_0/RT) \tag{5.48}$$

where g_m is the degeneracy of the most probably energy level. Expressing g_m by the entropy S according to:

$$S = R \ln g_m \tag{5.49}$$

one obtains

$$K^+ = \exp(-U_0/RT) \exp \Delta S/R \tag{5.50}$$

which is identical with the relation $\ln K^+ = -\Delta F^+/RT$. The rate of dissociation, k_b, then follows from K^+ [34]:

$$k_b = \kappa \frac{kT}{h} K^+ = \kappa \frac{kT}{h} \exp(-U_0/RT) \exp \Delta S/R \tag{5.51}$$

113

where κ is the transmission constant which indicates how many of the activated complexes actually disintegrate. By combining all front factors and $\exp \Delta S/R$ one may rewrite this equation as:

$$k_b = \omega_0 \exp(-U_0/RT). \tag{5.52}$$

The quantity ω_0 is proportional to T; this dependency is generally negligible against the T-dependency caused by the exponential term.

The rate k_c for the breakage of one out of n_c equal and independent bonds is n_c times the rate k_b for the breakage of an individual bond. If the bonds are not independent it will mean that more than one oscillator has to acquire enough vibrational energy for dissociation. Depending on the coupling such a *cooperative region* may count some 5 or more oscillators [35]. In this case an experimentally determined term U_0 stands for the total vibrational energy required by all (coupled) atoms which are involved in one chain rupture event. This could mean that the potential barrier between neighboring sites of an uncoupled atom is much less than U_0. This meaning of U_0 has to be kept in mind if the numerical value U_0 is being compared with other activation energies.

Another problem in connection with U_0 arises from the fact that in a kinetic experiment such as the determination of the rate of chain scission one derives the bond strength U_0, that is the maximum of the potential barrier (cf. Fig. 4.1). In calorimetric measurements the reaction enthalpy (dissociation energy D) is determined.

By application of external forces the free energy of the activated state is changed with respect to the ground state. One may say that ΔF^+ has to be replaced by $\Delta F^+ + \Delta F_{ext}$. The term ΔF_{ext} will be some function f of the stress ψ which is acting on the molecular chain along its axis; Equation (5.52) then becomes:

$$k_c = n_c \, \omega_0 \exp - \frac{U_0 + f(\psi)}{RT}. \tag{5.53}$$

The form of f is subject to discussion [35–37]. Linear and quadratic relationships have been proposed. A linear relationship follows if one assumes that the force acting on a particular chain is not changed by the displacement of the chain under action of the force (comparable to a dead weight loading). The correction term $f(\psi)$ will be quadratic in ψ if the stored elastic energy within a small chain section to either side of the point of chain scission is thought to be responsible for the rate of chain fracture. From the Morse potential one obtains a difference of square root expressions which, however, results in an almost linear relationship between force and activation energy in the force region of interest [36]. The effect of a *linear* stress potential on the intramolecular binding energy is shown in Figure 5.9.

Some information concerning the form of $f(\psi)$ and the relation between molecular stress ψ and macroscopic stress σ has been obtained by Zhurkov, Vettegren et al. [37, 32]. These authors have studied the effect of macroscopic stress σ on infrared absorption of oriented polymers. The frequencies of skeleton vibrations of polymer molecules show the effect of external axial forces on molecular binding force-constants. The shift and change of shape of suitable infrared absorption bands

114

Fig. 5.9. Effect of axial stress on intra-molecular binding potential (from [19]).

should reveal, therefore, the intensity and distribution of actual molecular stresses ψ. Experiments carried out with polypropylene and polyamide 6 have shown that for a stressed sample the maximum of the investigated absorption band decreases slightly and shifts towards lower frequencies [37, 32]. In addition a low frequency tail of the absorption band appears. In agreement with existing structural models the authors interpreted their results in assigning the shifted symmetrical part of the absorption band to the crystalline phase of the sample, the low frequency tail to the amorphous phase. The shift of the symmetrical part of the absorption band, $\Delta \nu_s$, turned out to be linear with macroscopic stress in all performed experiments:

$$\Delta \nu_s = \nu_0 - \nu_\sigma = \alpha \, \sigma. \tag{5.54}$$

Following a theory of Bernstein [38] the shift of the stretching frequency of a bond can be related to a shift in dissociation energy D of the same bond:

$$\nu_0^2 - \nu_\sigma^2 = c \, (D_0 - D_\sigma). \tag{5.55}$$

Since $\Delta \nu_s \ll \nu_0$ one may substitute $\nu_0^2 - \nu_\sigma^2$ by $\Delta \nu_s \cdot 2\nu_0$. A consequence of this condition is that also $(\nu_0^2 - \nu_\sigma^2)$ is proportional to σ. The choice of ν or ν^2 in Eq. (5.54) only affects α, not the power of σ. If one further assumes that the activation energy U_0 is changed by a similar amount as D_0 one obtains:

$$U_\sigma = U_0 - \frac{2\nu_0}{c} \, \alpha \, \sigma. \tag{5.56}$$

In order to derive the form of $f(\psi)$ the relation between σ and ψ for oriented crystalline phases has to be established. According to experimental and theoretical evidence it is assumed that a condition of homogeneous stress exists within the crystalline layers of a regular sandwich structure and that ψ is proportional to σ. Equation (5.56) thus yields that the activation energy for scission of an oriented chain molecule is linearly decreased by stresses acting upon this molecule. Equation (5.53) with (5.56) may be written as:

$$k_c = n_c \, \omega_0 \, \exp \left[(-U_0 + \beta \, \psi)/RT\right] \tag{5.57}$$

which is equivalent to Eq. (5.44). It is believed that the rate of fracture of chains *all subjected to the same local stress* ψ can correctly be described by Eq. (5.57).

On the basis of the foregoing considerations the response of two chains of different conformation to axial stresses has been evaluated in Figure 5.10. An extended chain segment ($L = L_0$) stretches elastically up to the time-dependent fracture stress. The calculations have been carried out employing a modulus of elasticity for the polyamide 6 chain of 200 GN/m^2. As indicated in Table 5.3 it is still open to discussion whether this value can be unambiguously accepted. The times to fracture have been calculated from Eq. (5.57) for a segment comprising 12 bonds of low strength ($U_0 = 188$ kJ/mol). The partially extended chain ($L = 1.1\,L_0$) contains initially four kinks (8 gauche bonds). It responds in region A (Fig. 5.10) mostly by the hindered rotation of its gauche bonds, which, in this region, accounts for 90% of the axial deformation. The resulting combined modulus amounts to 22 GN/m^2. If geometrically possible a transition to the fully extended conformation will occur in region B. Region C represents the energy elastic straining of the then extended chain segment of length $1.1\,L_0$.

The degradation of chains through ionic decomposition or chemical reactions under stress will be discussed where appropriate in the following chapters.

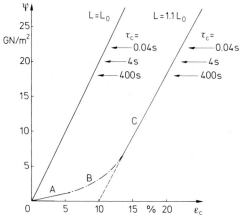

Fig. 5.10. Stress-strain diagram of polyamide 6 chain segments. L = contour length, 5.2 nm, 12 weak bonds, L_0 = distance between segment ends in the unstressed state, $\tau_c = 1/k_c$ = average lifetime according to Eq. (5.57), A = response due to hindered rotation of gauche bonds, B = intermediate region, C = energy-elastic response of extended chain.

References for Chapter 5

1. P. J. Flory: Principles of Polymer Chemistry, Ithaca, New York: Cornell University Press 1953.
2. L. R. G. Treloar: The Physics of Rubber Elasticity, 2nd Ed., London: Oxford University Press 1958, a) p. 61, b) p. 103ff, c) p. 108.
3. H. M. James, E. Guth: J. Chem. Phys. *11*, 455 (1943).
4. A. Abe, R. L. Jernigan, P. J. Flory: J. Am. Chem. Soc. *88*, 631 (1966).
5. J. P. Flory, A. D. Williams: J. Polymer Sci. A-2, *5*, 399 (1967).
6. P. R. Saunders: J. Polymer Sci. A-2, *2*, 3765 (1964).
7. W. Pechhold: Kolloid-Z. Z. Polymere *228*, 1 (1968).
8. H. G. Zachmann: Kolloid-Z. Z. Polymere *231*, 504 (1969).
9. J. A. Sauer, A. E. Woodward: Polymer Thermal Analysis, II., P. E. Slade, Jr., and L. T. Jenkins, Eds., New York: Dekker, 1970, p. 107.
10. P. B. Bowden: Polymer *9*, 449–454 (1968).
11. D. S. Boudreaux: J. Polymer Sci.: Polymer Phys. Ed. *11*, 1285 (1973).
12. W. J. Dulmage, L. E. Contois: J. Polymer Sci. *28*, 275 (1958).
13. T. R. Manley, C. G. Martin: Polymer *14*, 491–496 (1973) and ibid. 632–638 (1973).
14. H. Hahn, D. Richter: Colloid + Polymer Sci. *255*, 111–119 (1977).
15. E.g. J. Meixner: Handbuch der Physik, Vol. III/2, S. Flügge, Ed., Berlin: Springer-Verlag 1959, p. 413.
16. D. Langbein: Theory of Van der Waal's Attraction, Springer Tracts in Modern Physics, Vol. 72, Heidelberg: Springer-Verlag, 1974.
17. P. Lindenmeyer: Mechanical Behavior of Materials, Kyoto, Japan: Soc. Mat. Sci. 1972, Special Vol., p. 74.
18. H. Scherr: Dissertation Universität Ulm, Juli 1973.
19. H. H. Kausch: J. Polymer Sci. *C 32* (Polymer Symposia), 1 (1971).
20. A. D. Chevychelov: Polymer Science USSR *8*, 49 (1966).
21. H. H. Kausch, D. Langbein: J. Polymer Sci., Polymer Physics, Ed. *11*, 1201–1218 (1973).
22. H. H. Kausch, J. Becht: Deformation and Fracture of High Polymers, H. H. Kausch et al., Eds., New York: Plenum Press 1973, p. 317.
23. H. Baur, B. Wunderlich: Fortschr. Hochpolymeren Forsch. *7*, 388 (1966).
24. E. W. Fischer, G. Hinrichsen: Kolloid-Z. Z. Polymere *213*, 28 (1966).
25. J. D. Ferry: Viscoelastic Properties of Polymers, 2nd Edition, New York: J. Wiley and Sons, Inc. 1970.
26. A. Kadim: Thesis, Dept. of Chem. Engng., Technion, Israel Institute of Technology, Haifa, Israel 1968.
27. A. H. Abdel-Alim, A. E. Hamielec: J. Appl. Polymer Sci. *17*, 3769–3778 (1973).
28. J. W. Breitenbach, J. K. Rigler, B. A. Wolf: Makromol. Chemie *164*, 353–355 (1973).
29. A. M. Basedow, K. Ebert: Adv. Polymer Sci. (Fortschr. Hochpolymerenforsch.) *22*, 83–143 (1977).
30. E. M. Hagerman, C. C. Mentzer: J. Appl. Polymer Sci. *19*, 1507–1520 (1975).
31. Yu. Ya. Gotlib, A. A. Darinskii: Vysokomol. soyed. *A 16*, 2296–2302 (1974), Polymer Science USSR *16*, 2666 (1974).
32. S. N. Zhurkov, V. E. Korsukov: J. Polym. Sci., Polymer Physics Ed. *12*, 385–398 (1974).
33. E. A. Guggenheim: Handbuch der Physik, S. Flügge, Ed., Vol. III/2. Berlin: Springer-Verlag 1959.
34. S. Glasstone, K. J. Laidler, H. Eyring: The Theory of Rate Processes, New York: McGraw-Hill 1941.
35. H. H. Kausch: J. Macromol. Sci., Revs. Macromol. Chem., *C 4* (2), 243–280 (1970).
36. E. E. Tomashevskii: Soviet Physics – Solid State *12*, 2588–2592 (1971).
37. S. N. Zhurkov, V. I. Vettegren', V. E. Korsukov: 2nd Internat. Conf. Fracture, Brighton, England April 1969, paper 47.

38. H. I. Bernstein: Spectrochim. Acta *18*, 161 (1962).
39. P. J. Sheth, J. F. Johnson, R. S. Porter: Polymer *18*, 741–742 (1977), C. B. Wu, P. J. Sheth, J. F. Johnson: Polymer *18*, 822–824 (1977).
40. J. Skelton, W. D. Freeston, Jr., H. K. Ford: Appl. Polymer Symposia *12*, 111–135 (1969).
41. U. G. Gafurov, I. I. Novak: Polymer Mechanics 6/1, 160–161 (1972).
42. S. N. Zhurkov, V. I. Vettegren', V. E. Korsukov, I. I. Novak: Soviet Physics – Solid State *11*/2, 233–237 (1969).
43. B. E. Read, D. A. Hughes, D. C. Barnes, F. W. M. Drury: Polymer *13*, 485–494 (1972).
44. P. Bouriot: Premier Colloque du Groupe Français de Physique des Polymères, Strasbourg, 21–22 mai 1970.
45. S. Sikka: PhD Thesis: Application of FTIR to study external stress effects on the PETP backbone, University of Utah, Salt Lake City, December 1976. S. Sikka, K. Knutson: to be published.
46. V. S. Kuksenko, V. A. Ovchinnikov, A. I. Slutkser: Vysokomol. soyed A11 *9*, 1953–1957 (1969).
47. H. Hahn: Personal communication.

Chapter 6

Identification of ESR Spectra of Mechanically Formed Free Radicals

I. Formation

The action of axial tensile forces on a molecular bond R_1-R_2 results in a decrease of the apparent binding energy of the bond and thus in an increase of the probability for bond scission. If the reduction of the apparent binding energy is sizeable the mechanical action may be considered as the main cause of chain destruction. In as much as the scission of a chain molecule into organic radicals and the resulting appearance of unpaired free electrons is governed by mechanical forces a study of radical formation and of radical reactions will reveal information on the forces acting on a chain at a molecular level. The method of investigating free radicals by paramagnetic resonance techniques has been highly developed during the last thirty years [1, 2]. Since then it was successfully applied to elucidate the mechanism of the formation of free radicals in chemical reactions and under irradiation of visible and ultraviolet light, of x- and γ-rays, and of particles [1.3]. Also the value of the spectroscopic splitting factor g, the magnetic surrounding of the unpaired free electron spin, and the structure of the free radical have been studied. In all these cases the free electron spin acts as a probe, which, at least temporarily, is attached to a certain molecule, takes part in the motion of the molecule, and interacts with the surrounding magnetic field.

In investigations of the deformation and fracture of high polymer solids [4–67] by electron paramagnetic resonance techniques (EPR) similar experimental problems arise as during the above mentioned applications of EPR:
1. Interpretation of the form of the observed spectra, i.e. determination of the position of the free electron within the molecule.

119

2. Measurement of spectrum intensity, i.e. determination of the number of free electrons.

3. Observation of the changes displayed by the EPR signal and elucidation of the nature and kinetics of radical reactions.

From these primary data, information is sought to obtain with respect to the deformation behavior of chain molecules, their scission and their role in fracture initiation.

In the following a brief description of EPR spectroscopy will be given in order to provide an understanding of the main experimental technique which is used in these investigations. Subsequently reactions, identification, and concentration of mechanically formed free radicals will be discussed.

II. EPR Technique

A. Principles

The EPR spectroscopy of free electrons is based on the paramagnetism of free electron spins. For this reason one also speaks of electron spin resonance (ESR) spectroscopy. Electrons on fully occupied molecular orbitals generally do not give rise to a magnetic moment since — according to the Pauli principle — pairwise spins compensate each other. If, however, a bond is broken through homolytic scission, then free radicals with impaired electron spins are formed which are detectable. A free electron has the magnetic moment μ given by

$$\mu = g\beta s \tag{6.1}$$

where g is the spectroscopic splitting factor, β is the Bohr magneton, and s is the spin vector. In a magnetic field $\mathbf{H}$ the spin may assume one of two positions (spin quantum numbers m being either $+1/2$ or $-1/2$). The energy difference between these two states is

$$\Delta E = g\beta H. \tag{6.2}$$

The populations of the two energy levels at equilibrium are given by Boltzmann statistics which indicate that more of the free electrons will normally reside in the $-1/2$ "ground state". However, when electromagnetic energy of an appropriate (microwave) frequency ν is supplied, an electron can be "pumped" to the higher $+1/2$ state. The resonance condition is fulfilled when $h\nu$ is just equal to the energy difference between the two levels:

$$\Delta E = h\nu = g\beta H_{res} \tag{6.3}$$

where h is Planck's constant. In resonance, microwave energy is absorbed by the resonating sample if the upper energy level is constantly depopulated by thermal interaction between the spins and the molecular lattice. Generally the microwave

120

frequency employed by commercially available ESR spectrometers is fixed (at about 9.5 GHz) and the magnetic field is swept through the resonance (at about 3200 Oe).

B. Hyperfine Structure of ESR Spectra

Since the "free" electrons are attached to molecules, they will sense not only the presence of the applied external magnetic field but will show magnetic interactions with nuclei of surrounding atoms which have a magnetic moment. These interactions give rise to the "splitting" that can be used to identify particular radicals. Equation (6.3) is accordingly transformed into

$$H_{res} = h\nu/g\beta - \sum_{i=1}^{n} (A_i + B_i)\, m_{Ii}. \tag{6.4}$$

Here m_{Ii} is the spin quantum number of the interacting nucleus i, A_i the isotropic part of the magnetic coupling between radical electron and nucleus i, and B_i the anisotropic part which depends on the orientation of the radical within the external magnetic field.

For the important class of hydrocarbon radicals, B is different from zero only for the α-protons, i.e. for the nuclei of hydrogen atoms directly bound to the radical C atom. In well ordered systems (highly oriented polymers, single crystals) these radicals do have a strongly orientation dependent hyperfine structure.

In the simplest case of no interaction (A = 0, B = 0) there will be one line with its center at H_{res}. If only one proton interacts isotropically (B = 0) with the radical electron then the spectrum will show two lines displaced from H_{res} by $\pm A$. If two equivalent nuclei interact with the electron then m_I can assume the values +1, 0, and −1 so that a triplet of lines appears with distances of −A, 0, and +A from the center respectively. Since the state with $m_I = 0$ appears twice as often as the other ones the individual resonance lines have intensities as 1:2:1. It may generally be said that n equivalent protons give rise to a spectrum with (n + 1) lines which show an intensity distribution given by the binomial coefficients of exponent n. For illustration Figure 6.1 shows the spectra of alkyl radicals $-CH_2-\overset{.}{C}H-CH_2-$ in highly
    β   β

oriented polyethylene monofilaments which were produced by irradiation with fast electrons [37]. In one orientation (fiber axis parallel to H, spectrum at bottom) the total splitting parameter (A + B) of the α-protons is just equal to the isotropic parameter of the 4 equivalent β-protons. Therefore, a spectrum is observed which might be caused by 5 equivalent protons, i.e. a sextet with intensities 1:5:10:10:5:1 (some background including the small center line belongs to a second radical present within the sample). In the case of a perpendicular arrangement of fiber axis and magnetic field the spectrum shown in the upper portion of Figure 6.1 is observed. Now the α- and β-protons have different splitting parameters and a quintet of lines appears due to 4 equivalent β-protons with each quintet line split into a doublet by the α-proton. In all 10 lines result with intensity ratios of 1:1:4:4:6:6:4:4:1:1:. Again the small background of a superimposed second spectrum is observed. The orienta-

Fig. 6.1. Orientation dependency of the hyperfine structure of main chain (alkyl) free radicals: a. axis of highly oriented polyethylene fiber parallel to magnetic field H_0, b. axis of fiber perpendicular to H_0 (after [37, 38]).

tion dependency of the hyperfine structure of hydrocarbon radicals offers the opportunity to study the orientation of radical carrying chains.

C. Number of Spins

In most spectrometers lines are recorded which are the first derivative of the energy absorption peaks with respect to the magnetic field H_0. It is the plot of these first derivates that is usually called the ESR spectrum.

The total microwave energy absorption can be obtained by double integration of the ESR spectrum. In the absence of saturation effects the total power absorbed, $I(\infty)$, depends upon the number of spins N_0, the intensity of the oscillating magnetic (microwave) field H_1, and the frequency ν according to

$$I(\infty) = 2\,\pi^2\,\nu^2\,H_1^2\,\frac{1}{4\,kT}\,N_0\,g^2\,\beta^2. \tag{6.5}$$

To calculate absolute intensities of spins usually comparison of the unknown intensity N_0 is made with a standard sample with a known concentration N_s of spins [1]. Frequently a solution of the stable free radical 1,1-diphenyl-2-picryl hydrazyl (DPPH) in benzene is used as standard radical. If the "unknown" spectrum is sym-

122

metrical and narrow and the experimental conditions (sample size and shape, conditions of spectrometer operation) are the same in both cases the relation

$$N_0 = N_s \, I(\infty) \, / \, I_s \, (\infty) \tag{6.6}$$

is valid where N_s and $I_s(\infty)$ refer to the standard sample. If lines of identical shape and width are compared it is sufficient to measure the peak height h of the first derivative. The numbers N_1 and N_2 of unpaired electrons causing these peak heights are then related by:

$$N_1 / N_2 = h_1 / h_2 . \tag{6.7}$$

For most cases of stress induced free radical formation this relation has been used for evaluation of time-dependent intensities.

The number of free radicals necessary for detectable peak heights depends on the line shape and the effective volume of the spectrometer cavity. At the time covered by the referenced investigations (1960–1976) the lower limit of detectability of polymer radicals in the solid state can be given as 10^{13} spins/cm^3.

It may be mentioned that the number of free radicals present in a liquid or at the surface of a solid can also be determined by a non-spectroscopic method. This method exploits the color changes associated with the consumption of free radicals from a test solution e.g. of DPPH [39]. It should be emphasized again that great caution must be exercised in comparing absolute values of free radical concentrations obtained from differently shaped samples and/or in different cavities. According to general experience deviations of up to a factor of 2 must be expected.

III. Reactions and Means of Identification

The assignment of the specific peaks to specific types of free radicals is not always obvious; it sometimes involves highly intriguing techniques and the knowledge of related spectra and of chemical reactions to be expected [64, 67]. The main obstacles to be overcome are the broad line widths of resonance lines from solid specimens and the high rate of many radical reactions. It is evident that a large line width often prevents an effective resolution of nuclear hyperfine structure. The so-called "5 + 4 line" spectrum of mechanically destroyed methacrylic polymers [4] is an example of a spectrum which had only been identified after comparison with the 16-line-spectrum of an aqueous solution of the polymerization radical of methacrylic acid. Thus it was learned that the former spectrum is the unresolved form of the latter and has to be assigned to the same radical [40].

The difficulties in identifying an observed spectrum are well illustrated by the spectrum of 6 polyamide shown in Figure 6.2. It was obtained after stretching a fiber sample at room temperature in the ESR cavity by an automatic servo-control system to the point of radical detection, then scanning the magnetic field from 3071 to 3571 Oersted (mean, 3321 Oe) at 9.433 GHz and a time constant of 0.5 second.

Fig. 6.2. Apparent 5-line spectrum
as obtained after stretching of a
bundle of 6 polyamide fibers at
room temperature in an ESR cavity.

This apparent 5-line spectrum is obtained, under exclusion of oxygen, when highly stressed fibers or finely ground or milled material are investigated at room temperature [41]. A similar spectrum is observed after γ or electron irradiation of PA 6 [42, 43]. Two different interpretations of this 5-line spectrum have been given. Graves and Ormerod [42] suggested that an apparent quartet coming from the radical $-CH_2-CO-NH-\overset{\cdot}{C}H-CH_2-$ (radical I) is superimposed by a singlet which may be due to $-CH_2-\overset{\cdot}{C}HO-NH-CH_2-$ (II). Kashiwagi [43] deduced, after comparison with radicals from polyurea and succinamide, that the apparent 5-line spectrum from 6 polyamide and the 6-line spectra from 610, 66 and 57 polyamides are a triplet of doublets stemming from the radical I. It is, therefore, generally concluded that the radicals observed at room temperature in mechanically degraded polyamide material are mainly of type I, superimposed to a varying degree by a component of type II. The radicals I are secondary chain radicals which can only have been produced by the reaction of a hydrogen from a methylene group with an other — a primary — radical. Since the secondary radicals are reaction products they give no direct information on the nature and location of the primary process of mechanical degradation of the above solids.

The appearance of free radicals in the process of mechanical degradation reveals conclusively that under the action of local mechanical forces main chains have been broken and that the breakage of a main chain bond has led to the formation of a chain end free radical at either side of the broken bond. Knowledge of these facts can already be employed to study the relation between environmental parameters and the rate of chain scission. The kinetics of chain scission and its possible role in fracture initiation is investigated in detail in Chapters 7 and 8. Apart from the intensity which gives a quantitative measure of chain scission the form of free radical spectra has been used to determine the location of primary and secondary radicals within the molecule [64] and within morphological units (within crystals, on crystal surfaces, in amorphous regions).

If degradation and ESR investigation of a sample are carried out at liquid nitrogen temperature, the rate of radical reactions is slowed down sufficiently and a direct observation of the primary radicals produced by mechanical degradation becomes possible. In a thorough investigation Zakrevskii, Tomashevskii, and Baptizmanskii [10] have cleared up the scheme of radical reactions for 6 polyamide (caprolactam, capron). At 77 K they observed a complex spectrum showing the hyperfine structure of a sextet superimposed on a triplet. From the distance between the different lines of the sextet (splitting) and their intensity ratios the above authors established the presence of the radical $R-CH_2-\dot{C}H_2$ (III). This radical will form after rupture of any of the bonds 1 to 6 in the caprolactam unit:

$$^7CO^1CH_2\,^2CH_2\,^3CH_2\,^4CH_2\,^5CH_2\,^6NH^7CO-.$$

If the superimposed sextet is subtracted from the original spectrum a triplet remains which must be due to a scission product of the above molecule. The exclusive breakage of bonds 3 and 4 can be ruled out immediately since it would only lead to radicals of type III. Breakage of bonds 6 and 7 was eliminated after comparison with capron deuterated in the imino groups; breakage of bond 1 would not lead to a radical with a triplet spectrum. After inspection of all other possibilities, including secondary radicals, there remain as likely scission points only the bonds 2 and 5. After having studied α and ϵ methyl substituted caprolactam the authors [10] finally were able to state that in a stressed 6 polyamide molecule both the bonds 2 and 5 do break and with equal probability. The rupture of a PA 6 molecule, therefore, leads to three primary radicals:

$-CH_2-\dot{C}H_2$ (III, 50%), $-NH-\dot{C}H_2$ (IV, 25%) and $-CO-\dot{C}H_2$ (V, 25%).

Upon slowly rising the temperature from 77 K radical III begins to disappear leaving at 120 K only IV and V. At higher temperatures IV and V are gradually transformed into I which prevails above 240 K. Within the experimental limits mentioned above with respect to comparing the absolute numbers of radicals of different species there seems to be no loss of radicals during conversion of III, IV, and V into radical I.

In the following Sections and Tables the experimental conditions are reported which were employed during mechanical production and ESR investigation of free radicals. The assignments of primary and/or secondary radicals are given. For a discussion of the nature and kinetics of possible transfer reactions the reader is referred to the recent and comprehensive reviews of Rånby et al. [2] and of Sohma et al. [64]. Special problems in view of polymer morphology and of the reduction of chain strengths will be treated in Chapters 7 and 8.

IV. Assignment of Spectra

A. Free Radicals in Ground High Polymers

The appearance of macroradicals in mechanically destroyed polymers was first shown in 1959 [4—6]. Since then, fairly systematically natural and synthetic organic materials have been investigated with respect to formation of free radicals in mechanical degradation (see for instance the monograph by Ranby and Rabek [2] and the review papers by Butyagin et al. [7], Kausch [8], and Sohma et al. [64]). Owing to the originally limited sensitivity of ESR-spectrometers, the first experiments have been carried out with crushed polymers which have a high ratio of fracture surface to volume and, therefore, a comparatively large ESR signal.

The experimental equipment is described by various authors (referenced in 2). Grinding, milling, cutting or crushing of the polymer sample is done in vibration mills or mortars, under inert atmosphere, in vacuo, or in liquid nitrogen. Magnetic drills have been reported by which the shavings or grindings are produced within a sealed glass tube immediately at the ESR-cavity. Most equipment used in these experiments allows for a control of the gaseous environment and for temperature control from liquid nitrogen temperature (77 K) to above room temperature.

In Table 6.1 the presently available details of sample preparation (crushing technique, temperature, environment), the treatment of the crushed sample, the temperatures at which the ESR-spectra were taken, and the assignments of the observed spectra to primary and/or secondary free radicals are listed for 35 different polymeric materials. The common result of practically all of the reported ESR work [4–36] on comminuted polymers is that the mechanical action leads to the severance of a main chain bond and to the formation of chain end radicals as primary radicals. The only exception to this rule are the substituted polydimethylsiloxanes (No. 32 to 35) where there is a strong indication that the Si–O bond decomposes by an ionic mechanism rather than by homolytic bond scission [36]. In no case free radicals are produced by a mechanical stripping off of sidegroups or -atoms from the main chain. For this to occur stresses are required which simply cannot be transmitted onto the relatively small sidegroups present in the materials listed in Table 6.1. In fact the attempted degradation of low molecular weight compounds (paraffins, ethanol, benzene) with a molecular weight equal to or larger than that of those sidegroups failed, although mechanical means identical to those used successfully for macromolecules had been employed [13, 14, 62].

B. Free Radicals in Tensile Specimens

Mechanical degradation which leads to formation of free radicals can be achieved in various ways but generally two types of samples are being used in experiment: powders of finely ground polymers as discussed above or uniaxially stressed fibers, monofilaments, or strips of films. In samples of the first type all molecular fracture processes are completed before the ESR investigation begins. These samples can primarily be used to obtain information on the nature and absolute number of free radicals present within the degraded polymer.

Samples of the second type, tensile specimens, are in principle suitable for studying the time-dependent formation of free radicals and their effect on strength and on other macroscopic properties. It can be said a priori that the free radical intensities of tensile specimens — be it before or after rupture — will be very much lower than the intensities found in finely ground polymers with fracture surfaces of several thousand cm^2/g. Even with the increased sensitivity of presently available ESR spectrometers the investigation of stressed samples is limited to a few with favourable properties. The only experiments leading to a successfull investigation of time-dependent formation of free radicals were carried out with highly oriented, semi-crystalline fibers like 6 and 66 polyamide, polyoxamides, polyethylene, polyethylene terephthalate, or natural silk and with preoriented rubbers (natural rubber, polychloroprene) near liquid nitrogen temperatures.

In Table 6.2 experimental conditions and types of radicals observed are listed.

The experimental equipment necessary for stressing fibers within the resonant cavity has been extensively reviewed by Rånby et al. [2]. It is quite more elaborate than that for investigating powders, although the requirements for controlled temperature and atmosphere are nearly the same. Lever systems or servo-controlled hydraulic loading systems have been reported [29, 37, 44, 60] through which programmed load or strain functions can be applied to bundles of fibers (or other tensile specimens) in the resonant cavity. The stretching of the tensile specimens necessarily has to be carried out in that temperature regime within which free radical spectra can be well observed. For thermoplastic fibers this meant 200 to 320 K; for preoriented rubbers ambient temperatures of 93 to 123 K were necessary. At these temperatures the primary free radicals are sufficiently mobile to react rapidly with atom groups of the same or other chain molecules, with absorbed gases, or with additives or impurities acting as radical traps. For this reason the observed free radicals listed in Table 6.2 are secondary radicals — with the only exception of the highly stable primary radical of poly-p-(2-hydroxyethoxy) benzoic acid (PEOB) studied by Nagamura and Takayanagi [60]. The secondary radicals — if sufficiently resolved — have been assigned to either peroxy radicals ($R-O-\dot{O}$), main chain radicals ($R-\dot{C}H-R$), unknown impurities [55], or deliberately added radical traps as, for instance chloranyl [61].

Table 6.1. Condition of Degradation and Assignment of Resulting ESR Spectra of Polymers

Degraded Polymer		Sample preparation			(Heat) treatment time at temperature	
		process	temperature	environment	min	K
1. Polyethylene $-CH_2-CH_2-$	PE	milling	77	liqu. N_2 + 10% O_2		
		milling	240			
		milling	80–100	vacuum		
		milling	77	helium		
		sawing	77	liqu. N_2	none	
		milling	77	helium		
		sawing	77	liqu. N_2	5	132
		milling	77	helium		
		sawing	77	liqu. N_2	5	152
					5	233
		milling	77	helium	warming to 270	
2. Polypropylene $-CH_2-(CH_3)CH-$	PP	milling	80–100	vacuum		
		milling	80–100	vacuum	warming to 140	
		grinding	77			
		sawing	77	liqu. N_2		
		ball mill	77	vacuum		
		ball mill	77	vacuum	introducing air	
3. Polyisobutylene $-CH_2-(CH_3)_2C-$	PIB	milling dilute solution	77	hexane, carbon tetrachloride		
		cutting	77	liqu. N_2 + 10% O_2		
		milling	77	liqu. N_2 + 10% O_2		
Poly(1-butene) $-CH(CH_2-CH_3)-CH_2-$	PB	vibrating mill	77	inert		
		vibrating mill	77	inert		200
4. Polybutadiene $-CH_2-CH=CH-CH_2-$ cis-rich 54% trans	BR	sawing	77	liqu. N_2		
		sawing	77	liqu. N_2		
		ball mill	77	vacuum		
5. Polyisoprene $-CH_2-CH_3C=CH-CH_2-$		milling	77	liqu. N_2 + 10% O_2		
6. Polystyrene $-CH_2-(C_6H_5)CH-$	PS	milling dilute solution	77	various solvents plus monomers		
		milling	room	vacuum		

ESR spectra taken at K	Assignment of observed spectra to		Ref.
	primary radicals	secondary radicals	
1. 77			4, 9, 10
240			5
77		$-CH_2-\dot{C}H-CH_2-$	11
77	$-CH_2-\dot{C}H_2$	none	12
77		none	13, 14
135		$-CH_2-\dot{C}H-CH_3$	12
77		$-CH_2-\dot{C}H-CH_3$	13, 14
160	none	$-CH_2-\dot{C}H-CH_3$	12
77	none	$-CH_2-\dot{C}H-CH_3$ and $-CH_2-HCO\dot{O}-CH_3$ or $-CH_2-HCO\dot{O}-CH_2-$	13, 14
77	none	$-CH_2-HCO\dot{O}-CH_3$ or $-CH_2-HCO\dot{O}-CH_2-$	13, 14
77	none	$-CH_2-CH=CH-\dot{C}H-CH_2-$	12
2. 77		$-CH_2-CH_3\dot{C}-CH_2-$ (increasing with milling time)	11 11
140	$-CH_2-CH_3\dot{C}H$	$-CH_2-CH_3\dot{C}-CH_2-$ (Sextet)	9
77		none	13, 14
77		RO\dot{O}	
77	$-CH_2-CH_3\dot{C}H$ and $-CH_3CH-\dot{C}H_2$	none	15
77	$-CH_2-CH_3\dot{C}H$	$-CH_3CH-H_2CO\dot{O}$	15
3.			
77		$R_1-\dot{C}H-R_2$	16
77	$-CH_2-(CH_3)_2\dot{C}$	none	5
77		RO\dot{O}	4
77	$-CH_2-CH(CH_2-CH_3)-\dot{C}H_2$ and		65
77	$-CH_2-\dot{C}H(CH_2-CH_3)$	$-CH_2-\dot{C}(CH_2-CH_3)-CH_2-$	65
4.			
77	none	RO\dot{O}	13, 14
77	none	RO\dot{O}	13, 14
77	none	$R-\dot{C}H-CH=CH-CH_3$	62
5.	none	RO\dot{O}	4
6.			
77	$-CH_2-C_6H_5\dot{C}H$	reaction of $\dot{R}$ with monomer	17
room	none	RO\dot{O}	4

Table 6.1. (continued)

Degraded Polymer		Sample preparation			(Heat) treatment time at temperature	
		process	temperature	environment	min	K
6. (continued)		milling	77	liqu. $N_2 + O_2$		
		cutting	77	liqu. $N_2 + 10\% O_2$		
		shaving	77	vacuum	none	
		shaving	77	vacuum	none	
		shaving	77	vacuum	heating in air to 273	
7. Poly α methylstyrene $-CH_2-(C_6H_5)(CH_3)C-$		milling dilute solution	77 77 77	various solvents plus amines toluene toluene		
8. Polytetrafluorethylene $-CF_2-CF_2-$	PTFE	sawing milling ball mill	77 77 77	liqu. N_2 liqu. $N_2 + 10\% O_2$ vacuum	none	
9. Polyvinylalcohol $-CH_2-OHCH-$	PVAL	grinding dilute solution grinding	80 80	H_2O; D_2O vacuum		
10. Polyvinylacetate $-CH_2-(OCOCH_3)CH-$	PVAC	shaving shaving cutting milling dilute solution	77 77 77? 77	vacuum vacuum liqu. $N_2 + 10\% O_2$ various solvents and monomers	heat. in vac. to 300	
11. Polymethacrylate $-CH_2-(CO_2CH_3)CH-$	PMA	cutting milling dilute solution	77 77	liqu. $N_2 + 10\% O_2$ various solvents and monomers		
12. Polymethylmethacrylate $-CH_2-(CO_2CH)(CH_3)C-$	PMMA	milling milling sawing shaving shaving shaving grinding grinding grinding grinding dilute solution	77 77 77 77 room 77 77 77 77 77	liqu. $N_2 + 10\% O_2$ liqu. $N_2 + 10\% O_2$ liqu. N_2 vacuum vacuum liqu. $N_2 + 10\% O_2$ vacuum vacuum vacuum benzene	heat. in air to 300 none	
13. Polyacrylonitrile $-CH_2-CNCH-$	PAN	vibro mill vibro mill	77 77	air		

ESR spectra taken at K	Assignment of observed spectra to		Ref.
	primary radicals	secondary radicals	
77	none	ROȮ	4
77	none	ROȮ	5
77	$-CH_2-C_6H_5\dot{C}H$		18, 19, 20
300	none	$-CH_2-\dot{C}_6H_4CH_2$	6
273	none	ROȮ	19
7. 77 to 300	$-CH_2-(CH_3)(C_6H_5)\dot{C}$	reaction of $\dot{R}$ with amines	21
77		$R_1-\dot{C}H-R_2$	16
177	none	$R_1-\dot{C}H-R_2$	16
8. 77	none	$-CF_2-CF_2O\dot{O}$	13, 14
77	none	$-CF_2-CF_2O\dot{O}$	4
77 and 243	$CF_2-\dot{C}F_2$	none	62
9. 80–170	$-CH_2-OH\dot{C}H$ and $-OHCH-\dot{C}H_2$	none	22
80–350	$-CH_2-OH\dot{C}H$	$-CH_2-OH\dot{C}-CH_2-$	22, 23
10. 77	$-CH_2-(OCOCH_3)\dot{C}H$	none	19
room	none	none	19
77?			5
	$-CH_2-(OCOCH_3)\dot{C}H$		
77		reaction of $\dot{R}$ with monomer	17, 21
11. 77			5, 9
	$-CH_2-(COOCH_3\dot{C}H$		
77		reaction of $\dot{R}$ with monomer	17
12. room		ROȮ	4
77		ROȮ	4
77		slight ROȮ	13,14
77		R_1-CH-R_2	19
room		none	19
300	$-CH_2-(COOCH_3)(CH_3)\dot{C}$	none	5
77		$R_1-\dot{C}H-R_2$	16, 62, 63
273		no change of $R_1-\dot{C}H-R_2$	16
283		$R_1-\dot{C}H-R_2$ disappears	16
77		$R_1-\dot{C}H-R_2$	16
13. 77			23
195	$-CH_2-CN\dot{C}H$		23
195		ROȮ	24

Table 6.1. (continued)

Degraded Polymer		Sample preparation			(Heat) treatment time at temperature	
		process	temperature	environment	min	K
14. Poly(oxymethylene) $-CH_2-O-$	POM	sawing grinding	77 80	liqu. N_2 vacuum		
15. Polyethyleneoxide $-CH_2-CH_2-O$	PEO	grinding grinding dilute solution	80 80	vacuum $H_2O + CH_3COOH$		
16. Polypropyleneoxide $-CH_2-CH_3CH-O-$		grinding	80	vacuum		
17. Polyethyleneterephthalate $-O-CH_2-CH_2-O-CO-C_6H_4-CO-$	PETP	vibro mill	77			
18. Polycarbonate $-O-C_6H_4-C(CH_3)_2-C_6H_4-O-CO-$	PC		77			
19. Poly(2,6-dimethyl-p-phenylene oxide) $-(CH_3)_2C_6H_2-O-$	PPO	shaving	77 77 room	liqu. $N_2 + O_2$ vacuum air		
20. Polycaprolactam $-CO-(CH_2)_5-NH-$	6-PA	cutting milling grinding milling milling milling milling	77 77 77 77 77 77	liqu. $N_2 + 10\% O_2$ various (liqu. N_2, helium) vacuum helium vacuum helium	heated to 295	
21. Poly α methylcaprolactam $-CO-CH(CH_3)-(CH_2)_4-NH-$ Poly ϵ methylcaprolactam $-CO-(CH_2)_4-CH(CH_3)-NH-$		milling milling	77 77	vacuum vacuum		
22. Poly(hexamethyleneadipamide) $-NH-(CH_2)_6-NH-CO-(CH_2)_4-CO-$	66-PA	grinding	77 77	liqu. N_2 liqu. N_2	O_2 introduced	
23. Polyurethane (Solithane 113) $-O-CO-NH-(CH_2)_n-$	PU	grinding grinding	77 room	liqu. N_2	quenched in liqu. N_2	
24. Natural Silk		milling milling milling	77 77 77	helium helium helium		

ESR spectra taken at K	Assignment of observed spectra to		Ref.
	primary radicals	secondary radicals	
14. 77 88–350	$-CH_2\dot{O}$	$-CH_2O\dot{O}$ $-O-\dot{C}H-O-$ or $-O-\dot{C}H-OH$	13, 14 22
15. 80–230 80–200	$-O-\dot{C}H_2$ or → $-O-\dot{C}H_2$	$-O-\dot{C}H-CH_2-$	22 22
16. 80–200	$-O-CH_3\dot{C}H$ or →	$-O-CH_3\dot{C}-CH_2$	22
17. 77	$R-CO-O-\dot{C}H_2$		25
18. 77	$-O-\dot{C}_6H_4$ and $-OC-O-H_4\dot{C}_6$		26
19. 77 77 room	$-(CH_3)_2C_6H_2-\dot{O}$	$RO\dot{O}$ none none	27 27 27
20. 77 77 77 155 195 113 77	none $-CH_2-\dot{C}H_2$; $-CO-\dot{C}H_2$ $-NH-\dot{C}H_2$ $-CO-\dot{C}H_2$; $-NH-\dot{C}H_2$ none none	mostly $RO\dot{O}$ $-CO-NH-\dot{C}H-CH_2-$ $-CO-NH-\dot{C}H-CH_2-$ and singlet	5 9, 10, 12, 28 10, 66 12 22 9, 29 12
21. 77 77	$-CO-\dot{C}H(CH_3)$ $-NH-\dot{C}H(CH_3)$		10 10
22. 77 233	$-CO-\dot{C}H_2$; $-NH-\dot{C}H_2$	$-CO-NH-\dot{C}H-CH_2$ $RO\dot{O}$	30, 31, 32 30, 32
23. 77 77		singlet singlet	30, 33 33
24. 77 120 295	$-NH-\dot{C}H_2$; $-CO-\dot{C}H_2$ $-NH-\dot{C}H-CH_3$; $-CO-\dot{C}H-CH_3$ none	none $-NH-\dot{C}H-CO-$	12 12 12

133

Table 6.1. (continued)

Degraded Polymer	Sample preparation			(Heat) treatment time at temperature	
	process	temper- ature	environment	min	K
25. Polyethylene-Sulfide $-CH_2-CH_2-S_n-$	milling	77	liqu. N_2		
26. Polypropylene-Sulfide $-CH_2-CH_3CH-S_n-$	milling	77	liqu. N_2		
27. Ebonite, Thiokol, and other polymers containing sulfide bonds in the main chain	shaving compression shaving	room room 77	air air liqu. N_2 + 10% O_2		
28. Polydimethylsilmethylene $-Si(CH_3)_2-CH_2-$	drilling	77	liqu. N_2	none	
29. Polyoxymethylenedimethylsiloxane $-Si(CH_3)_2-CH_2-O-CH_2-O-$ $-CH_2-Si(CH_3)_2-O-$	drilling	77	liqu. N_2	none	
30. Polydimethylsilphenylene-dimethyl siloxane $-Si(CH_3)_2-C_6H_4-Si(CH_3)_2-O-$	drilling	77	liqu. N_2	none	
31. Polydimethylsildiphenylene- oxide dimethylsiloxane $-Si(CH_3)_2- C_6H_4-O-C_6H_4Si(CH_3)_2-O-$	drilling	77	liqu. N_2	none	
32. Polydimethylsiloxane $-Si(CH_3)_2-O-$	drilling	77	liqu. N_2	none	
33. Polydimethyl-(methylvinyl)- siloxane $-Si(CH_3)_2-O-$ 99.9%	drilling	77	liqu. N_2	none	
34. Polydimethyl-(methylethyl)- siloxane $-Si(CH_3)_2-O-$ 92 % $-Si(C_2H_5)(CH_3)-O-$ 8%	drilling	77	liqu. N_2	none	
35. Polydimethylsiloxane (crosslinked polymer) $-Si(CH_3)_2-O-$ 98.5% $-Si(O-)(CH_3)-O-$ 1.5%	drilling	77	liqu. N_2	none	

ESR spectra taken at K	Assignment of observed spectra to		Ref.
	primary radicals	secondary radicals	
25. 77	$-S-\dot{C}H_2$	$-CH_2-S-\dot{C}H-CH_2-$	34
26. (?)	no R$-\dot{S}$		34
27. room 77 77	R$-\dot{S}$		5, 34, 35
28. 77	traces of $\equiv Si-\dot{C}H_2$	RO$\dot{O}$	36
29. 77	traces of $\equiv Si-\dot{C}H_2$	RO$\dot{O}$	36
30. 77	none	RO$\dot{O}$ and traces of $\equiv Si-\dot{C}H_2$	36
31. 77	none	RO$\dot{O}$ and traces of $\equiv Si-\dot{C}H_2$	36
32. 77	none	none (ionic degradation)	36
33. 77	none	none (ionic degradation)	36
34. 77	none	none (ionic degradation)	36
35. 77	none	none (ionic degradation)	36

Table 6.2. Form of Spectrum and Concentration of Free Radicals Formed in Tension

Polymer	Sample and treatment	Environment
1. Polyethylene	bundle of filaments	hexane at 243 K
	yarn, 300 filaments, 0.15 mm diam.	N_2
	moulded sheet, 3.5 mm thick	$N_2 + O_2$, 160 to 294 K
2. Polypropylene	yarn, 200 filaments, 0.20 mm diameter	N_2
3. Acrylonitrile-butadiene	cut sheets	2.8 mg O_3/l
	100% preoriented at 300 K	N_2 at 118 to 168 K
4. cis-Polyisoprene (natural rubber)	vulcanized sheet, 2 mm thick prestretched; sulfur-cured;	Ar + N_2 at 93 K
	dicumyl-peroxide cured	Ar + N_2 at 93 K
5. Polychloroprene	dicumyl-peroxide cured, 100% preoriented at 300 K	Ar + N_2 at 93 K
6. Polystyrene	stack of strips	hexane at 243 K
	film, 1.3 mm thick	(air?)
7. Styrene copolymers	film, 1.3 mm thick	(air?)
8. Polymethylmethacrylate	bundle of filaments	hexane at 243 K
9. Polyethylene terephthalate	bundle of filaments	hexane at 243 K
	yarn, 7000 filaments, 0.02 diameter	N_2
	(mono filament?)	air at room temperature
10. Polycarbonate	bulk, drawn to necking	air at room temperature
11. Polycaprolactam	bundle of fibers	hexane at 243 K
	bundle of fibers	vacuum
	bulk material	vacuum
	yarn, 7000 filaments, 0.02 diameter	N_2
	(monofilament?)	air at room temperature
	monofilament	air at room temperature
12. Poly(hexamethylene adip-amide) (66 polyamide)	preoriented film, 0.5 mm thick	air at room temperature
13. Poly(laurolactam) (12 polyamide)	yarn, 6800 filaments, 0.02 diameter	N_2
14. Polyoxamide	bundle of fibers	air at room temperature
15. Natural silk	bundle of fibers	hexane at 243 K
16. Silicone rubber	cut sheet	2.8 mg O_3/l
	cut sheet, 100% prestrain at 300 K	N_2 at 194 K
17. Poly(p[2 hydroxyethoxy] benzoic acid)	bundle of fibers	vacuum

References for Chapter 6

1. P. Ayscough: Electron Spin Resonance in Chemistry: Methuen, London 1967.
2. B. Rånby, J. Rabek: ESR Spectroscopy in Polymer Research. Heidelberg: Springer-Verlag 1977.
3. H. Fischer: Magnetische Eigenschaften freier Radikale: Landolt-Börnstein-Tabellen, Neue Serie II/1, K.-H. Hellwege, A. M. Hellwege, Eds., Berlin: Springer-Verlag 1965.

ESR spectra taken at K		Observed radical	Concentration spins/cm^3	Ref.
1.	77	peroxy (R–O–$\dot{\text{O}}$)	$5 \cdot 10^{16}$	45
	291	singulet	5 to $50 \cdot 10^{14}$	37, 46
	103	peroxy	1.6 to $48 \cdot 10^{15}$	47
2.	291	none	10^{14}	37, 46
3.	room	asymmetric triplet	up to 10^{18}	48
	118 to 168	asymmetric triplet	up to $2 \cdot 10^{16}$	49
4.		asymmetric quintet		
	113	($-CH_3C=CH-\dot{C}H_2$?)	4 to $11 \cdot 10^{17}$	49, 50
	113	peroxy	well observable	50
5.	113	peroxy		51
6.	77	none	10^{14}	45
	(room)	none	10^{11}*	52
7.	(room)	none	10^{13}	52
8.	77	none	10^{14}	45
9.	77	peroxy	$8 \cdot 10^{15}$	45
	291	none	10^{14}	37, 46, 53
	113	peroxy	up to $15 \cdot 10^{16}$	53
10.	room	ill resolved quartet	up to $1.6 \cdot 10^{16}$	54
11.	77	quintet ($-CO-NH-\dot{C}H-CH_2-$)	$5 \cdot 10^{17}$	44, 45
	298	quintet	$1 \cdot 10^{17}$	55
	298	none		55
	291	quintet	$1 \cdot 10^{18}$	28, 32, 37, 46
	113	peroxy	up to $7 \cdot 10^{16}$	53
	77	quintet	$5 \cdot 10^{16}$	56
12.	300	triplet (unresolved quintet?)		57
13.	291	quintet	$5 \cdot 10^{14}$ to $5 \cdot 10^{15}$	37, 46
14.	300	quintet	$1 \cdot 10^{16}$	58
15.	77	doublet	$7 \cdot 10^{17}$	44, 45
16.	room	multiplet	up to $2 \cdot 10^{16}$	48
	194	multiplet	up to $2 \cdot 10^{16}$	59
17.	room	slightly asymmetric triplet	well observable	60

* (indirectly estimated)

4. S. E. Bresler, S. N. Zhurkov, E. N. Kazbekov, E. M. Saminskii, E. E. Tomashevskii: Zh. tekhn. fiz. *29*, 358 (1959), Soviet Physics – Techn. Physics *4*, 321–325 (1959).
5. S. E. Bresler, E. N. Kazbekov, E. M. Saminskii: Polymer Sci. USSR *1*, 540 (1959).
6. P. Yu. Butyagin, A. A. Berlin, A. E. Kalmanson, L. A. Blumenfeld: Vysokomol. Soyed. *1*, 865 (1959).
7. P. Yu. Butyagin, A. M. Dubinskaya, V. A. Radtsig: Uspekhi khimii *38*, 593 (1969). Russian Chem. Rev. *38*, 290–305 (1969).

8. H. H. Kausch: Rev. Macromol. Chem. *5(2)*, 97 (1970).
9. K. L. DeVries, D. K. Roylance, M. L. Williams: Report from the College of Engng., Univ. of Utah, Salt Lake City, UTEC DO 68–056, July 1968.
10. V. A. Zakrevskii, E. E. Tomashevskii, V. V. Baptizmanskii: Fizika Tverdogo Tela *9*, 1434–1439 (1967); Soviet Physics – Solid State *9*, 1118–1122 (1967).
11. V. A. Radtsig, P. Yu. Butyagin: Vysokomol. Soyed. *A 9*, 2549 (1967); Polymer Sci. USSR *9*, 2883 (1967).
12. V. A. Zakrevskii, V. V. Baptizmanskii, E. E. Tomashevskii: Fizika Tverdogo Tela *10*, 1699 (1968), Soviet Physics – Solid State *10*, 1341–1345 (1968).
13. J. Sohma, T. Kawashima, S. Shimada, H. Kashiwabara, M. Sakaguchi: Nobel Symposium 22, Almquist & Wicksell, Stockholm, 225–233 (1972).
14. T. Kawashima, S. Shimada, H. Kawashibara, J. Sohma: Polym. J. (Japan) 5, 135–143 (1973).
15. H. Yamakawa, M. Sakaguchi, J. Sohma: Rep. Progr. Polymer Physics, Japan, XVI, 543 (1973). M. Sakaguchi and J. Sohma: Rep. Progr. Polymer Physics, Japan, XVI, 547 (1973). M. Sakaguchi, H. Yamakawa and J. Sohma: Polymer Letters *12*, 193 (1974).
16. P. Yu. Butyagin, I. V. Kolbanev, V. A. Radtsig: Fizika Tverdogo Tela *5*, 2257–2260, Soviet Physics – Solid State *5*, 1642–1644 (1964).
17. A. M. Dubinskaya, P. Yu. Butyagin: Vysokomol Soyed. *A 10*, 240–244 (1968), Polymer Sci. USSR *10*, 283–288 (1968).
18. A. M. Dubinskaya, P. Yu. Butyagin: Vysokomol. Soyed. *B 9*, 525 (1967).
19. S. N. Zhurkov, E. E. Tomashevskii, V. A. Zakrevskii: Fizika Tverdogo Tela *3*, 2841–2847 (1961), Soviet Physics – Solid State *3*, 2074 (1962).
20. J. Tino, M. Capla, F. Szöcs: European Polymer J. *6*, 397 (1970).
21. A. M. Dubinskaya, P. Yu. Butyagin, R. R. Odintsova, A. A. Berlin: Vysokomol. Soyed. *A 10*, 410–415 (1968), Polymer Sci. USSR *10*, 478–485 (1968).
22. V. A. Radtsig and P. Yu. Butyagin: Vysokomol. Soyed. *A 7*, 922 (1965), Polymer Sci. USSR *7*, 1018 (1965).
23. P. Yu. Butyagin: Dokl. Akad. Nauk *140*, 145 (1961).
24. P. Yu. Butyagin, I. V. Kolbanev, A. M. Dubinskaya, M. Yu. Kisljuk: Vysokomol. Soyed. *A 10*, 2265–2277 (1968).
25. S. A. Kommissarov, N. K. Baramboim: Vysokomol. Soyed. *A 11*, 1050–1058 (1969), Polymer Sci. USSR *11*, 1189–1199 (1969).
26. V. A. Zakrevskii, E. E. Tomashevskii: Vysokomol. Soyed. *B 10*, 193 (1968).
27. T. Nagamura, M. Takayanagi: J. Polymer Sci., Polymer Phys. Ed. *13*, 567 (1975).
28. D. K. Roylance, K. L. De Vries, M. L. Williams: P. II. Internat. Conf. Fracture, Brighton, April 13–18, 1969.
29. D. K. Roylance: Ph. D. Dissertation, Dept. of Mech. Engng., Univ. of Utah, Aug. 1968.
30. D. K. Backman: Ph. D. Thesis, Univ. of Utah, May 1969.
31. D. K. Backman, K. L. De Vries: J. Polymer Sci. A-1 *7*, 2125 (1969).
32. M. L. Williams, K. L. De Vries: 14th Sagamore Army Materials Res. Conf. Aug. 23 (1967).
33. K. L. De Vries, D. K. Roylance, M. L. Williams: J. Polymer Sci. A-2, *10*, 599 (1972).
34. V. A. Zakrevskii, E. E. Tomashevskii: Polymer Sci. USSR *8*, 1424 (1966).
35. S. N. Zhurkov, V. A. Zakrevskii, E. E. Tomashevskii: Fizika Tverdogo Tela *6*, 1912–1914 (1964), Soviet Physics – Solid State *6*, 1508–1510 (1964).
36. K. A. Akhmed-Zade, V. V. Baptizmanskii, V. A. Zakrevskii, S. Misrov and E. E. Tomashevskii: Vysokomol. Soyed. *A 14*, 1360–1364 (1972), Polymer Sci. USSR *14*, 1524–1530 (1972).
37. J. Becht: Dissertation, Technische Hochschule, Darmstadt (1971).
38. H. H. Kausch, J. Becht: Kolloid-Z. u. Z. Polymere, *250*, 1048 (1972).
39. T. Pazonyi, F. Tudos, M. Dimitrov: Angew. Makrom. Chemie *10*, 75–82 (1970).
40. H. Fischer: Polymer Letters *2*, 529 (1964).
41. J. B. Park: Ph. D. Thesis, Univ. of Utah, June 1972.
42. C. T. Graves, M. G. Ormerod: Polymer *4*, 81 (1963).
43. M. Kashiwagi: J. Polymer Sci. *A 1*, 189 (1963).
44. S. N. Zhurkov, A. Ya. Savostin, E. E. Tomashevskii: Dokl. Akad. Nauk. SSR, *159*, 303 (1964), Soviet Physics – Doklady 9, 986 (1964).

45. S. N. Zhurkov, E. E. Tomashevskii: Proc. Conf. Physical Basis of Yield and Fracture, Oxford 1966, p. 200.
46. J. Becht, H. Fischer: Kolloid-Z. u. Z. Polymere *240*, 766 (1970).
47. L. A. Davis, C. A. Pampillo, T. C. Chiang: Bull. Amer. Phys. Soc. *18*, 433 (1973), J. Polymer Sci., Polymer Phys. Ed. *11*, 841–854 (1973).
48. K. L. De Vries, E. R. Simonson, M. L. Williams: J. Macromol. Sci. – Phys. *B-3*, 671 (1970).
49. R. T. Brown, K. L. De Vries, M. L. Williams: Polymer Preprints *11/2*, 428 (1970).
50. R. Natarajan, P. E. Reed: J. Polymer Sci. A-2, *10*, 585 (1972).
51. E. H. Andrews, P. E. Reed: In: Deformation and Fracture of High Polymers: H. H. Kausch, J. A. Hassell and R. I. Jaffee, Eds., Plenum Press, New York 1973, p. 259.
52. L. E. Nielsen, D. J. Dahm, P. A. Berger, V. S. Murty, J. L. Kardos: J. Polymer Sci., Polymer Phys. Ed. *12*, 1239 (1974).
53. T. C. Chiang, J. P. Sibilia: J. Polymer Sci., Polymer Phys. Ed. *10*, 2249 (1972).
54. B. Yu. Zaks, M. L. Lebedinskaya and V. N. Chalidze: Vysokomol. Soyed. *A 12*, 2669 (1970), Polymer Sci. USSR *12*, 3025 (1971).
55. D. Campbell, A. Peterlin: Polymer Letters *6*, 481 (1968).
56. B. Crist, A. Peterlin: Makromol. Chem. *171*, 211–227 (1973).
57. G. S. P. Verma, A. Peterlin: Polymer Preprints *10*, 1051 (1969).
58. P. Matthies, J. Schlag, E. Schwartz: Angew. Chem. *77*, 323 (1965).
59. R. T. Brown, K. L. De Vries, M. L. Williams: Polymer Letters *10*, 327 (1972).
60. T. Nagamura, M. Takayanagi: J. Polymer Sci., Polymer Phys. Ed. *12*, 2019–2034 (1974).
61. J. Ham, M. K. Davis, J. H. Song: J. Polymer Sci., Polymer Phys. Ed. *11*, 217 (1973).
62. M. Sakaguchi, J. Sohma: J. Polymer Sci., Polymer Phys. Ed. *13*, 1233–1245 (1975).
63. M. Sakaguchi, S. Kodama, O. Edlund, J. Sohma: J. Polymer Sci., Polymer Letters Ed. *12*, 609 (1974).
64. J. Sohma, M. Sakaguchi, in: Adv. Polymer Sci., Vol. 20, Berlin, Heidelberg: Springer-Verlag 1976.
65. V. A. Radtzig: Vysokomol. Soyed *A 17*, 154–162 (1975). Polymer Sci. USSR *17*, 179–190 (1975).
66. J. Tino, J. Placek, F. Szöcs: European Polymer J. *11*, 609–611 (1975).
67. D. Campbell: J. Polymer Sci, Part D, Macromol. Rev. *4*, 91–181 (1971).

Chapter 7

Phenomenology of Free Radical Formation and of Relevant Radical Reactions (Dependence on Strain, Time, and Sample Treatment)

I. Radical Formation in Thermoplastics

A. Constant Rate and Stepwise Loading of Fibers

Historically, the first ESR experiments with mechanically produced free radicals have been made in the Ioffe-Institute in Leningrad in 1959 [1] with ground or milled polymers, the samples being investigated after completion of the fracture process. To elucidate the effect of structural or environmental parameters on the kinetics of stress-induced free radical formation it is necessary to study — by ESR — highly stressed chains *during* the loading process. As outlined in Chapter 5 a notable energy elastic straining of a chain can only be achieved if the chain cannot relieve stresses internally by a change of conformation or externally by slippage against the field exerting the uniaxial forces. *Vice versa the mechanical scission of a chain must be taken as an indication that at the time of chain rupture axial stresses ψ equal to the chain strength ψ_c had not only been reached but also maintained during the average lifetime period τ_c of the chain segment.*

In view of the fact that the slippage of chains, microfibrils, or fibrils reduces or prevents the mechanical scission of chains highly oriented thermoplastic fibers with little potential for plastic deformation are the most suitable objects for a study of chain scission kinetics. ESR investigations using fibers of polyamide 6, 66, and 12, PE, PP, PETP and of other materials have been carried out in a limited number of laboratories, at first in the USSR, later in the USA, Germany, Great Britain, and Japan (see Table 6.2). Practically all researchers worked with highly oriented monofilaments, stacks of strips, or mostly with commercial yarns made up of a few hundred filaments each having a diameter of about 20 μm. As discussed in Chapter 2 the filaments consist of fibrils and these in turn of microfibrils with a diameter of 20 to 40 nm. The experimental set-up allowing uniaxial straining of these samples under controlled atmosphere and directly within the ESR cavity has been described and referenced in Chapter 6, Section IV B. It has also been seen that the primary radicals formed in tensile straining can — with one exception — be assigned to the breakage of main chains. In the following discussion, concentrations of secondary — rather than of primary — radicals are usually analyzed. It can be understood that the conversion factor from the number of primary radicals to the number of secondary radicals is constant (and, if not stated otherwise, close to unity).

If a semicrystalline microfibril is subjected to stresses, the resulting deformation will be non-homogeneous at the molecular level. The bulk of the deformation will be born by the amorphous regions. As discussed in Chapter 5 the largest stresses are transferred upon extended chain segments which share the strain imparted onto an amorphous region. The stress induced scission of chains must, therefore, be expected to occur within the amorphous regions.

Becht and Fischer [2] have demonstrated directly that free radicals are formed in the amorphous regions. These authors observed that polycaprolactam samples swollen in methacrylic acid did not show in subsequent stressing the usual ESR spectrum of the polyamide radical but the polymerization radical of the methacrylic acid. Based on the logical assumption that only amorphous regions had been swollen it was proved, therefore, that free radicals had formed in the amorphous regions only. Verma et al. [3] reached at the same repeatedly confirmed conclusion studying irradiation-produced radicals in semicrystalline polymers. Those radicals had been created under γ-irradiation within the whole volume of a PA 66 film i.e. as well within amorphous as within crystalline regions. At room temperature Verma observed — depending on sample orientation within the ESR cavity — a three, four, or six line spectrum. He assigned the pronounced anisotropy of the spectrum to the fact that the majority of the remaining radicals is located in the well oriented crystalline blocks. If the free radicals were introduced into the same material by stretching, no conspicuous anisotropy of the ESR spectrum could be observed. In this case the radicals were obviously located in sites showing rather small local chain orientation. In terms of fiber morphology the chain scission events must have taken place, therefore, in the amorphous regions between the crystal lamellae within the microfibrils (Fig. 2.11, 2.17b, 2.18b), at microfibril surfaces, or in interfibrillar regions (Fig. 2.11). The question of the location of chain scission points will be touched upon repeatedly within this chapter.

At first the phenomenology of free radical production in preoriented yarns will be discussed taking 6 and 66 polyamide as examples. In constant load rate tests a very steep increase in concentration of fairly stable secondary radicals is observed in a strain regime of between 8% and rupture strain (16 to 25%) which corresponds to a stress regime of between 500 and 900 MN/m^2 (Fig. 7.1). Load rate tests are a suitable means to demonstrate the effect of mechanical action on radical formation and accumulation, but they are not sufficient to discriminate between the simultaneous effects of stress, strain, and time on the rate of radical formation. From constant load tests it was learned that, under constant load, the rate of radical formation decreases quite rapidly whereas the amount of radicals formed in virgin samples reaches a constant equilibrium value (Fig. 7.2). If a sample is strained repeatedly to a strain level ϵ_1 then free radicals will be formed, i.e. chains will break, only during the first straining cycle. If, however, in subsequent straining the level ϵ_1 is exceeded, more chains will be seen to break. In a step-strain test the radical population increases stepwise (Figs. 7.3 and 7.4).

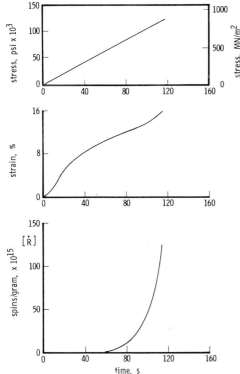

Fig. 7.1. Typical stress, strain, and free radical formation curves in constant load-rate straining of 6 polyamide fibers [4].

In the first paper on this subject [7] Kausch and Becht gave a detailed discussion of the notable fact that within each strain step the radical concentration increases concurrently with a decrease of the macroscopic tensile stress. The conclusion was drawn that the local molecular stresses are unevenly distributed. They are not simply a constant multiple of the macroscopic stresses. The chain segments breaking towards the end of a strain interval quite obviously do not participate in the macro-

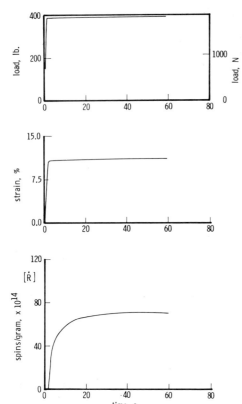

Fig. 7.2. Load, strain, and free radical production in a constant-load test of 6 polyamide fibers [4].

Fig. 7.3. ESR spectra of secondary radicals produced after stepwise loading of 6 polyamide fibers. The numbers correspond to stress levels of 0, 690, 760, 794, 828, 876, and 897 MN/m² respectively [5].

Fig. 7.4. Concentration of free radicals formed in stepwise straining of 6 polyamide fibers and measured uniaxial stress as a function of strain and time [6].

scopic stress relaxation (a decrease of the axial chain stress to 0.9 ψ_c increases the lifetime by two orders of magnitude and virtually stops the chain scission process). It was furthermore concluded that the microfibrils do not unload through slippage and that the lateral rigidity of the crystal blocks within the sandwich-structured microfibrils must be large enough so as to permit the build-up of large stress concentrations. If the lateral rigidity of the crystal blocks were insufficient the fibril would break up into submicrofibrils. This situation is schematically shown by Figure 7.5. The

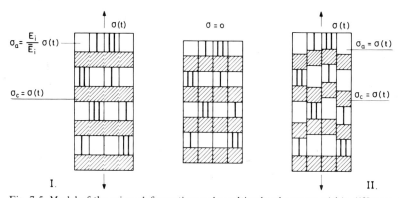

Fig. 7.5. Model of the micro-deformation and resulting local stresses within different amorphous (σ_a) and crystalline (σ_c) regions of semicrystalline fibers; case I. large, case II. small lateral rigidity of crystalline phases; E_i: Young's moduli of amorphous subregions, supposedly depending on the concentrations of highly extended tie molecules which are schematically indicated by the number of heavy lines between crystalline regions.

145

central diagram represents the alledged submicrofibrillar structure within an un-stressed microfibril. If the crystal blocks break up while being strained the tensile stresses in every part of the structure will equilibrate and be equal to the external stress (right hand side). The external stresses are, as one knows from Chapters 1 and 5 al-ways much smaller than the chain strength. If, however, the layered structure is also preserved at large strains stresses proportional to local strain and tensile modulus E_i of the amorphous subregions can be exerted (left hand side of Fig. 7.5). This means that within a subregion with large E_i correspondingly large — although locally well confined — stresses $\epsilon_a E_i$ — can be reached. For this reason the left hand side model is proposed to describe the mechanical response of an axially strained microfibril.

The observed fact that chains break even after (20 minutes of) stress relaxation not only demands the integrity of the crystal blocks but also an intimate and per-sistent lateral cohesion between microfibrils within a fibril and between fibrils within a filament. In the same way as studied in detail in Chapter 5 for single chains the shear displacement of the ends of microfibrils against the interfibrillar cohesion per-mits the transfer onto the microfibrils of those forces which accumulate within the stress-transfer lengths to the axial stress σ. This stress relaxes at constant filament elongation. The continued rupture of chains indicates that the axial *strains* of the microfibrils are maintained during such stress relaxation. Those strains, however, can only be maintained if large scale slip of microfibrils or fibrils does not occur.

In the described experiments with PA 6 fibers the cause of stress relaxation must obviously be sought within the microfibrils. Following the model shown on the left hand side of Figure 7.5 decrease of $\sigma(t)$ at constant microfibril elongation can be caused.

— either by the gradual and fairly general decrease of the relaxation moduli $\overline{E}_i$ of *practically all* amorphous regions within the microfibril
— or by the preferential and large decrease of the $\overline{E}_i$ of just a few amorphous regions, accompanied by their large deformation and by the retraction of others.

The previously discussed intimate lateral cohesion would favor the more homo-geneous decrease of all relaxation moduli. The same conclusion was drawn by Geza-low et al. [8] from the uniformity of stress induced long period changes.

The following statements will serve, therefore, as the basis for a mathematical analysis of the (static) chain scission experiments such as step-strain or small loading rate tests at sufficiently low temperatures:

— stress relaxation or creep is not caused by slippage of fibrils or microfibrils;
— chain scission occurs in tie-segments, i.e. in those segments of a chain molecule which connect two different crystal blocks traversing in a fairly extended con-formation the amorphous region in-between (e.g. Fig. 5.2); most frequently tie-segments interconnect adjacent crystal blocks within a microfibril, but also more distant or interfibrillar ties have to be considered;
— eventually breaking tie-segments are *solidly held,* most probably by the crystal blocks; although the *clamping* extends over 2 to 5 nm the tie-segments are treated as if they had a well defined end-to-end distance L_0 and contour length L;
— chain scission occurs if the axial chain stress ψ reaches chain strength ψ_c;
— the large value of molecular stress concentration (ratio of ψ to σ of the order of 50) follows from the fact that individual (highly rigid and nearly extended) chains

are subjected to the comparatively large strains of the intra- or interfibrillar amorphous regions they span.

At this point only the results of the mathematical analysis with respect to the kinetics of the formation of free radicals are presented. A discussion of the possible role of chain scission in fracture will be resumed in Chapter 8.

The analysis concerns the general case of fairly extended tie-segments of contour length L_i, end-to-end distance L_0, and elastic modulus E_k. Each segment spans an arbitrary number n_{ai} of amorphous regions (width L_a) and $n_{ci} = n_a - 1$ of crystalline regions (width L_c). As discussed in Chapter 5 a segment i will experience ultimate stresses ψ_i only after it has assumed the most extended conformation accessible to it. The elastic stress then is:

$$\psi_i = E_k \left[\frac{L_a}{L_i} (1 + \epsilon_a) - 1 + \frac{L_0 - L_a}{L_i} (1 + \epsilon) \right]. \tag{7.1}$$

At a given average axial microfibril strain ϵ the stresses ψ_i are largest for tie-segments interconnecting adjacent crystal blocks ($L_0 = L_a$). The third term in Eq. (7.1) then disappears. If the tie-segments extend over several sandwich layers and crystal blocks of the same or different microfibrils the first term in Eq. (7.1) becomes small. In that case the average *tensile* strain ϵ determines the stress level ψ_i. It is important to point out again that in this analysis no loading of interfibrillar tie-segments through *shear strains* (slippage between microfibrils) will be considered. At a degree of crystallinity of 50% ϵ is by a factor of ($2-E/E_c$) smaller than ϵ_a, where E and E_c designate the relaxation modulus of the microfibril as a whole and that of the crystalline regions.

The mathematical analysis of the free radical production curves aimes at a determination of the distribution $N_0(L_i/L_0)$ and of the kinetic parameters U_0, ω_0, and β. It can be carried out taking into consideration only tie-segments between adjacent crystal blocks ($L_0 = L_a$). If only long interfibrillar tie-segments were present the resulting distribution $N_0'(L_i/L_0)$ would be *narrower* and approximately obtainable through the transformation $N_0' \left(\frac{L_i}{L_0} \right) = N_0 \left(\frac{[1 + \epsilon_a] L_i}{[1 + \epsilon] L_0} \right)$. The following analysis also assumes a homogeneous distribution of amorphous strains within a microfibril. Consideration of a non-homogeneous distribution of strains ϵ_{ai} would again result in *narrowing* $N_0(L_i/L_0)$. From Eqs. (5.57) and (7.1) one obtains straight forwardly the expectation value of the number of broken chains or formed free radicals:

$$[R \cdot] = 2 \sum_i N_0 (L_i/L_0) \left\{ 1 - \exp \left[- \int_0^t \bar{\omega}_b \exp (\beta \psi_i/RT) \, d\tau \right] \right\} \tag{7.2}$$

where $\bar{\omega}_b = \omega_0 \exp (-U_0/RT)$.

Through a numerical iteration procedure, a distribution of L_i/L_0 can be determined [7] in such a way that Eq. (7.2) describes the experimental points (e.g. those radical concentrations shown in Fig. 7.4). The length distribution of the intrafibrillar tie-segments obtained in that case is shown in Figure 7.6, it ranges from $L/L_0 = 1.06$

Fig. 7.6. Distribution of relative chain lengths of tie molecules in the amorphous regions of 6 polyamide fibers as obtained from numerical evaluation of step-strain ESR data [4, 7]. I chains broken until fracture, II chains fully extended and contributing with chain modulus E_k to sample modulus E, III chains in non-extended conformation.

to 1.18. If the same experimental radical concentration points were to derive from the rupture of fairly long interfibrillar tie-segments then their relative lengths must have a distribution of L/L_0 between 0.97 (!) and 1.05. Since a given distribution of segment lengths is essentially reflected by the step heights of radical equilibrium concentrations a histogram of these step heights serves to characterize the apparent segment length distribution without introducing any assumptions on the nature of the tie-segments. In Figure 7.7 histograms from step strain tests carried out at different temperatures are reproduced.

The association of the observed free radical distributions with apparent chain length distributions requires the smallest number of additional assumptions. If it were to be assumed that the microfibrils contained a monodisperse distribution of

Fig. 7.7. Distributions of relative chain lengths shown as histograms of free radical concentrations obtained in step-straining of 6 polyamide fibers at (a) $-25\,°C$, (b) room temperature, (c) $+50\,°C$, and (d) $+100\,°C$ [4, 5].

148

relative segment lengths (e.g. $L/L_0 = 1.0$) it were also necessary to assume that

— the amorphous strains are unevenly distributed throughout the fiber
— the molecular extensibility is comparable to the amorphous strains, i.e. reaches 25 to 35%
— there is a rapid strain hardening of an amorphous region after breakage of the tie molecules (otherwise collapse of that region would lead to complete stress relaxation in the whole microfibril).

Particularly the latter two conditions seemed to be unrealistic. So the early proponents [3—7] were led to their interpretation that a fiber contains tie chains of different conformations ("tautness").

A third interpretation of step-strain data was advanced by Crist and Peterlin [9]. They suggested that there is an uneven distribution of strains due to differences in the lengths of the many thousand filaments strained simultaneously in any of the above experiments. The uneven-length effect undoubtedly broadens the observed distributions of relative chain lengths. But the premature breakages of individual filaments, and the formation of their fracture surfaces, cannot account for the number of free radicals formed. To further disprove this contention Hassell and DeVries examined the free radicals formed in straining of highly oriented bulk nylon 66 banding material [10]. They obtained quite similar and even slightly wider histograms in straining bulk strip material than bundles of nylon 66 fibers. A scanning electron micrograph of the strip material fracture surface reveals that in banding material — as postulated for yarns — defects are introduced throughout the stressed sample volume (Figs. 7.8 and 7.9). The final fracture surface propagates along the path of least resistance which includes the previously developed defect zones. It is only at a level of deformation close to the rupture strain that a difference between the straining of a single fiber and of a bundle thereof becomes noticeable. A statistical explanation of this fact has been given in Chapter 3.

Fig. 7.8. Fracture surface of dry 66 polyamide banding material [110].

Fig. 7.9. Fracture surface of 66 polyamide banding material, before fracture stored to equilibrium at 65% relative humidity [10].

Convincing evidence against the assumption that segment length distributions are macroscopic artefacts was derived from step-temperature tests carried out by Johnsen and Klinkenberg [11] which will be discussed in Section C.

B. Effect of Strain Rate on Radical Production

Local molecular stresses can partly be relieved by slip or uncoiling of molecules. In thermoplastics the relaxation times governing those viscoelastic deformations at room temperature are of the order of milliseconds to minutes, i.e. smaller or comparable to the duration of the mechanical excitation. Rapid loading, then, can bring more nonextended chain segments to high stresses and eventually to chain scission than slow loading during the same strain interval. But their number should be small. On the other hand, the higher rigidity of a rapidly loaded viscoelastic body leads to a decrease of the strain at fracture and, therefore, also to decrease of the largest strain, ϵ_a, imposed on the amorphous regions.

The effect of strain rate on chain scission of rigidly clamped segments follows from Eq. (7.2). The higher the loading rate the smaller the time t necessary to reach a certain strain, the shorter also the time interval a stressed chain spends at the stress level $\sigma(t)$ and the smaller the probability for chain scission at that level [7]. In Figure 7.10 the decay curves of polyamide 6 segments of *monodisperse* contour lengths $L = 1.09 \, L_0$ have been plotted. At 20% *amorphous* strain ϵ_a a segment of length L will assumedly have a fully extended conformation and experience a *chain* strain of 10%. At this strain level chain scission becomes noticeable at small strain rates $(0.1 \cdot 10^{-6} \, s^{-1})$. At larger strain rates an equivalent decay is observed later. For the given system an increase of strain rate by a factor of 10 will shift the decay curve to higher strain values by $\Delta\epsilon_a = 0.9\%$.

The effect of a change of strain rate in the course of a tensile experiment can be predicted from Figure 7.10. An instantaneous increase in strain rate is equivalent

Fig. 7.10. Effect of rate of strain on the scission of polyamide 6 segments of contour length $L = 1.09 \, L_0$ (after 12.7).

to changing from a certain level of depletion (e.g. 40% at point A) to a level where the depletion at that rate should have been much lower (6% at B). In other words a fraction of 34% of those chains which are supposed to scission beyond B are already gone. Consequently the rate of chain scission following an increase in strain rate should be reduced. This has been quantitatively verified by Klinkenberg [13].

Only few experiments are known where the concentration of free radicals as a function of strain rate has been investigated [12–16]. In Figure 7.11 the total number of *spins at fracture* of a 6 polyamide fiber sample is shown to increase in the range of clamp displacement rates between 1 and $10 \cdot 10^{-3}$ cm/s. Above $20 \cdot 10^{-3}$ cm/s (which corresponds to a strain rate of $1.43 \cdot 10^{-3}$ s^{-1}) a drop in free radical production is indicated [14]. The total rate effect seems to be not very large, however. For the present experiment, it did not exceed 50%. Similar values have been reported [16] from an investigation of the cold drawing of polycarbonate [15]. Here the total concentration of free radicals obtained increased by 140% with strain rates varying from 0.017 s^{-1} to 1.7 s^{-1} (Fig. 7.12). The effect of strain rate on the formation of free radicals in polychloroprene will be discussed in Section II A.

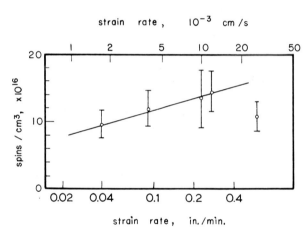

Fig. 7.11. The effect of strain rate on the radical concentration at break of 6 polyamide fibers [14].

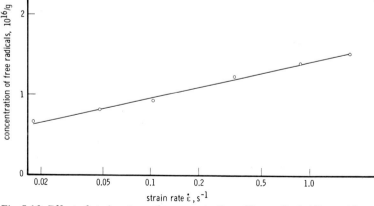

Fig. 7.12. Effect of strain rate on the concentration of free radicals after necking of polycarbonate [15, 16].

In all of these cases the variation of free radical concentration seems to reflect more the changes of material response than the kinetics of decay of stressed chains. The effect of strain rate on radical production in comminution (grinding, milling) would be very difficult to determine and no results have been reported.

C. Effect of Temperature

1. Apparent Energy of Bond Scission

It is presumed that the stress-induced scission of chain molecules is achieved through the cooperative action of mechanical forces (lowering the bond's binding potential barrier) and of statistically fluctuating thermal vibrations supplying the remaining increment of energy which is necessary to split the loaded bond. It is also believed that Eq. (5.57) provides an adequate description of how mechanical and thermal energy affect the rate k_c of the thermo-mechanical chain scission process. If this presumption is correct, the lack of thermal vibrational energy should increase the stability of a stressed bond. Vice versa an increase in thermal energy should make previously stable bonds reach the critical level of excitation and should break them. It is of interest to quantitatively analyze this aspect of interaction of thermal and mechanical energy terms in the scission kinetics of 6-polyamide chains.

The largest energy available to one $RCONHCH_2-CH_2R'$ bond within a time period of $1/k_b$ is (Eq. 5.57 for $n_c = 1$):

$$U(T) = \beta \psi + RT \ln \omega_0/k_b. \tag{7.3}$$

Using as before a value of ω_0 equal to 10^{12} s^{-1}, a value of β of $5.53 \cdot 10^{-6}$ m^3/mol, and a rate $k_b = 1/\tau_b = 1/48$ s^{-1} the thermal term in Eq. (7.3) has, at a temperature of 20 °C, a value of 77.2 kJ/mol. The stress term $\beta \psi$ must reach, therefore, a value of 110.8 kJ/mol before U(T) is larger than the U_0 of PA 6 and *bond rupture* at room temperature can occur. At 200 K the thermal term has decreased to 52.7 kJ/mol. In order to provide a total energy of 188 kJ/mol, the mechanical term must now reach 135.3 before bond breakage will occur within 48 s. For a *chain segment* containing n_c weak bonds the rate of scission, $1/\tau_c$, employed in the above calculations will be $1/\tau_c = n_c/\tau_b$.

If one assumes that neither β nor U_0 depend on temperature, one has to conclude that at 200 K it is possible to raise the molecular stress to a level which is 17% higher than the largest molecular stress attainable at room temperature. In other words, a polyamide fiber highly strained at 200 K will quite stable contain many bonds with $\beta \psi$ values between 110 and 135 kJ/mol. Those bonds will break, of course, if the temperature is raised to room temperature provided that no other means of stress relaxation (slip, uncoiling) become available to the strained chain segment.

The search for chain scission events during an upward temperature scan can be called a critical experiment for the validity of both the kinetic equation (Equation 5.57) and the morphological model (Fig. 7.5). These investigations have been carried out in the Deutsches Kunststoff-Institut in Darmstadt. Following earlier experiments by Becht [5] Johnsen and Klinkenberg [11, 13] step-strained 6 polyamide fibers at 206, 227, 248, 269 and 290 K. These tests showed the same stepwise increase of radical concentration and a relaxing stress as reported earlier by them [6, 7, 11, 17] and other authors [3, 4, 10, 14] and as described in Section A. The result of a step-temperature experiment is shown in Figure 7.13. Whereas the stress relaxation is much more pronounced here than it is in a step-strain test, the radical accumulation

Fig. 7.13. Concentration of free radicals and uniaxial stress in step-temperature test as a function of temperature and time for 6 polyamide fibers [11].

appears to be very similar in both experiments. The authors [11] established that the final radical concentration [$\dot{R}$ (T, ϵ)] practically did not depend on the way T and ϵ were attained. Their experiment has proven the contention made earlier about the kinetics of thermomechanical chain scission. It has also proven that the highly loaded chains do show in the region T < 260 K essentially no slip or anelastic deformation during the time they are held at the high load, despite the considerable amount of overall stress relaxation. A slight loss of stressed chains becomes apparent at T > 270 K.

The equivalence of thermal and mechanical action on a given network of chains is further illustrated by calculating the relative length L/L_0 of those intrafibrillar chains which are just about to break. One considers that condition to be fulfilled if the ψ (L), calculated from Eq. (7.1) for $L_a = L_0$ and introduced into Eq. (7.3), yields U_0:

$$\left(\frac{L}{L_0}\right) = \frac{(1 - xE/E_c)\epsilon/(1-x) + 1}{1 + (U_0 - RT\ell n\omega_0/k_c)/\beta\, E_k} \, . \tag{7.4}$$

E and E_c are respectively the (temperature-dependent) Young's moduli of the sample as a whole and of the crystal lamellae in chain axis direction, x is the degree of sample crystallinity.

Equation (7.4) relates sample strain ϵ, temperature T, and relative length L/L_0 of critically loaded chain segments, but subject to the condition that no chain slip occurs simultaneously. Under this condition the following equivalence between incremental changes of sample strain and temperature is obtained from Eq. (7.4):

$$\Delta\epsilon = -\frac{(1-x)\,\Delta T}{(1-xE/E_c)}\left[\frac{L\,R\,\ell n\,\omega_0/k_c}{L_0\,\beta\,E_k} - \left\{\left(1 + \frac{U_0 - RT\,\ell n\,\omega_0/k_c}{\beta\,E_k}\right)\frac{L}{L_0} - 1\right\}\right.$$
$$\left.\cdot\frac{d(E/E_c)/dT}{1-xE/E_c}\right].$$

$$(7.5)$$

The two terms in the square brackets refer respectively to the change of chain strength and of strain distribution with temperature. Using the kinetic parameters employed in evaluating Eq. (7.3) for PA 6 chains, values of $E_c = 25$ GN/m^2 and of $E_k = 200$ GN/m^2 as previously, and an estimated temperature coefficient $d(E/E_c)dT$ of $-0.8 \cdot 10^{-3}$ K^{-1} one obtains at a temperature of $-20\,°$C the following quantitative evaluation of Eq. (7.5):

$$\Delta\epsilon = -\,[0.12 \cdot 10^{-3}\,L/L_0 + 0.48 \cdot 10^{-3}\,(1.1\,L/L_0 -1)]\,\Delta T.$$

$$(7.6)$$

This means that for a critically loaded PA 6 chain segment with $L/L_0 = 1.1$ a temperature increase of 10 K can be compensated by a decrease of strain of only 0.23%. The result of this calculation agrees fairly well with the average value of 0.3%/10 K derivable from the experiments of Johnsen and Klinkenberg [11]. For preoriented PA 6 fibers they observed that at $-60\,°$C a sample strain of 14.6% led to the same number of (10^{17}) free radicals as a strain of 12.2% at $+20\,°$C.

The loading of possibly present long interfibrillar tie-segments is essentially determined by $1 + \epsilon$ rather than $1 + \epsilon_a$. Accordingly in Eq. (7.5) the first term in square brackets and a front factor equal to ΔT must be retained. For PA 6 this leads to a $\Delta\epsilon$ equal to $-0.24 \cdot 10^{-3}\,\Delta T\,L/L_0$, i.e. to 0.26%/10 K at $L/L_0 = 1.1$. This comparison does not permit, therefore, to rule out the existence of interfibrillar tie-segments.

2. Rate of Bond Scission

A *rate of bond scission* can be defined in terms of time as $d[\dot{R}]/dt$ or in terms of strain as $d[\dot{R}]/d\epsilon$. The relation between the two rate expressions is defined by the identity

$$\left(\frac{d[\dot{R}]}{dt}\right)_{\epsilon_1,\,\dot{\epsilon},\,T} = \dot{\epsilon}\left(\frac{d[\dot{R}]}{d\epsilon}\right)_{\epsilon_1,\,\dot{\epsilon},\,T}$$

$$(7.7)$$

which has to be taken at constant $\dot{\epsilon}$ and T at corresponding sample strains ϵ_1.

In the absence of plastic flow (chain scission at strains of 10 to 30%) the rate can be derived numerically from Eq. (7.2); since the rate will depend on the shape of the tie-segment length distribution, N_0 (L_i), no general analytical function can be given for $d[\dot{R}]/d\epsilon$. The effect, however, of a change of $\dot{\epsilon}$ on $d[\dot{R}]/d\epsilon$ is the same as that illustrated by Figure 7.10. The effect of the strain ϵ_1 at which the rate of bond scission is determined, depends on the shape of the segment length distribution, i.e. on the number $\Delta N(L/L_0)$ of chains having a relative length L/L_0 suitable to be broken by an increment $\Delta\epsilon$ of sample strain:

$$\left(\frac{d[\dot{R}]}{d\epsilon}\right)_{\epsilon_1,T} = \frac{2\Delta(L/L_0)}{\Delta\epsilon}\left(\frac{dN}{dL/L_0}\right)_{\epsilon_1,T}. \tag{7.8}$$

In Figures 7.6 and 7.7 typical segment length distributions are given.

A change of temperature will affect the rate of radical production in three ways. Firstly the number N of highly extended chains will decrease if thermally activated slip becomes possible. Secondly the activation energy U(T) decreases with increasing temperature. Thirdly the sample modulus E and the crystalline modulus E_c are affected by temperature.

If in a region of sufficiently low temperatures chain slip can be excluded i.e. if

$\frac{dN}{dL/L_0}$ is constant with T the temperature affects solely the width $\Delta(L/L_0)$ of the

chain segment population which is critically loaded by a strain increase:

$$\frac{\Delta(L/L_0)}{\Delta\epsilon} = \frac{(1 - xE/E_c)/(1 - x)}{[1 + (U_0 - RT\ell n\omega_0/k_c)/\beta E_k]}. \tag{7.9}$$

Employing the same values as used in the evaluation of Eq. (7.3) one arrives at $\Delta(L/L_0) = 1.7\,\Delta\epsilon$ at 20 °C and at $\Delta(L/L_0) = 1.6\,\Delta\epsilon$ at −60 °C. This means for example that a given distribution of segments with relative lengths L/L_0 between 1.1 and 1.3 break at room temperature within the strain interval from 12 to 13.5%. At a temperature of −60 °C sample strains of between 13.5 and 15.1% are required.

A completely different chain loading mechanism prevails during the plastic deformation of polymers at strains of between 30 and several hundred per cent. In that case a chain will rupture as a consequence of the frictional forces encountered by the chain itself or by other morphological units in the dynamic shearing (cf. Chapter 5 II E). The obtainable axial chain stresses are proportional to the molecular or fibrillar friction coefficients and to the strain rate $\dot{\epsilon}$. The number of critically loaded chains will reflect, therefore, the strong increase of the friction coefficient with decreasing temperature. Davis et al. [19] strained sheets of high molecular weight polyethylene in air and observed the formation of peroxy radicals. At a level of true strain $\ell n\, \ell/\ell_0$ of e.g. 1.1 − which corresponds to a conventional strain of 200% − the concentration of the peroxy radicals increased from $5 \cdot 10^{14}$ cm^{-3} at 294 K to 10^{16} cm^{-3} at 160 K. The rate of radical accumulation, $d[\dot{R}]/d\ell n(\ell/\ell_0)$, revealed two transition regions, one between 180 and 200 K and another beginning at 250 K.

The low temperature transition was assigned to the transition of amorphous PE to the glassy state, the other to slippage of chains within crystal lamellae [19].

3. Concentration at Break

The increase of temperature has a fourfold effect on the number of free radicals observed at the point of macroscopic fracture. Firstly as seen previously the bond strength is decreased and thus chain scission at a given molecular stress is facilitated. Secondly the decrease in intermolecular attraction and the increase in molecular mobility lead to more rapid relaxation of molecular stresses. For the same reason – thirdly – the density of stored elastic energy at a given strain level is decreased which will in turn affect crack stability and propagation. Fourthly, the increased reactivity of the free radicals may lead to an increased discrepancy between the concentrations of radicals formed and of radicals observed at the point of failure.

Whenever possible the fourth point has been taken into consideration and radical concentrations have been corrected for losses due to recombination or other decay reactions of the (secondary) radicals under consideration. In the case of 6 polyamide fibers [5, 11, 17, 18] the radical decay is a second order recombination reaction with the rate constant depending on type, location, and mobility of the radicals (see Section III below).

In Figure 7.14 the concentration of free radicals at fracture of 6 polyamide fibers is shown as observed and also corrected for radical decay during and after sample fracture [18]. The indicated decrease of the radical concentration at break towards

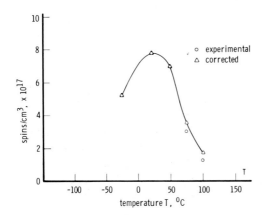

Fig. 7.14. Concentration of free radicals at break of 6 polyamide fibers as a function of breaking temperature [4, 5], ○ observed concentrations, △ concentration corrected for decay of free radicals.

lower temperatures was well confirmed by Johnsen and Klinkenberg [11] who extended their measurements to −67 °C. At that temperature they obtained a concentration which was by a factor of 8 smaller than that at room temperature.

The first three of the initially mentioned temperature effects can be better understood if one makes use of the segment length distribution model. If, in fact, at low temperatures the changing bond strength is the only temperature dependent parameter then a chain length distribution derived at about $-25\,^{\circ}C$ should exactly include the (narrower) distributions at still lower temperatures. Vice versa from one measured step-strain free-radical curve all narrower ones must be predictable. Johnsen and Klinkenberg [11, 13] have confirmed that the chain length distributions obtained at temperatures between -67 and $-25\,^{\circ}C$ are derived from one and the same "master" distribution which is skewed to a different extent at the right hand side. At higher temperatures (-4 and $17\,^{\circ}C$) discrepancies between the apparent distributions and the master distribution are observed. They are interpreted as indication of structural changes leading to a relaxation of some overstressed chains. These chains may, however, break at a later stage of the loading process [11, 13]. The relaxation and slip effect is stronger at higher temperatures and at $50\,^{\circ}C$ it clearly outweighs the effect of diminished bond strength, i.e. the total number of broken bonds at fracture decreases. Beyond $50\,^{\circ}C$ most of the chain segments belonging to the master distribution escape chain scission because the crystal blocks are no longer able to exert and maintain large enough molecular stresses. At this point it may remain open which mechanisms are responsible for the stress relieving structural changes.

In the above sections were discussed the scission of tie chains in highly oriented fibers. Judging from the absence of any scale effect the molecular fracture events are evenly distributed throughout the amorphous volume of the stressed sample. Those "volume" concentrations of 10^{15} to 10^{18} spins/cm^3 accounted for almost the total radical intensity whereas the final fracture surfaces (with estimated radical populations of 10^{12} to 10^{13} spins/cm^2) did not noticeably contribute towards the total number of radicals observed in tensile straining of the fiber samples. This is different if the mechanical degradation is achieved through milling, grinding and/or cutting. In all these techniques comminution is achieved through failure of virgin material resulting from the application of multiaxial stress particularly involving compressive components. In impact loading of platelike or granular samples compressive waves may be generated. Upon reflection on boundaries or discontinuities, compressive waves become dilational waves giving rise to tensile fracture. All forces combine to overcome the cohesive forces and to create new surface area. Those chains which break in this process are either intersecting weak cross-sections and/or are contained in shear zones along a suitable direction. It must be assumed, therefore, that chains are broken preferentially — but not exclusively — in the zones of new surface.

The surface concentration of free radicals in ground polymers has been determined by different investigators [20–23]. As to be expected the mobility of the chains within the degrading material strongly effects the obtainable radical concentrations. Backman and DeVries [21, 22] sliced 66 PA, PP, and PE in a nitrogen atmosphere and transferred the slices within less than 0.5 s into liquid nitrogen and determined the number of free radicals by ESR. The degradation temperature was varied between $-20\,^{\circ}C$ and $90\,^{\circ}C$. The authors observed a rapid decrease of surface concentrations at the respective glass transitions (Fig. 7.15). It should be noted that the rate of decay of the radicals formed also increases beyond T_g, but that this loss of radicals amounts to less than 30% [18]. From the level of the low temperature radical concentrations it can be concluded that the average fracture plane breaks between 5 and 10% of all chains intersecting the plane. Pazonyi et al. [22] investigated

Fig. 7.15. Surface densities of spins observed on comminuted polymers as a function of the degradation temperature T plotted as fraction of the corresponding glass transition temperature [20].

polyvinylchloride with differing contents of plasticizer (30, 40, 50, and 60% respectively of dioctylphthalate, DOP). They estimated that cutting at 30 °C led to the scission of (128, 105, 55 and 34% respectively of all) chains intersecting the fracture plane. They also investigated low and high density polyethylene and found 1.39 and $1.17 \cdot 10^{15}$ spins/cm^2 which means they had to cut 120 and 104% respectively of the intersecting chains. Pazonyi et al. did their cutting in an ethanolic solution of diphenyl picryl hydrazyl the color change of which was used as a measure for the number of free radicals formed. The unusually large molecular cutting ratios given by Pazonyi et al. and their discrepancy with the values cited earlier are noted and discussed by Salloum and Eckert [23]. They point out that reactions of DPPH with ethanol and with polymer surfaces not containing free radicals may take place thus accounting for the excessively high radical numbers determined by Pazonyi. Salloum and Eckert drilled PE, Nylon, and PP in a solution of DPPH in benzene. They obtained 0.1—2 (PE), 0.5—4 (6 PA) and $7—9 \cdot 10^{13}$ spins/cm^2 (PP) respectively. The first two values are in agreement with the results of Backman and DeVries, the latter is not.

Both laboratories [20—22] conclude that free radicals are formed within a surface layer of finite thickness. From a study of oxygen diffusion rates and of the conversion of the original ESR signal from secondary 6 nylon radicals to that of peroxy radicals Backman and DeVries determine that 50% of all "damage" occurred within a layer of less than 0.6 μm from the surface. The remaining 50% of chain radicals were found in a depth of up to 3 μm from the surface. Considering the morphology of the degraded polymers, the mechanics of the comminution process, and the mobility of the primary formed free radicals the spatial distribution of the secondary radicals can be understood. In this case the pull out and breakage of *individual PA chains* appears unlikely in view of the crystal strength. As we have discussed in Chapter 5, a 6 PA chain embedded within a crystallite with more than 1.7 nm of its length will rather break than be pulled away from the crystallite. The pull out of *whole microfibrils* from a fracture surface very probably will occur and will be accompanied by breakage of interfibrillar tie chains thus introducing damage within a surface layer of up to 1 μm depth. Even more important will be the heavy plastic deformation of the material ahead of a growing fracture surface. The propagation of free radicals certainly contributes to deepen the layer of apparent damage. The layer depths observed in these experiments do coincide, however, very well with the lower limits of particle sizes obtainable in mechanical degradation [24] indicating, that the damage may indeed be introduced mechanically up to the observed depths.

D. Effect of Sample Treatment

The effect of heat treatment on structure, fracture strength and chain scission has been very extensively investigated by Statton, Park and DeVries [25–27] and by Lloyd [5]. With respect to our understanding of chain breakages in annealed fibers the following morphological changes seem to be particularly noteworthy. Relaxed heat treatments result in [25]:
1. Increases of the degree of crystallinity, perfection of crystallites, density, chain folding, Young's modulus, and intensity of the fluid-like component of the NMR-signal.
2. Decreases of sample length (shrinkage), breaking strength, sonic modulus, and T_g.

The changes are the more intense the higher the annealing temperature. Tension during annealing tends to modify or reduce the above listed effects. Relaxed and tension annealing do, however, increase the long period.

The morphological changes certainly have a bearing on the molecular stress distribution and chain breakage. In Figure 7.16 the histogram from an ESR step-strain test is reproduced which was obtained from a tension-free annealed polyamide sample at 199 °C [27]. The histogram is broader, indicating a less homogeneous distribution of molecular strains, and it is shifted towards higher strains by about two to four

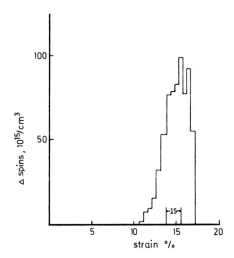

Fig. 7.16. Histogram of radical concentration from 6 polyamide fibers slack annealed at 199 °C and subsequently step-strained at room temperature [4. 5].

per cent of strain. This may be seen from Figure 7.17 where the chain length distribution of the annealed sample is compared with that of an unannealed reference sample. The used probability plot had been calculated under the assumption that the histogram in Figure 7.16 represents a normal distribution of chain lengths skewed at the right hand side.

Fig. 7.17. Probability plot of histogram of radical concentration of 1. reference sample and 2. sample slack annealed at 199 °C [4, 5].

The observed conformational changes in slack annealing have been interpreted [25–27] as an increase in relative length of (the formerly) extended tie chain segments due to defect migration out of the crystal blocks. The number of regular chain folds also increases. Shrinkage of the yarn appears to be governed by the number of folds introduced. The structural changes occurring during annealing are mechanically stable and not simply reverted under tensile stress. In Figure 7.18 a model representation of the conformational changes in annealing are given [4, 5]. Because of defect migration one obtains in tension annealing a relaxation of local molecular strains and thus a larger number of extended chain segments. On the basis of this experimental evidence Statton has constructed a string model of slack annealed, tension annealed, and reference fibers which is shown in Figure 7.19 [28]. The effect of the annealing process on the total number of free radicals formed in subsequent straining at room temperature and on fracture strength is shown in Figures 7.20 and 7.21. The annealing at maximum tension (11.7% elongation) does lead to a smaller number of spins at fracture and may be explained by the breakage of chains during tension annealing. Intermediate positive annealing strains (not shown in Fig. 7.20) do slightly increase the number of spins. The opposite effects of annealing temperature and of constraint of the sample during annealing on the homogeneity of the chain length distribution are shown in Figure 7.22. These effects are discussed in more detail in Chapter 8.

160

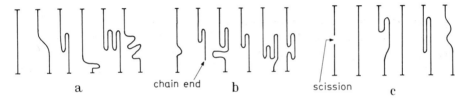

a   chain end   b   scission   c

Fig. 7.18. Model of tie chain conformation in (a) reference sample, (b) fiber slack annealed, and (c) fiber annealed under tension [4, 5].

SLACK ANNEAL at 199 °C   CONTROL   HIGHEST TENSION ANNEAL at 199 °C

Fig. 7.19. String-model by Statton [27, 28] representing effect of annealing on fiber morphology and chain conformation.

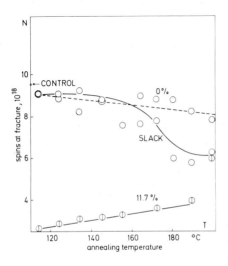

Fig. 7.20. Concentration of free radicals at fracture (at room temperature) as function of annealing temperature [27].

161

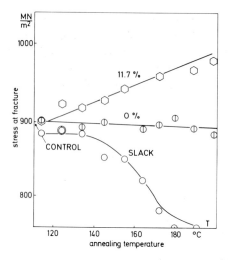

Fig. 7.21 Effect of annealing temperature and annealing condition on room temperature fracture stress of 6 polyamide fibers [27].

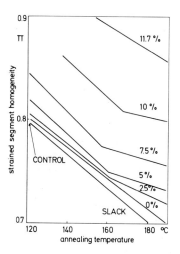

Fig. 7.22. Effect of annealing temperature and constraint during annealing on the strained segment homogeneity π which is defined as the ratio of that strain where the first radicals are observed to the strain at break [27].

II. Free Radicals in Stressed Rubbers

A. Preorientation, Ductility, and Chain Scission

In Section I we have discussed the stress-induced chain scission of thermoplastic polymers. As we have seen chain scission could occur whenever the intermolecular forces acting on tightly embedded sections of extended (tie) molecules were strong enough to resist segment slip during deformation in such a manner as to give rise to axial forces equal to or larger than the mechanical strength of the molecular chain. Exactly the same considerations apply to chain scission in elastomers which are in a temperature region below T_g. Andrews, Reed, and Natarajan [29–31] have studied under these aspects the deformation behavior of sulphur cured natural rubber (cis-polyisoprene), dicumyl peroxide-cured natural rubber, and polychloroprene. They observed that (see Fig. 7.23) in the temperature region I from glass transition (200 K) to 160 K the deformation behavior of natural rubber was very similar to that of a thermoplastic polymer, with a clearly discernible yield point and a yield stress which

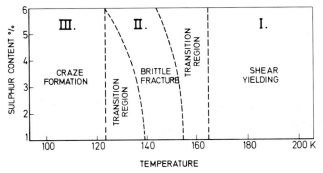

Fig. 7.23. Deformation and fracture behavior of preoriented polyisoprenes and polychloroprene (after Andrews and Reed, [31].

increased with decreasing temperature. Following a transition region there was a narrow temperature band (II) between 150 and 130 K within which brittle fracture occurred at strains of less than 5%. For still lower temperatures – and only for these temperatures – the authors [29–31] made the very interesting observation that the deformation behavior was determined by the state or orientation of the stressed network. Unoriented rubber specimens behaved brittle just as they do within the temperature band above. A second group of samples was prepared by first preorienting them at room temperature by uniaxial straining to 100 to 200%, and then cooling them down to 90 to 130 K (region III). These samples showed a ductile deformation behavior (Fig. 7.23). In this case the ductile low temperature deformation was accompanied by microvoiding in narrow striations, the formation of detectable amounts of free radicals, and by absorption of environmental gases within the microvoids. These gases expanded upon warming of the deformed specimens and gave rise to a peculiar foaming effect. The observed free radical spectra of sulphur-cured polyisoprene have been assigned to allyl main chain radicals stemming from the scission of the (weakest) bond located between the two α-methylene groups. These radicals appeared together with a contribution of possibly RṠ radicals. The spectra of dicumyl peroxide-cured polyisoprene and polychloroprene were assigned to secondary peroxy radicals ROȮ. The concentration of free radicals (in polychloroprene) varied with strain rate and yielded a maximum of $6 \cdot 10^{16}$ spins/g at a strain rate of 0.01 s^{-1} [31].

Taking all these observations together, the following model of the deformation behavior of elastomers below their glass transition temperature is proposed: In region I intermolecular attraction is sufficiently strong to permit energy elastic deformation of chain segments. Gradually at first, and exclusively beyond the yield point, slip and reorientation of chain segments occur. Chain scission is negligible because chains will rather slip than break. In temperature region II where brittle fracture occurs irrespective of the state of preorientation the intermolecular attraction seems to be just large enough to permit the axial loading of chain segments up to their breaking stress. In the absence of local strain hardening the largest crack initiated within the specimen during deformation to the 5% level will rapidly expand and prevent the growth of any other crack nuclei. As discussed for thermoplastics, the formation of essentially one plane of fracture is hardly sufficient to produce a detectable amount of free radicals. Unoriented rubber specimens (samples frozen in the relaxed state) show the same brittle behavior also in temperature region III (90 to 130 K). Specimens preoriented by extensions of 100 to 200%, however, obviously have gained the property of local strain hardening. The nature of this effect can only be speculated on and probably is related to the build up of partially oriented microfibrils within the preoriented elastomers. The result of local strain hardening is that the flaws and defects which are still present develop into a system of interconnected microvoids and extended fibrils forming the large and clearly visible striations. Environmental gases are adsorbed to the newly formed surfaces. Since the microfibrils are strong enough to carry the load transferred upon them the strained sample will not break but more and more defects will develop into crazes and striations. The largest number of striations appears at medium extension rates (0.01 s^{-1} in polychloroprene). The large local deformation (up to more than 100% strain) which is found throughout the total volume of the sample together with the strong intermolecular attraction between different chain segments leads to a large axial excitation -- and breakage -- of those chain segments which play the role of tie molecules. As the discussed results show, the spin concentration at break of elastomers in temperature region III is comparable to that obtained from 6 polyamide. In case of a *large* preorientation of 300% (in polychloroprene) a ductile deformation behavior and the formation of free radicals are observed [31]. The deformation is macroscopically homogeneous. No striations or crazes are visible indicating that the strain hardening is more effective than at lesser preorientations. The growth of microvoids obviously is stopped before they coalesce and form crazes. The absence of large scale gas adsorption and of foaming upon subsequent heating together with differences in the observed ESR spectrum support this view [31].

Brown, DeVries, and Williams duplicated these experiments and extended them to Hycar 1043, an acrylonitrile-butadiene rubber [32] and to Silastic E RTV, a silicone elastomer [33]. They confirmed for these polymers the above described effect of prestrain on the stress-strain behavior at low temperatures (118 to 193 K) the formation of free radicals with increasing sample strain, and the effect of strain rate on radical concentration with a maximum at a rate of $9 \cdot 10^{-4}$ s^{-1}. For these polymers they did not observe, however, the three temperature zones mentioned above which are characterized by distinctly different types of mechanical response. They also propose that during preorientation numerous regions of higher strength and degree of orientation are created which later serve to arrest developing microcracks before those can attain a critical size. The fact that for the latter two polymers [32, 33] only one region of ductile and one region of brittle fracture were observed may be attributed to different physical and chemical properties of those networks (configuration, cross-link density, glass transition temperature). Some of these structural parameters will be studied in the following section.

B. Cross-Link Density, Impurities, Fillers

We have seen above that in low temperature ductile deformation of rubbers a large number of chain segments break which indicates that the local axial stresses exceed the strengths of the chain segments. We have shown earlier that the stresses required for bond breakage are more than two orders of magnitude larger than those applied macroscopically to deform the rubber specimens. The appearance of free radicals, therefore, indicates the highly inhomogeneous distribution of local stresses. The radical intensity gives an account of the number N of segments which have been stressed beyond their strength; N will depend, therefore, on all parameters which raise the local stresses like cross-link density or content of reinforcing fillers, which decrease the bond strength like presence of hetero-groups in the main chain, or which affect the number of observed radicals through radical reactions.

Andrews and Reed [31] observed an increase in radical intensity with increasing cross-link density of sulphur cured natural rubber (Fig. 7.24) using the preorientation-technique described above. This observation is well in keeping with the fact that

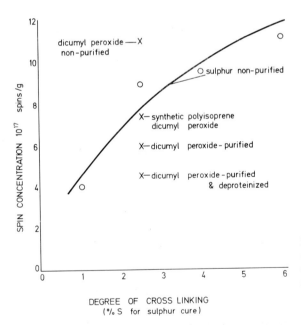

Fig. 7.24. Effect of crosslink density and crosslinking agent on the concentration of free radicals at fracture at 100 to 120 K (after Andrews and Reed [31]).

tensile stresses of identically strained rubbery specimens increase with increasing cross-link density i.e. decreasing chain length between cross-links. The effect of impurities on the concentration of formed free radicals although clearly demonstrated by the data of Figure 7.24 is not fully understood. In the absence of conspicuous

165

radical reactions or a direct effect on bond strengths it is believed that the impurities constitute stress raising inhomogeneities [31].

It is not always necessary to prestrain rubbers to induce the development of a large number of microcracks distributed throughout the bulk of the rubber sample. Wilde et al. [34, 35] have shown that certain granular fillers can promote bond rupture during sample fracture. Again the studies were by necessity conducted at temperatures below 210 K where the radicals were sufficiently stable. Fillers in elastomers are commonly used for a variety of reasons. In any case the presence of such fillers increases the complexity of the micromechanics of deformation and fracture. The presence of the filler results in very non-homogeneous localized stress fields ranging from levels of high strain concentration to complete separation between the matrix and filler (dewetting). The filler particles may serve as crack arrestors and/or initiators and thereby alter the fracture strength, resilience, and/or toughness. Wilde studied four systems all filled to approximately 50% by weight. These were:

1. 29 μm glass beads in EPDM,
2. silane treated 15 μm glass bead filled polyisoprene,
3. NaCl filled polyisoprene and
4. HiSil 233 (silica) filled polyisoprene.

The first system provided very weak interaction between filler and matrix, the last very strong interaction, and the other two demonstrated intermediate filler-matrix interaction. All systems were tested in temperature region I (170 to 200 K), a region within which an unfilled rubber without preorientation produces no free radicals in tensile straining. Three of the four filled systems, however, behaved differently. It was found that the untreated glass filled elastomer fractured with no detectable production of free radicals, fracture in the rubbers filled with silane treated glass and NaCl produced easily detectable concentrations of radicals $(3.21 \cdot 10^{14}$ spins/cm^3). The HiSil system produced at some temperatures and strain rates, up to an order of magnitude more radicals than NaCl or treated glass $(7.86 \cdot 10^{14}$ spins/cm^3). In his PhD thesis Wilde [35] details a rather comprehensive comparison of scanning electron microscope photographs with the ESR results. The photographs indicate that at room temperature dewetting started to occur in the untreated glass system at strains of less than 10 to 20%, in the treated glass beads and NaCl systems it occurred at strains of between 50 and 100%, and in the HiSil system at strains greater than 200%. Scanning electron micrographs of the fracture surfaces at all temperatures between 150 and 300 K revealed the same findings and showed 1. that untreated glass beads were sitting essentially loose in the "smooth" vacuoles or voids on the fracture surface, 2. that treated glass surfaces and NaCl behaved similar to untreated glass except that here the vacuoles were not so smooth and that there was material adhering to the filler particles, and 3. that the HiSil particles were embedded in the matrix at the fracture surface with the rubber adhering completely to the particle surfaces.

A slightly different two-phase system with strong bonds at phase boundaries is that obtained from triblock copolymers such as styrene-butadiene. As discussed in Chapter 2 such a copolymer molecule is composed of rigid (styrene) end blocks joined by elastomeric (butadiene) center blocks. The styrene blocks aggregate to form small domains which act as cross-links leading to a *rubberlike elasticity* of the block copolymer at *ambient temperatures* but to the possibility of *plastic deformation* at *high temperatures*. For an understanding of the fracture behavior of these systems it would be useful to identify in which of the phases molecular rupture most often occurs. The direct means of accomplishing this goal would be to fracture the material and analyze the hyperfine structure of the resulting ESR spectra. In the temperature range between liquid nitrogen and room temperature, however, tensile straining does not lead to sufficiently observable accumulations of free radicals. As a consequence, De Vries, Roylance, and Williams [36] adapted the less conclusive but more straight-forward method of comparing the spectrum of styrene-butadiene block copolymers (SBS) with those of styrene and butadiene alone. These studies were done on grindings to accumulate large amounts of fracture surface and at liquid nitrogen temperatures. The low temperature made the radicals more stable and hopefully "froze" the primary radicals. Comparison of the spectra of the three materials showed that the spectrum in the SBS contained all the lines of the butadiene radical but not of the styrene radical. The radical of the SBS system was, therefore, assigned to the butadiene phase. Unfortunately the study leaves open the questions as to whether or not the radical formed by grinding at low temperatures is the same as that presumably formed under normal conditions at room temperature and whether the observed radical is the primary radical or a secondary one.

Taking together the experimentally observed effects of the five parameters *impurity content, degree of cross-linking, ease of crystallite formation, strength of filler-matrix adhesion*, and *presence of different phases* on the nature and intensity of free radicals formed we derive at the following conclusions. All five parameters tend to increase the apparent cross-link density and decrease the extensibility of chain segments between cross-links. They thus enhance the forces acting at a given strain and also the ultimate forces that can be transmitted onto chain segments. Both conditions are necessary before chain stresses can exceed chain strengths. Furthermore all five parameters increase the number of actual or potential defect sites and cause a better spatial distribution of chain scission events. If chains fail they will always be located in the amorphous or more flexible regions of the (hetero-phase) elastomers.

III. Mechanically Relevant Radical Reactions

In Sections I and II of this Chapter primary and secondary free radicals have been treated as microprobes which characterize occurrence and molecular environment of chain breakages. As shown in Chapter 6 the primary mechano-radicals are always chain end radicals which are mostly unstable. At a rate depending on temperature these radicals will transfer the free electrons and thus "convert" to secondary radicals. This reaction and also subsequent conversion and decay reactions including recombination are relevant with respect to an interpretation of the fracture process in two ways. Firstly these reactions interfere with a determination of the concentration and molecular environment of the original chain scission points. Secondly they change the physical properties of other network chains through the introduction of unpaired electrons and the formation of crosslinks. For a discussion of *spectro-scopic details* and of the *stability* and *conformation* of the free radicals the reader is referred to the comprehensive monograph of Rånby and Rabek [37] and to the review articles of Campbell [38] and of Sohma et al. [39]. In the following the *mechanically* relevant aspects of radical transfer and decay are reported.

A. Transfer Reactions

At this point the geometrical aspect of transfer reactions is to be investigated. Through breakage of an elastically strained chain the primary radicals are formed at either side of the breakage site. Because of the axial chain stresses the new chain end radicals rapidly retract from each other. They are thus being prevented from recombination. The primary radicals then convert to secondary ones mostly by transfer of a hydrogen atom (see Table 6.1). If hydrogen transfer occurs *along* one and the same chain no effect on the physical network properties is to be foreseen. If, however, a hydrogen is abstracted from a *neighboring* chain the radical is transferred to a mid-chain position at a hitherto uneffected chain segment. There is strong evidence for the occurrence of both processes, the radical migration *along* a chain [39] and the hydrogen hopping *accross* chains [40]. The hydrogen hopping process was demonstrated by Dole et al. [40] who had molecular deuterium gas, D_2, diffuse into irradiated polyethylene; D_2 was then converted to HD by the following reactions:

$$R + D_2 \longrightarrow R_1 D + \dot{D}$$
$$\dot{D} + R'H \longrightarrow \dot{R}' + HD.$$

Although in this experiment a gaseous tracer had been used it is believed that hydrogen abstraction from a hydrocarbon chain proceeds in the same manner. In view of the mobility of the chain end radical the hydrogen abstraction seems to be a logical vehicle for the transfer of a free electron to a central position within another chain segment.

The radical migration along a chain was found by Sohma et al. [39] to give the only plausible explanation for his observation that in PE and PMMA chain end radicals convert to a near-end radical:

```
  H   H           H   H   H
- C - C·  ⟶   - C - C - C - H
  H   H           H   ·   H
 (77 K)        (5 min. at 152 K)
```

```
  CH₃   H   CH₃   H              CH₃   H   CH₃   H
- C   - C - C   - C·  ⟶       - C   - C - C   - CH.
  R     H   R     H              R     ·   R     H
 (after short time              (after 0.5 to 24 h.
  milling at 77 K)               milling at 77 K)
```

Sohma indicates that also in these polymers radical migration along the chain and successive hydrogen abstraction are both important mechanisms of radical migration.

Transfer reactions are, of course, influenced by the presence of impurities or additives which are active as radical acceptors or radical scavengers. Their mechanical relevance lies in the fact that they reduce or prevent transfer of free radicals to central chain positions, radical recombination, and/or cross-linking. The latter aspect is discussed in Chapter 8. For a review of radical reactions with H_2, D_2, O_2, SO_2, NO and NO_2, NH_3, H_2S, and various halogens the reader is referred to the works of Campbell [38] and of Rånby and Rabek [37], the latter being especially concerned with the reactions of oxygen in its different forms (triplet and singlet molecular oxygen, atomic oxygen).

B. Recombination and Decay

The *normal decay* of macroradicals by recombination in the solid phase is a second order reaction [5, 11, 17–18, 37–39, 41–43]

$$[R(t)] = [R_0]/(1 + k[R_0]t). \tag{7.10}$$

The decay rate term kR_0 depends — as expected — on temperature: for unloaded fibers under He atmosphere values of $0.0014 \ s^{-1}$ (75 °C) and $0.0110 \ s^{-1}$ (100 °C) have been determined [18]. This value ties in with the $0.001 \ s^{-1}$ measured at 50 and 60 °C under N_2 [11, 41]. It should be noted that the recombination reaction is slowed down by the application of tensile forces as for instance during the pre-fracture loading period: under near maximum tensile stress the decay rate terms were only $0.00035 \ s^{-1}$ (75 °C) and $0.00075 \ s^{-1}$ (100 °C) respectively [18]. The kinetics of decay is also affected by conversion reactions (e.g. of alkyl into allyl radicals) or catalytic effects [37].

The effect of the mobility of the radical site on the recombination rate has been clearly shown by the pressure [44–46] and temperature dependence [41–43, 47–49] of radical decay in e.g. PE, PP, PA 6, 66, and 12, and PPO. The decay curves of free radical populations in (irradiated) polymers as a function of increasing temperature reveal a number of plateaus each corresponding to radicals trapped in a particular morphological site. A transition from one plateau to the next indicates that one *species* of the trapped free radicals has become sufficiently mobile for recombination (chemically identical radicals in crystalline and amorphous regions may differ in steric configuration which also effects the rate of recombination [37, 42, 47]). The transitions are hence associated with the corresponding mechanical γ, β, and α dispersions [41–43].

As already indicated earlier apparent steps in those radical concentration curves which are obtained by straining fibers by equal increments at different temperatures [11, 19], are not solely related to dispersion regions of the modulus but reflect details of the relative segment length distribution as well.

169

For chemically different radical species generally different decay constants with different activation energies are observed. Tino et al. [42] determined for the central polyamide radical $-NH-\dot{C}H-CH_2-$ in the temperature region of 15 to 50 °C an activation energy for decay of 10 kcal/mol, for the chain end radical $-CONH-\dot{C}H_2$ a value of 5.2 kcal/mol.

Studies of the effect of hydrostatic pressure on free radical decay have the aim of elucidating the mobility and the distance of migration. A series of experiments has been carried out at the Polymer Institute of the Slovak Academy of Sciences in Bratislava. The rate constants of free radical decay decrease exponentially with increasing pressure [44–46]. At low temperatures the rate constant varies only slightly with pressure. The retarding effect becomes stronger at higher temperatures; the closer the decay temperature to the glass transition temperature T_g the more pronounced the stabilizing effect of pressure. Of course the pressure effect shows a saturation once the pressure is so large as to inhibit the molecular motion in question, 8 kb for the α relaxation in PE and PVAC at 80–110 °C [44], 15 kb for the α relaxation in 6 PA [44].

C. Anomalous Decay

Sohma et al. [39] investigated the thermal decay of mechano-radicals. They observed for PE, PP, and PTFE an increase in the concentration of free radicals while increasing the temperature from 77 to 170 K. They termed this behavior which is not found for irradiation-produced radicals, anomalous. The anomalous increase is enhanced by excess triboelectric charges of the sawed samples and by the presence of oxygen [39]. On the basis of their extensive investigations the authors advance a mechanism for the creation of free radicals through heat treatment at the rather low temperatures of 77 to 170 K. They postulate the formation of (spectroscopically neutral) carbanions during the degradation process at 77 K, here illustrated for PP:

$$-CH_2-(CH_3)\dot{C}H + e^- \quad \rightarrow \quad -CH_2-(CH_3)\dot{C}H^-$$
$$\uparrow$$
$$\text{(carbanions, no ESR spectrum)}$$
$$\downarrow$$
$$-(CH_3)CH-\dot{C}H_2 + e^- \quad \rightarrow \quad -(CH_3)CH-\dot{C}H_2^- .$$

During the "heat" treatment (100–173 K) the carbanions react with O_2 to form spectroscopically active peroxy radicals. The released electrons leak to the ground:

$$-CH_2-(CH_3)\dot{C}H^- \;\; + O_2 \;\; \rightarrow \;\; -CH_2-(CH_3)HCO\dot{O} \;\; + e^-$$
$$-(CH_3)CH-\dot{C}H_2^- \;\; + O_2 \;\; \rightarrow \;\; -(CH_3)CH-H_2CO\dot{O} \;\; + e^- .$$

D. Radical Trapping Sites

Mechanically relevant information has been obtained from an analysis of radical trapping sites. Thus it has been concluded in previous sections of this chapter that

170

it is the amorphous regions of a semicrystalline polymer where mechano-radicals are formed. Apart from that changes in sample morphology during mechanical working have been investigated by ESR technique. Kusumoto, Takayanagi et al. [50–51] studied PE and PP films by successively removing the amorphous material through nitric acid etching. They then analyzed the ESR spectra obtained by γ irradiation of the so treated films. Thus they were able to correlate the octet observed in PP with radicals trapped at defect sites within the crystallites, the nonet with radicals in the amorphous fold surfaces. The latter are especially effective radical trapping sites. The authors analyzed the effect of quenching, annealing, and cold-drawing on the mosaic-block structure of their films.

From a change in line shape parameter of the PE sextet spectrum and of the temperature dependence of radical recombination rates the authors determined the increase in unoriented material content during mechanical fatigue [51]. For the study of the surfaces of as-grown and annealed PE single crystals Kusumoto et al. [52] also employed the spin-probe method.

The ESR technique has permitted to identify initial rupture points in mechanically degraded copolymers. Lazar and Szöcs [53] investigated random, block, and graft copolymers of methylmethacrylate (MMA) and styrene (S). They found in a degraded mixture of equal parts MMA and S homopolymers equal concentrations of MMA and S radicals. In a random copolymer (MMA : S = 1 : 1) the styrene radical prevailed clearly. By comparing the reactions of mechano-radicals in ball-milled PP with those of "γ-radicals" Kurokawa et al. [54] concluded that the mechano-radicals were produced and trapped on the fresh surfaces created by the breaking-up of solid polypropylene.

References for Chapter 7

1. S. E. Bresler, S. N. Zhurkov, E. N. Kazbekov, E. M. Saminskii, E. E. Tomashevskii: Zh. tekhn. fiz. *29,* 358 (1959); Sov. Phys.-Techn. Physics *4,* 321–325 (1959).
2. J. Becht, H. Fischer: Kolloid-Z. u. Z. Polymere *229,* 167 (1969).
3. G. S. P. Verma, A. Peterlin: Polymer Preprints *10,* 1051 (1969).
4. H. H. Kausch, K. L. De Vries: Int. Fracture *11,* 727–759 (1975).
5. B. A. Lloyd: Ph. D. Thesis, Department Mechanical Engng. Univers. of Utah (June 1972); B. A. Lloyd, K. L. DeVries, M. L. Williams: J. Polymer Sci. A-2, *10,* 1415–1445 (1972).
6. J. Becht: Dissertation, Technische Hochschule, Darmstadt (1971).
7. H. H. Kausch, J. Becht: Rheologica Acta *9,* 137 (1970).
8. M. A. Gezalov, V. S. Kuksenko, A. I. Slutsker: Mechanica Polymerov *8,* 51 (1972); Polymer Mechanics USSR *8,* 41 (1972).
9. B. Crist, A. Peterlin: Makromol. Chem. *171,* 211–227 (1973).
10. W. H. Hassell: M. S. Thesis, Department Mechanical Engng. Univers. of Utah (June 1973).
11. U. Johnsen, D. Klinkenberg: Kolloid-Z. u. Z. Polymere *251,* 843 (1973).
12. H. H. Kausch: J. Polymer Sci. C *32,* 1–44 (1971).
13. D. Klinkenberg: 38. Dtsch. Physikertagung, Nürnberg, 23.–27. Sept. 1974. D. Klinkenberg, Dissertation, Technische Hochschule, Darmstadt 1978.
14. M. Williams, K. L. De Vries: Proc. Fifth Internat. Congr. Rheology, S. Onogy, Ed., Vol. 3, University of Tokyo Press, Tokyo 1970, 139.
15. B. Yu. Zaks, M. L. Lebedinskaya, V. N. Chalide: Vysokomol. soyed *A 12,* 2669 (1970); Polymer Sci. USSR *12,* 3025 (1971).

16. H. H. Kausch, J. Becht: Kolloid-Z. u. Z. Polymere *250*, 1048 (1972).
17. J. Becht, H. Fischer: Kolloid-Z. u. Z. Polymere *240*, 766 (1970).
18. K. L. DeVries, B. A. Lloyd: Univers. of Utah, personal communication.
19. L. A. Davis, C. A. Pampillo: J. Polymer Sci., Polymer Phys. Ed. *11*, 841–854 (1973).
20. D. K. Beckman, K. L. DeVries: Department Mechanical Engng. Univers. of Utah, June (1969).
21. D. K. Backman, K. L. DeVries: J. Polymer Sci. A-1, *7*, 2125 (1969).
22. T. Pazonyi, F. Tudos, M. Dimitrov: Angew. Makrom. Chem. *10*, 75–82 (1970).
23. R. J. Salloum, R. E. Eckert: J. Appl. Polymer Sci. *17*, 509–526 (1973).
24. T. Pazonyi, F. Tudos, M. Dimitrov: Plaste u. Kautschuk *16*, 577–581 (1969).
25. W. O. Statton: J. Polymer Sci., (Polymer Symposia) *32*, 219 (1971).
26. J. B. Park: Ph. D. Thesis, Department Mechanical Engng. Univers. of Utah, (June 1972).
27. W. O. Statton, J. B. Park: to be published J. Polymer Sci., Polymer Phys. Ed.
28. W. O. Statton: Univers. of Utah, personal communication.
29. E. H. Andrews, P. E. Reed: J. Polymer Sci. B-*5*, 317 (1967).
30. R. Natarajan, P. E. Reed: J. Polymer Sci. A-2, *10*, 585 (1972).
31. E. H. Andrews, P. E. Reed in: Deformation and Fracture of High Polymers, H. H. Kausch, J. A. Hassell, and R. I. Jaffee, Eds. New York: Plenum Press 1973, 259.
32. R. Brown, K. L. DeVries, M. L. Williams: Polymer Preprints *11/2*, 428 (1970).
33. R. T. Brown, K. L. DeVries, M. L. Williams: Polymer Letters *10*, 327 (1972).
34. K. L. DeVries, T. B. Wilde, M. L. Williams: J. Macromol. Sci-Phys. B7 *(4)*, 633 (1973).
35. T. B. Wilde: Ph. D. Dissertation, Department of Mechanical Engng. Univers. of Utah (1970).
36. K. L. DeVries, D. K. Roylance, M. L. Williams: J. Polymer Sci. A-2, *10*, 599 (1972).
37. B. Rånby, J. F. Rabek: ESR-Spectroscopy in Polymer Research, Heidelberg: Springer-Verlag 1977.
38. D. Campbell: J. Polymer Sci., Part D (Macromol. Rev., Vol. 4) 91–181 (1970).
39. J. Sohma, M. Sakaguchi: Advances in Polymer Sci. *20*, 109–158 (1976).
40. M. Dole, D. R. Johnson, S. Kubo, W. Y. Wen: Department of Chemistry, Baylor University, Waco, Texas.
41. F. Szöcs, J. Becht, H. Fischer: Europ. Polymer J. *7*, 173–179 (1971).
42. J. Tino, J. Placek, F. Szöcs: Europ. Polymer J. *11*, 609–611 (1975).
43. T. Nagamura, N. Kusumoto, M. Takayanagi: J. Polymer Sci. (Polymer Phys. Ed.) *11*, 2357–2369 (1973).
44. F. Szöcs, J. Tino, J. Placek: Europ. Polymer J. *9*, 251–255 (1973).
45. F. Szöcs, O. Rostasova, J. Tino, J. Placek: Europ. Polymer J. *10*, 725–727 (1974).
46. F. Szöcs, O. Rostasova: J. Appl. Polymer Sci. *18*, 2529–2532 (1974).
47. H. Kashiwabara, S. Shimada, J. Sohma: ESR Applications to Polymer Research (Nobel Symposium 22), Stockholm: Almqvist & Wiksell 1973.
48. T. Nagamura, M. Takayanagi: J. Polymer Sci. (Polymer Phys. Ed.) *13*, 567–578 (1975).
49. J. Tino, J. Placek, P. Slama, E. Borsig: Polymer *19/2*, 212–214 (1977).
50. N. Kusumoto, K. Matsumoto, M. Takayanagi: J. of Appl. Polymer Sci. A-1, *7*, 1773–1787 (1969).
51. T. Nagamura, N. Kusumoto, M. Takayanagi: Intern. Conf. on Mech. Behavior of Materials, Kyoto, Japan, Aug. 1971.
52. N. Kusumoto, M. Yonezawa, Y. Motozato: Polymer *15/12*, 793–798 (1974).
53. M. Lazar, F. Szöcs: J. of Polymer Sci. Part C *16*, 461–468 (1967).
54. N. Kurokawa, M. Sakaguchi, J. Sohma: Polymer J. *10/1*, 93–98 (1978).

Chapter 8

The Role of Chain Scission in Homogeneous Deformation and Fracture

Section II B of Chapter 2 gave a description of the uniaxial deformation behavior of an unoriented thermoplastic polymer. It was indicated that — depending on experimental and material parameters — failure could occur at any of the different stages of a tensile loading process:

— as *brittle fracture* due to crack initiation and propagation during anelastic deformation
— as a consequence of plastic flow after *necking* or *homogeneous yielding*
— as *ultimate failure* following plastic deformation with strain hardening.

It is useful, therefore, to distinguish between *brittle strength, yield strength,* and *ultimate strength* of a polymer.

Quite obviously the role of chain loading and scission within the three strength determining mechanisms will be quite different. It has been stated repeatedly throughout this book that the load carrying capabilities of chain molecules are utilized most effectively if chain orientation and intermolecular attraction permit the gradual accumulation of large axial chain stresses and oppose slip and

173

void formation. It is for this reason that highly oriented polymer fibers are the most suitable objects to study chain loading and scission. In Chapter 7 the experimental observations concerning the formation of mechano-radicals and of their conversion have been reported. With respect to the phenomenology there is little disagreement in the literature. In the first section of this chapter the possible effects of chain scission and of radical reactions on ultimate strength will be discussed, which is a more controversical subject.

The second section concerns unoriented polymers and failure mechanisms, which are homogeneous on a macroscopic scale. Thus the conditions of impact loading (large loading rates), creep (constant load), yielding, necking, and drawing (constant strain rate), and fatigue (intermittent or cyclic loading) will be investigated. Elastomers also belong to the group of unoriented polymers although their failure may occur after considerable elongation and orientation.

The chapter closes with a discussion of chain scission caused or assisted by the simultaneous action of stress and aggressive environment.

I. Small-Strain Deformation and Fracture of Highly Oriented Polymers

A. Underlying Problems

The undisputed fact of *the rupture of chains in large numbers during mechanical action* [1] *is, in itself, neither proof nor even indication* that macroscopic stress relaxation, deformation and fracture *are a consequence* of the breakage of chains. As was pointed out by Kausch and Becht [2] the observed total number of broken chains is much too small to account (by virtue of their load carrying capability) for the measured reductions in macroscopic stress. As shown in Figure 7.4 the stress relaxation within a strain step (of 0.65%) amounts to 60 to 100 MN/m^2. The load having been carried by the $0.7 \cdot 10^{17}$ chain segments broken during such strain step, however, is calculated to be 2.4 MN/m^2 if the tie-segments had traversed just one amorphous region; it is n · 2.4 MN/m^2 if they had spanned n such regions. For these and most of the following calculations a sandwich-type model of the fiber structure will be used such as represented in Figure 7.5 I. In the case of n = 1 the macroscopic stress relaxation obviously has an intensity 25 to 40 times larger than the decrease of the accumulated *molecular* stresses derived from the number of the observed chain scission events. However, here also one must say that *this discrepancy* in *itself is neither proof nor even indication* that macroscopic stress relaxation, deformation and fracture are *not* a consequence of the breakage of chains.

A basic problem of microstructural fracture theories involving chain scission is clearly an elucidation of this descrepancy. Such a clarification can only be attempted by widening the scope and by including other experimental observations. Three principal questions must be raised before supporting or discarding a chain scission fracture hypothesis:

— Does the number of chain scission events *observed* by any one method correspond to the number of chains *actually breaking*?

- Does chain scission effect the mechanical properties of a polymer network *in an indirect way,* e.g. through the release of stored elastic energy or ensuing radical reactions?
- At what point and why does the phase of spatially homogeneous defect development turn into local crack growth? Is chain scission *a necessary or an accidental precursor* to mechanical instability?

The discussion of the underlying problem of the nature of the primary molecular processes of damage, their interaction and possible propagation has been initiated and constantly stimulated through the various detailed studies carried out in the A. F. Ioffe Physicotechnical Institute in Leningrad by S. N. Zhurkov and his colleagues [3—33]. The individual problems attacked concern

- investigation of the loading of chain segments before chain scission by IR technique [4—16]
- determination of the absolute numbers [16—18] of chain scission events using ESR [19—21] and IR [22—26] spectroscopy
- formation of microcracks studied by X-ray diffraction [27—29]
- kinetics of thermal, mechanical and molecular degradation [30—33]
- analysis of the structure, mechanical behavior and fatigue under various environmental conditions in view of the foregoing considerations. These are reviewed in [3] and will be discussed later in this Section.

A general remark must be made concerning the comparison and transfer of results obtained from different specimens (e.g. films or fibers), methods (e.g. IR or ESR), or laboratories. Whereas *relative data* such as the effect of time, temperature or stress on the behavior of one specimen (in one spectrometer at one laboratory) are generally in good agreement inconsistencies still persist with respect to *absolute* figures and to the possible *structural* and/or *chemical identity* of different specimens. The following discussion of the fracture work and of fracture theories involving chain scission will be centered around the three principal questions raised earlier even though they will not always be answered.

B. Loading of Chains before Scission

To date two methods have been employed for the determination of the distribution $N(\psi)$ of (large) axial chain stresses:
- the indirect method of observing the formation of free radicals after a lowering of chain strengths through increasing temperatures
- the direct method of evaluating the stress-induced shift of infrared (IR) adsorption bands.

The indirect method of obtaining stress-distributions is based on the fact that the strength of a chain segment decreases with increasing temperature. During an upward temperature scan, therefore, those chain segments will be broken successively whose strength $\psi_b(T)$ becomes smaller than their axial load. The method has been discussed in Section IC1 of Chapter 7 for groups of chain segments with different relative lengths L/L_0. The ensuing relative length distribution $N(L/L_0)$ can

Fig. 8.1. Dichroism of polypropylene;
parallel dichroic vibrations at 842, 975, 998, 1045, 1168, 1254, 1304 cm^{-1}
perpendicular dichroic vibrations at 809, 899, 941, 1330 cm^{-1}.

easily be transformed into a stress distribution $N[\psi(L/L_0)]$ with the aid of Eq. (7.1).

This indirect method has a rather restricted applicability. It is limited to a temperature region $T < T_1$ where interfibrillar slip is still negligible and to molecular stresses $\psi > \psi_b(T_1)$. Furthermore the method does not permit the solution of a problem inherent to all ESR measurements; namely to determine whether the absolute concentration of broken bonds is equal to twice the observed concentration of free radicals or whether it is (much) larger.

To overcome this difficulty Zhurkov and his collaborators began in 1965 to study by an independant method, the IR technique, the effect of axial stress on molecular chains [4–16]. Subsequently similar [35, 37–40] infra-red studies were carried out in the USA. All these investigations are based on the fact that frequency and intensity of skeletal or backbone vibrations of chain segments respond to a superimposed deformation of the vibrating skeleton. In polypropylene for example C–C stretch vibration modes have been assigned [16, 36, 41, 42] to the parallel dichroic bands (Fig. 8.1) at 1326, 1168, 1045, 975, 842, 456, and 398 cm^{-1}. All of these bands have been found to be stress sensitive [36]. The stress effect manifests itself as a shift of the frequency of maximum absorption and in the occurrence of a "tail" at longer wavelengths (Fig. 8.2).

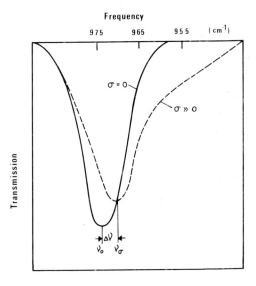

Fig. 8.2. Schematic effect of stress on a skeletal vibration of a semicrystalline, highly oriented polymer after (15, 36).

The mechanisms of the distortion of IR absorption bands of stressed polymers have been discussed especially by Gubanov [7–9], Kosobukin [13], Vettegren and Novak [15] and Wool [36]. There is a general agreement that the distorted IR absorption band profile $D(\nu)$ can be related to a large number of separate oscillators with strongly overlapping absorption bands whose maxima are shifted by different amounts. The possible causes of the frequency shift of an individual stressed oscillator have been considered to be the quasielastic deformation of the harmonic oscillator (reduction of force constant under stress), the elastic increase of bond angles, changes of segments conformation, and the creation of defects. For small deformations one predicts [4—16, 36], that for the first three mechanisms there is a quite linear change of frequency with *molecular* stress ψ:

$$\Delta \nu' = \nu_\psi - \nu_0 = \alpha' \psi. \tag{8.1}$$

As discussed previously molecular stresses are a function of molecular strains which depend on sample morphology and chain orientation. Any real sample, therefore, contains a variety of differently stressed oscillators. The profile of the deformed band, $D(\nu)$, representing such a system of oscillators can be expressed by a convolution integral:

$$D(\nu) = \int_{-\infty}^{\infty} F(n) U(\nu - n) \, dn. \tag{8.2}$$

Here $F(n)$ is the distribution of stressed oscillators, $U(\nu)$ the shape of the normalized undeformed band, and n an integration variable. The distribution $F(\nu)$ can be obtained by deconvolution and transformed into a molecular stress distribution by Eq. (8.1).

As already indicated and shown in Figure 8.2 the distribution of molecular stresses in semicrystalline polymers contains a *peak of apparently homogeneously stressed chain segments* — related to the crystalline regions — and a *tail of very differently stressed chains* — related to highly stressed tie segments. For a *regular* sandwich structure of crystalline regions having a modulus E_c and a chain axis orientation described by $\cos \theta$ one observes a frequency shift $\Delta \nu$ of the *crystalline peak:*

$$\Delta \nu = \alpha' \frac{E_k}{E_c} \sigma \cos^2 \theta = \alpha \sigma. \tag{8.3}$$

Thus the macroscopic stress sensitivity factor α depends on sample-morphology and orientation. As shown in Figure 8.3 quite different stress sensitivities have been observed for identical bands in different samples [36]. In any real sample the macroscopic stress sensitivity factor α is an average value. For the 975 cm^{-1} band of polypropylene it ranges from 2 to 6 cm^{-1} per GN/m^2 (Fig. 8.3). The 1168 cm^{-1}

Fig. 8.3. Stress-induced shift of the peak of the 975 cm^{-1} absorption band of polypropylene as found by different investigators [6, 15, 34, 36].

band of the same polymer has a larger stress sensitivity of 10 cm^{-1} per GN/m^2 [15]. Vettegren and Novak [15, 16] report an α of 2.2 cm^{-1} per GN/m^2 for the 930 cm^{-1} band of PA 6 and of 3.6–6 cm^{-1} per GN/m^2 for the 974 cm^{-1} band of PETP. For the latter band depending on sample treatment Mocherla and Statton find stress sensitivity values of 12 to 20 cm^{-1} per GN/m^2 [40]. More recently Vettegren et al. [16] have obtained a stress sensitivity value of 8 cm^{-1} per GN/m^2 at 1350 cm^{-1} for polyacrylonitrile.

In the tail region of the "stressed 975 cm^{-1} band" of PP frequency shifts of up to 35 cm^{-1} are discernible which correspond to molecular stresses up to 12 GN/m^2. Although the significance and dynamics of "stressed-IR" investigations are not yet understood in detail a number of interesting observations have been made:

– the integrated area of a deformed band (975 cm^{-1} PP) remained constant with σ increasing from 0 to 700 MN/m^2 [5] indicating that no new oscillators had been created under stress;

– the area fraction in that band related to overstressed segments increased with stress to 15% at $\sigma = 400$ MN/m^2 and 18% at 490 MN/m^2 [12]; for the 1170 cm^{-1} band an area fraction of 6% is reported [16] together with 15% for PETP (976 cm^{-1}), and 7% for PA 6 (930 cm^{-1}), all at a stress of 500 MN/m^2;

– it is interesting to note that the segment density l/q L for PP segments of 5 nm length amounts to 56 · 10^{19} segments per cm^3 of which 15% or 8.5 · 10^{19} segments/cm^3 would be overstressed at $\sigma = 400$ MN/m^2;

– at a constant stress level this tail area is directly proportional to the content of non-crystalline material, i.e. the overloaded segments seem to be contained in the amorphous regions [5];

– the tail regions of the deformed bands (930 cm^{-1} PA 6; 970 cm^{-1} PETP; 975 cm^{-1} PP) show fairly well defined limits at the lower frequency end corresponding to *fairly well defined limits of sustained molecular stress* ψ; these limits are determined [6, 15, 16] to be 11–13 GN/m^2 in PP, 15–20 GN/m^2 in PETP (lavsan), 8 GN/m^2 in PAN and 16 to 20 GN/m^2 in PA 6;

– with some samples [12, 24, 25] these limits are already reached at moderate macroscopic stresses smaller than 50% of sample strength; in other samples the limits are only attained at a stress level close to the breaking stresses [16];

Fig. 8.4. Maximum stress on molecular segment, ψ_m, as a function of applied uniaxial stress σ for PP (a) and PETP (b) for samples of different strengths σ_b; σ_b indicated by ↓ (after [16]). T = 26 °C.

Figure 8.4 reproduces the results of Vettegren et al. [16] on the maximum stresses on molecular segments for three morphologically different samples each of PP and PETP; the — non-specified — morphological differences do not seem to effect the limiting stress value; in samples a) the largest molecular stress concentration (ψ_m/σ) is larger than in b) or c) but in all three samples it is apparently extremely homogeneous as indicated by the fact that the maximum value of ψ_m is only reached at the maximum of σ; this behavior ties in with the observation [5] that the deformation of a band (975 cm^{-1} PP) of a highly drawn sample ($\lambda = 10$) is much less than that of a sample drawn to $\lambda = 5$;

— the observed molecular stress limit is also a function of temperature; in PETP Zhurkov et al. [6] determined a decrease from 22 GN/m^2 at 100 K to 15 GN/m^2 at 400 K (on the basis that the observed molecular stress limits correspond to the chain strengths ψ_b and using $\omega_0/k_b = 48 \cdot 10^{12}$ s^{-1} as for PA 6 segments one derives an activation volume of $11.7 \cdot 10^{-6}$ m^3/mol and an activation energy $U_0 = 316$ kJ/mol for the PETP segments); from later experiments Vettegren et al. [16] obtained activation volums of $13.2 \cdot 10^{-6}$ m^3/mol for PETP, $15.0 \cdot 10^{-6}$ m^3/mol for PA 6 and $15.6 \cdot 10^{-6}$ m^3/mol for PP;

— the dichroism of a band permits the determination of the average degree of orientation $(\overline{\cos^2 \theta})$ of loaded chain segments; Zhurkov et al. [5] found for a PETP with $\overline{\cos^2 \theta} = 0.75$ that the most highly stressed segments approached $\overline{\cos^2 \theta} = 1$, i.e. complete uniaxial orientation (Fig. 8.5);

— the distribution of molecular stresses in PP varies during deformation (creep) and stress relaxation as indicated by the different time dependent intensity changes of fixed frequencies in the deformed part of the spectrum (dynamic polarized IR studies [6, 35])

Fig. 8.5. Average orientation $(\overline{\cos^2 \theta})$ of highly stressed PETP segments as a function of stress; as abscissa the frequency shifts of the 974 cm^{-1} band and the corresponding axial chain stresses (using a stress sensitivity factor $\alpha = 3.6$ cm^{-1} per GN/m^2) have been employed (after [5]).

— creep deformation tends to increase the highly loaded chain area of the 975 cm^{-1} band in PP at the expense of intermediately stressed segments [6, 35] further supporting the hypothesis that the band deformation is indeed due to a *shift of frequencies* and not to a *creation of new oscillators*

— stress relaxation after shock loading of highly oriented PP showed an initial 10 to 70 s period of fast decay at practically constant degree of crystal orientation [35] as evidenced by the absorption of the 899 cm^{-1} band which is a highly dichroic, crystalline or "helix" band [43]; at later times a "steady state" relaxation takes place which is accompanied by a decrease of absorption of the 899 cm^{-1} band indicating a decrease of crystallinity or of helical regularity [35];

— in shock loading PP (e.g. up to a strain of 10.5% in less than 0.1 s) the largest absorption at 955 cm^{-1} is observed after t = 69 s when considerable stress relaxation has occurred whereas in ramp loading at a rate of 10%/min the largest absorption is attained at the maximum of stress at 10.5% strain; the largest increase in intensity at 955 cm^{-1} is (3.2 times) larger in shock loading than in ramp loading [38]; the molecular stress transfer in this highly oriented PP, therefore, is a viscoelastic process involving the deformation of the amorphous regions and the resistance to uncoiling of helices; Wool [39] has carried out a detailed experimental and computational analysis of stress relaxation, dynamic IR behavior and bond rupture; he concluded that different stress sensitivity values for crystalline regions (2.1 cm^{-1} per GN/m^2) and individual chains (8 cm^{-1} per GN/cm^2) should be employed, that the most highly stressed chains (952 cm^{-1}) relax first thus contributing to the increase in intensity at higher frequencies (e.g. 955 and 960 cm^{-1}), and that bond scission will be negligible if its activation energy U_0 is equal to or larger than 121 kJ/mol (29 kcal/mol); with U_0 = 105 kJ/mol very little chain scission occurs (affecting IR tail intensities by less than 0.3%), with 84 kJ/mol the majority and with U_0 = 63 kJ/mol (15 kcal/mol) all of the overloaded chains are expected to break; in his PP material chain scission did not seem to contribute measurably to the relaxation phenomena, this argument was backed by the observation that the absorption bands characteristic for chain-end groups remained constant within 0.5% [39]; on the other hand in the PP material of Vettegren et al. [16] stress relaxation at 100 °C was accompanied by the measurable appearance of chain-end groups.

The dynamic IR behavior of PETP has recently been investigated by Mocherla et al. [40]. Apart from the already discussed stress effects the authors observed the existence of an initial region of small stresses (up to 70 MN/m^2) within which stress effects (on the 973 cm^{-1} band of PETP) were absent. They concluded from extended studies of differently heat-treated films that this initial region coincides with the range of elastic response of the sample; obviously small elastic stresses are predominantly transmitted by secondary forces not giving rise to axial distortion of the chains.

Summarizing the above results from this important new method one may say that amorphous chain segments exist in appreciable numbers which carry up to about twenty times the average stress. There are upper limits of molecular stress which in the case of polyamide 6 seem to be given by the chain strength (21 GN/m^2). With polypropylene the upper limit of molecular stresses in the highly, uniaxially oriented material of Wool is due to the inception of a distortion of helical chain conformations and of crystal lamellae. No bond breakages could be traced.

In loading their oriented PP films Zhurkov, Vettegren et al. [6—16] observed, however, that an initially present concentration of $1-10 \cdot 10^{18}$ carbonyl groups per cm^3 increased [24]. They found this increase and the formation of other end groups (see below) related to the decrease of the number of overloaded chains [16]. The Russian authors concluded from the equality of the activation energies of thermal and mechanical destruction of their PP films (29 kcal/mol) with that of end-group accumulation (30 kcal/mol) that the kinetics of all three processes is controlled by the same molecular process, the thermomechanical scission of chain segments. According to the calculations of Wool, chain scission should practically never occur in PP if U_0 is equal to 29 kcal/mol. Those calculations, however, most probably refer to a constant temperature in the network of loaded and overloaded chains. Consideration of the large strain energy release at chain breakage would lead to the adjustment that local temperatures T are higher than the ambient temperature T_0 and that the effective chain strengths, $\psi_b(T)$, are smaller than $\psi_b(T_0)$. It seems likely, therefore, that the phenomenological observations do not contradict each other especially taking into account possible structural differences of the PP samples of the laboratories.

The upper limits of molecular stresses deduced for PETP ($15-20\, GN/m^2$) and PAN ($8\, GN/m^2$) were also related to chain breakage by the Russian authors [16]. With these materials, however, no comparable results on scission products (end groups, free radicals) have been reported. In these cases as with the PP material of Wool the upper stress limit seems to derive from the limited resistance of the crystal lamellae against distortion and break-up.

The latter conclusion is supported by the investigations of Becht and Kausch [44—48] concerning the deformation of highly oriented semicrystalline fibers. In a regular sandwich structure critical axial forces can be exerted onto tie segments if, and only if, the crystal lamellae can withstand stresses up to the chain strength. Otherwise crystal disintegration would precede chain scission. Using the spin probe technique, calorimetry and molecular weight measurements Becht [44—47] demonstrated the strain effect on crystal integrity. He irradiated samples of highly oriented 6 PA, 12 PA, PP, PETP, and PE by 1-Mev electrons at liquid nitrogen temperature. Subsequently, all samples were heated to their glass-transition-temperature (or above) for at least 5 minutes; thus, all radicals in the amorphous phase disappeared, and radicals remained within the crystallites only. These samples were then strained within the cavity of an ESR spectrometer at room temperature.

The rate-determining steps of the radical-decay reactions in these cases are transfer reactions and oxygen diffusion. They are both very strongly correlated with the mobility of the radical-carrying molecular segments with respect to their surroundings, i.e., the crystal lattice. An increase in these rate constants, therefore, is a measure of an increase in mobility, which may be due to a decrease in intermolecular attraction and/or to the introduction of new crystal defects.

In Figure 8.6, the concentration of free radicals is plotted versus a reduced time $K_0 \Delta t$, where K_0 is the rate of radical decay a t zero strain and Δt is the time elapsed

Fig. 8.6. Effect of sample strain ϵ on crystal distortion as evidenced by the rate of radical decay in irradiated and annealed highly oriented fibers (after [48]).

from the inception of sample straining. It should be noted that in PE, PP, PETP, and 12 PA, the application of sample strain has a twofold result: an immediate and irreversible decrease of the (crystalline) radical population and a stress-dependent reversible increase in the rate of decay. In 6 PA neither of these two effects can be observed. It was concluded from this experiment that in strained 6 PA, the tie molecules fail before axial chain stresses are reached which are sufficient to distort the crystal lamellae. In PE, PP, PETP, and 12 PA, however, forces can be transmitted which clearly affect the crystal lamellae. It is logical to assume that these forces are primarily transmitted by the tie segments. One would have to conclude that the axial load bearing capability of microfibrils of PE, PP, PETP and 12 PA is limited through the beginning crystal distortion. Thus chain strengths are not fully utilized in these polymers.

Still another combination of chain and crystal properties seems to exist in the PE films used by Vettegren et al. [15–18] for their end-group studies. Under stress vinyl, methyl, and oxygen-containing end groups formed readily and in large concentrations while no stress effects on IR bands were reported.

C. Spatially Homogeneously Distributed Chain Scissions

The scission of molecular chains generally leads to highly reactive chain end radicals and eventually to the formation of new end groups (cf. Chapters 6 and 7). Unless oxygen is excluded the following end groups are formed most frequently: methyl, aldehyde, ester, carbonyl, carboxyl, or vinyl groups. All of these groups have characteristic IR absorption bands (Table 8.1). Based on this consideration Zhurkov and his collaborators quantitatively investigated the build-up of new end groups in loaded polymer films [16–18, 22–26]. They employed a double beam technique the principle of which is illustrated in Figure 8.7. The optical densities $D = \ell n \, I_0/I$ of a stressed (σ) and an unstressed reference film (r) are compared:

$$\Delta D = D_\sigma - D_r = k(C_\sigma d_\sigma - C_r d_r). \tag{8.4}$$

Here k is the absorption coefficient, d the thickness of the film, and C the concentration of end-groups. Figure 8.8 shows such a difference spectrum recorded by Zhurkov et al. [17, 18] from a polyethylene film ruptured at room temperature.

Table 8.1. IR Absorption Bands Suitable for End-Group Analysis [17, 18, 24, 39].

(End) Group	Structure	Absorption Band ν (cm^{-1})	Absorption Coefficient [24] k (10^{-19} cm^2)
Vinyl	$-CH_2-CH=CH_2$	910	2.52
Vinylene	$-CH_2-CH=CH-CH_2-$	965	1.46
Aldehyde	$-CH_2-CHO$	1070	2.52
Methyl	$-CH_2-CH_3$	1379	1.02
Vinyl	$-CH_2-CH=CH_2$	1645	0.68
Carboxyl	$-CH_2-COOH$	1710	8.92
Aldehyde	$-CH_2-CHO$	1735	2.47
Ester	$-CH_2-COO-CH_2-$	1745	6.36

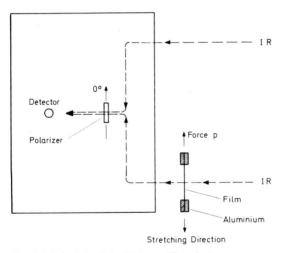

Fig. 8.7. Principle of double-beam IR technique.

The spectra have been evaluated with respect to the kinetics of end-group accumulation and to the absolute number of chain scission events. The results will be reported although it seems as if a full understanding has not yet been obtained in terms of sample morphology, molecular weight changes and ESR results. So far only three polymer samples (PE-LD, PE-HD and PP investigated in the Joffe-Institute) have yielded end-group concentrations sufficient for detection. Thus the initial remarks concerning comparison and transfer of results apply here.

Fig. 8.8. Difference IR spectrum of unstressed and stressed oriented polyethylene (after [16–18]).

Fig. 8.9. Accumulation of aldehyde-end groups formed in high density polyethylene (after [16–18]). (a) T = 20 °C; σ (MN/m²): 1. 360, 2. 350, 3. 330, 4. 280; (b) σ = 280 MN/m²; T (°C): 1. 67, 2. 46, 3. 20.

Subjecting uniaxially oriented films to uniaxial loads Zhurkov et al. [16–18, 22–26] measured by such a technique the accumulation of aldehyde groups in high pressure and low pressure polyethylene and in polypropylene after predetermined loading intervals. Typical results are reproduced in Figure 8.9. The concentration C(t) of aldehyde-end groups increased with loading time according to a first-order kinetic equation:

$$C(t) = C^* (1\text{-exp-}Kt) \qquad (8.5)$$

with rate constant K and apparent concentration C* of mechanically excitable – and thus breakable – bonds. The authors found that the concentration C* depended only slightly on temperature and stress (changes from 1 to $4 \cdot 10^{19}$ cm^{-3} for stresses between 180 and 500 MN/m² at temperatures between −70 °C and +20 °C) whereas the rate "constants" K varied by several orders of magnitude (Fig. 8.10). The authors noted that the rate constant K for accumulation of end groups was governed by an effective, stress-dependent activation energy E_* which is plotted in Figure 8.11. Using the apparent linear relation between E_* and σ one obtains:

185

$$K = K_0 \exp\left[-(E_A - \alpha_* \sigma)/RT\right]. \qquad\qquad (8.6)$$

The activation energies E_A and activation volumes α_* for the three polymer films investigated are listed in Table 8.2 together with the preexponential factors K_0.

In these experiments the authors also observed that the concentration of end groups at break for a series of specimens did not depend on stress, temperature, or type of loading (Fig. 8.12). The apparent "critical concentration" of *all* newly formed end groups in PE-HD amounted to $3.9 \cdot 10^{+19}$ cm^{-3} or to $\simeq 2 \cdot 10^{+19}$ bond ruptures

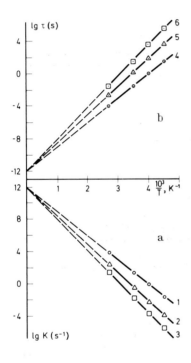

Fig. 8.10. Rate constants K for end-group accumulation and times to fracture τ for high density polyethylene (after [16–18]); σ(MN/m): 1. 400, 2. 300, 3. 200, 4. 400, 5. 300, 6. 200.

Fig. 8.11. Apparent activation energy. U_* for macroscopic breakage under uniaxial stress σ; E_* for accumulation of end groups under stress σ (after [18]).

Table 8.2. Kinetic Parameters of End Group Formation in Certain Polymer Films (after Zhurkov, [18]).

Polymer	Preexponential factor K_0 (s^{-1})	Activation energy E_A (kcal/mol)	Activation volume α	
			$\dfrac{kcal\ mm^2}{N\ mol}$	$\dfrac{10^{-6}\ m^3}{mol}$
Polyethylene (low-pressure)	10^{11}	27	0.026	110
Polyethylene (high-pressure)	10^{12}	28	0.059	250
Polypropylene ˘	10^{12}	30	0.051	210

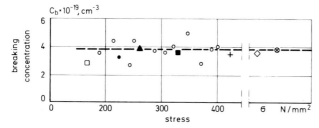

Fig. 8.12. Concentration C_b of end groups just prior to macroscopic fracture of high density polyethylene (after [18]); conditions of loading: samples broken in creep at constant stress at (+) 200 K; (■) 273 K; (○) 293 K; (□) 373 K; (●) 333 K; (▲) 323 K; (◊) samples broken under linearly increasing deformation with time; (⊕) samples broken under stress increasing linearly with time.

per cm^3. Vettegren and Chmel [70] estimated that in a *surface layer* of 1 μm thickness the concentration of end groups may even be higher by a factor of 20 to 60. For polypropylene they derived concentrations of $6 \cdot 10^{19}$ cm^{-3} broken bonds in the surface layer – as compared to $0.25 - 0.40 \cdot 10^{19}$ cm^{-3} reported by Zhurkov et al. [17, 18] for the interior of the film. The latter value should be compared with the number of overloaded 5 nm segments. In Section I B a concentration of $8.5 \cdot 10^{19}$ cm^{-3} overloaded segments had been determined from the area of the tail region of the "stressed IR band". Such a comparison indicates that of $N_0 = 56 \cdot 10^{19}$ cm^{-3} segments present in the interior material of the PP film $0.15\,N_0$ are initially overloaded and $0.0045\,N_0$ will have been broken at the point of macroscopic fracture. In another study on PP [25] a concentration of broken bonds of $3 \cdot 10^{19}$ cm^{-3} is reported, i.e. a fraction of 35% of the overloaded bonds is broken throughout the total volume of the sample. In the latter case wholesale destruction of the polymer must have occurred since on the average *each molecule* in a (monodisperse) $M_n = 50000$ g/mol sample *would have been broken 2.5 times*. In the former case the effect is ten times smaller, but M_n is still reduced to 40000 g/mol and M_w to 43750 g/mol. The effect is even more drastic with higher initial molecular weights. The *fracture criterion* derived at the Leningrad Institute for the three polymer films on the basis of the kinetics of thermal and mechanical destruction and of end-group formation is that of a *critical concentration of newly broken chains* [16–18].

In view of the fact that so far the end-group analysis has only been successful with these three polymer films one would welcome further studies on the extent to which the end-group formation depends on the initial concentration of carbonyl groups – or of other oxygen-containing groups. More extensive molecular weight measurements, especially by fractionating methods would also be useful to corroborate the absolute concentrations of end groups [220].

If the number of breaking chain segments is as large as indicated by the *end-group analysis* and if each segment breaks at the limiting stress value derived from the *analysis of deformed-IR bands* then the accumulated molecular stresses would be comparable within an order of magnitude to the applied macroscopic stress. In that case it must be assumed that apart from conformational rearrangements and chain slippage chain scission events noticeably influence observed stress-strain curves. So far the PP material from Leningrad is the only polymer which has lent itself successfully to both of the above mentioned IR analysis. In the cited literature [4–33] there is no reference to an attempt to explain stress-strain or stress-time data of this PP material on the basis of its end group-formation kinetics.

As discussed frequently within this text *stress-relaxation* data can not and must not be related solely to chain scission [2–52]. Attempts have been forwarded, however, to explain the *stress-strain* curves of polyamide-6 [49–51] and PEOB fibers [52] on the basis of free radical production kinetics. In these models the concept was employed that tie segments show a distribution of relative lengths. The *(tautness)* distribution, $N(L/L_0)$, was assumed to remain unchanged in a wide time-temperature regime. As detailed in Chapter 5, a tie-chain segment will respond to a strain ϵ_a predominantly rubber-elastically if $\epsilon_a < (L-L_0)/L_0$, it will break if $\epsilon_a > (1 + \psi_b/E_k)L/L_0 - 1$, and it will respond energy-elastically in the intermediate region where it happens to be fully extended and subcritically loaded. From these considerations a homogeneous four-phase model can be derived (Fig. 8.13). It is based on the assump-

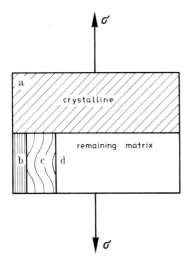

Fig. 8.13. Four-phase model of highly oriented semicrystalline fiber. (a) crystalline lamellae, (b) fully extended tie segments, (c) non-extended tie segments, (d) matrix (folds, ciliae, broken chains).

tion that a highly oriented semi-crystalline fiber can be represented by crystalline and amorphous regions in series with the latter consisting of three fractions, i.e. of the fully extended segments, of the non-extended tie-chains, and of the remaining matrix (folds, ciliae, broken chains). Once the strain-dependent widths of the fractions are obtained from a relative segment length distribution and appropriate moduli are assigned to the individual phases a *stress-strain* curve can be constructed. Such a procedure was proposed in 1971 by DeVries and Williams [49]. It has been executed by Lloyd [50] and by Klinkenberg [51] for PA 6 fibers and by Nagamura et al. [52] for PEOB multi-filaments. The calculated stress-strain curves can be made to fit the stress-strain response of virgin material; they contain as possible sources of uncertainty the *choice of the "elastic" moduli* and of the *number and distribution of tie-segment lengths,* i.e. of $N(L/L_0)$. The choice of the form of $N(L/L_0)$ determines which number of segment lengths is assigned to the fraction of non-extended chain segments by way of extrapolation. Lloyd [50] and Nagamura et al. [52] have chosen a normal distribution to represent $N(L/L_0)$.

A fit of observed and calculated stress-strain curves of first and second stretching cycles was obtained if the calculated number of radicals formed (or of chains broken) was taken as considerably larger than the observed number. Nagamura et al. [52] report a factor f_c of 40 for PEOB fibers, Klinkenberg [51] of 20 for PA 6 fibers. If one wishes to explain such a discrepancy while maintaining the basic model one is forced to make additional assumptions:

— either the number of broken chains is systematically larger by f_c than the number of observed radicals, e.g. due to a Zakrewskii-mechanism [20] to be discussed later;
— or the breakage of N_1 chains in a particular volume element leads to the preferential unloading of $f_c N_1$ extended chains; this amounts to the consideration of a microfibrillar substructure;
— in any event provision must be made to account for the anelastic deformation of the non-extended chains; the stress-induced change of the conformation of a tie-segment profoundly reduces the axial chain stress while not effecting $N(L/L_0)$ (cf. Chapter 5 IB and III); in subjecting a fiber to several stress-strain cycles a large fraction of the more extended relaxed, conformations formed during the first cycle are present at the beginning of the second cycle; these stress-induced, temporarily stable conformational changes are considered to be mostly responsible for the differences between first and second loading cycles [2].

It is the common observation of the above authors [49–52] that *qualitatively* the slopes of stress-strain and of radical concentration-strain curves correspond. For a *quantitative* correspondance one would have to assume that 20 to 40 times as many chains break as free radicals are observed. Such an assumption appears to be rather large in view of the facts that no correspondingly strong decrease of the molecular weight has been reported and that the load bearing capability of the fiber material outside of the immediate fracture zone does not necessarily suffer. Prevorsek [53] showed that the strengths of fiber segments subjected repeatedly to breaking loads did not decrease. The rupture segments obtained in a first loading exhibited a larger strength than the original fiber and maintained this strength in subsequent loadings (Fig. 8.14). It seems more likely, therefore, that the concentration of chain scission

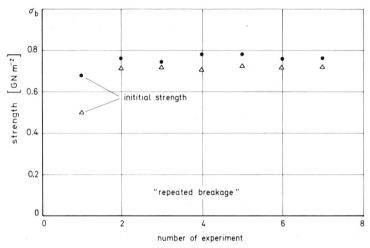

Fig. 8.14. Load-to-break of PA-6 fiber segments strained repeatedly (after [53]).

events is of the order indicated by the ESR investigations, i.e. 10^{16} to $5 \cdot 10^{17} \, cm^{-3}$. Such concentrations in themselves do not weaken the polymer network decisively since they amount to the breakage of only 0.002 to 0.10% of *all amorphous* chain segments [2]. Unless the qualitative agreement between the slopes of stress-strain and free radical-strain curves is coincidental it can be sought in an amplification effect accompanying the rapid release of stored elastic energy. The ensuing local temperature rise will facilitate the extension of kinked chains under annihilation of kinks which is practically equivalent to unloading those chains *without breaking them* (cf. Chapter 5). Within the framework of this model one would have to conclude that the breakage of one amorphous segment at maximum load $q\psi_b$ leads to conformational changes in the *surrounding* segments of the same microfibrillar region resulting in local load decreases of $40 \, q \, \psi_b$. In view of the fact that on the average there are more than 2000 surrounding segments (of 5 nm length in any amorphous region 400 nm^2 in cross-section) the value of $40 \, q \, \psi_b$ appears very reasonable.

One word of caution must be said with respect to the extrapolation of the segment length distribution $N(L/L_0)$ into the unknown regions of L/L_0. For this purpose DeVries et al. [49, 50] and Nagamura et al. [51] used the assumption that the "visible" part of the segment length distribution (Fig. 7.16) constitutes the skewed part of a normal distribution. A plot of the skewed distribution on probability coordinates (Fig. 7.17) apparently does not disprove the contention made. They had derived the parameters of the normal distribution (mean, variance) using the existence and location of an inflection point in the — always only partially known — cumulative distribution. It cannot be excluded that such inflection points result from the increased unloading of microfibril ends at larger strains since the stress-transfer length L_ϵ of a micro-fibril (with modulus E_f, diameter d) increases in case of perfect inter-fibrillar adhesion (of strength τ) with $L_\epsilon = E_f d_f \epsilon / 2\tau$. The eventual error of an extrapolation of $N(L/L_0)$ into the range of not fully — or even slightly — loaded segments is of not too much importance in calculating accumulated molecular stresses. It is of importance, however, if the distribution $N(L/L_0)$ is used to derive a fracture criterion. A fracture criterion in which macroscopic failure coincides with breakage of the last tie segment of the total distribution must be ruled out. Against such a fracture criterion the above mentioned argument holds that at the point of macroscopic failure 99.7% of all tie-segments are still unbroken (Fig. 7.6).

Although DeVries, Lloyd, and Williams [49, 50] work with a normal distribution of relative chain lengths they do not put it to a full and questionable use. Their assumptions may be reformulated on the basis of the above considerations. It is assumed 1. that the segment length distribution $N(L/L_0)$ — derived from spatially homogeneous chain scission events of virgin material — be representative for the homogeneity of molecular stress distribution in any average amorphous region of the fiber, 2. that an initially narrow distribution continues to rise sharply with segment length and 3. that the *complete* break-down of tie segments within microfibrillar amorphous regions be restricted to a comparatively small fracture zone which does not contribute measurably to the free radical population. They use the widths of their chain length distributions $N(L/L_0)$ and correlate it with macroscopic strength (Fig. 8.15). In other words they correlate the non-homogeneity of the molecular stress distribution with macroscopic strength — and find a negative correlation.

Fig. 8.15. Correlation of the homogeneity of segment lengths (narrow relative length distribution) with uniaxial strength σ_b [49, 50, 54].

It has been stated frequently in this work that the tie segments which eventually break are not the main source of strength of highly oriented fibers. The total number of radicals formed at macroscopic fracture does not *measure*, therefore, the strength of the sample. The wide differences in radical concentrations $N(\dot{R})$ at break of comparably strong fibers (cf. Table 6.2) illustrate this statement. Again these concentrations should be taken only as an indication of the *homogeneity* of *stress distribution between* different microfibrils and of *stress response within* different microfibrils. The later a possible defect zone starts its accelerated growth the longer the phase of spatially homogeneously distributed chain scission and the larger the number of chains which are loaded up to breakage. If one compares, under this aspect, members from one family of samples strength-related conclusions seem to be justified.

In Chapter 7 the morphological changes in PA 6 fibers due to annealing and their effect on chain scission have been discussed. In Figure 7.18 and 7.19 it had been indicated that slack annealing is accompanied by a relative lengthening of tie-segments and to a widening of the length distribution. The indicated losses of homogeneity lead to accelerated flaw growth already at stresses lower than those exerted on the control sample, i.e. to a loss of strength (Fig. 7.20, "slack"). Annealing at fixed fiber ends leads to a certain loss in homogeneity while maintaining the average relative segment length. The loss in strength (Fig. 7.21) and in the number N of radicals at break (Fig. 7.20) as compared to the control sample is small. This result confirms the findings of Levin et al. [21] with highly oriented tensile Kapron annealed at 200 °C. Annealing under a 11.7% prestrain increases the relative segment length (Figs. 7.18 and 7.19), segment homogeneity (Fig. 7.22), and sample strength (Fig. 7.21).

The *opposing effects* of increasing annealing temperature and increasing tensile prestrain on segment homogeneity are clearly indicated on Figure 7.22. The concentration $N_{\dot{R}}$ of radicals at break formed after annealing initially increases with pre-

Fig. 8.16. Effect of prestrain at annealing at 164 °C on strength and number of broken bonds of 6 polyamide fibers (after [55–57]).

strain (Fig. 8.16), showing a maximum at a 5% prestrain. At that strain radical formation in prestraining (N_p) starts. The total number of radicals ($N_p + N_{\dot{R}}$) is approximately constant in that region. The sample strength increases by 20% up to a maximum at 7.5% prestrain and decreases slightly thereafter (Fig. 8.16).

D. Formation of Microcracks

All of the foregoing considerations have been concerned with a distribution of chain
scission events homogeneous on a macroscopic scale. The accelerating effects of
stress concentrations at microstructural inhomogeneities and of mutual interaction
of scission points have so far been neglected. Such effects were not apparent in the
discussed IR and ESR investigations of chain rupture; in any event they did not
prevent tremendous concentrations of scission points being accumulated before
final fracture. This fact can certainly be taken as an indication that during a considerable period of straining, chain scission points remain isolated "defects" and
do not give rise to unstable cracks.

On the other hand the appearance of *submicrocracks* in loaded polymers has
been known for many years since the Leningrad school [17, 18, 27, 28] employed
X-ray scattering techniques to search for it. Such submicrocracks were found in PE,
PP, PVC, PVB, PMMA, and PA 6. The authors noted two important regularities of
the submicrocrack formation [28]. First of all, the submicroscopic cracks appear at
finite sizes with their transverse dimensions practically independent of load duration,
stress value, and temperature (Table 8.3). Secondly the transverse size of the submicrocracks is determined by the polymer structure. For oriented crystalline polymers the transverse size coincides with the microfibril diameter; for non-oriented,
amorphous polymers having a globular structure it coincides with the diameter of
the globules [28].

Two explanations have been given for the formation of these submicrocracks:
Zakrewskii [20] proposes a radical chain reaction and Peterlin [58] suggests that the
ends of microfibrils are the crack nuclei.

Table 8.3. Characteristics of Incipient Cracks in Polymers (after Zhurkov Kuksenko
et al., [27, 29])

Polymer	Micro crack diameter		N_{cr}	Density change $\Delta\rho/\rho$	
	longitudinal nm	transverse nm	cm^{-3}	calculated	measured
Polyethylene	15	17	$6 \cdot 10^{15}$	–	–
Polypropylene	20	32–35	$7 \cdot 10^{14}$	–	–
Polyvinyl chloride	–	60	$8 \cdot 10^{14}$	–	–
Polyvinyl butyral	–	50	$3 \cdot 10^{14}$	–	–
Polymethylmethacrylate	80	170	$4 \cdot 10^{12}$	$0.4 \cdot 10^{-2}$	$0.3 \cdot 10^{-2}$
Polycaproamide	5	9–25	$5 \cdot 10^{16}$	$1,8 \cdot 10^{-2}$	$2.0 \cdot 10^{-2}$

The Zakrewskii-mechanism was proposed by the Russian school following their studies of the formation of free radicals and of end-groups. They had noted that the concentration of free radicals was either by a factor of 4000 (PE-HD) smaller than the concentration of newly formed end-groups or even unmeasurably small (PP). This led them to search for a reaction by which chains might break without increasing the concentration of free radicals.

The mechanism of the proposed mechano-chemical reaction chain is illustrated in Figure 8.17. The thermo-mechanical scission of a regular covalent bond (a) leads to the formation of two chain-end radicals (b). As amply discussed in Chapter 6 the latter radicals are highly reactive. They will transform by hydrogen abstraction into stable end groups at the same time creating main-chain radicals within neighboring chains (c). It is then assumed by Zakrevskii and Zhurkov [17, 20] that the decreased energy of activation for the thermal scission reaction of radicalized chains (cf. Chapter 4) also leads to a lower tensile strength of these chains causing them to rupture (d). Repetition of the steps c and d would not increase the number of free radicals present but would multiply the number of chains broken. In Figure 8.18 the energy levels of the proposed steps of the reaction chain are indicated. It should be noted, however, that the lowering of the activation energy for scission of a radicalized chain is effective only where thermal equilibrium can be established (cf. Chapter 4 II B).

Fig. 8.17. Mechano-chemical reaction chain after Zakrevskii and Zhurkov [17, 20, 55]: (a) stress-induced chain scission, (b) formation of chain-end radicals, (c) radical reaction leading to main-chain radicals, (d) scission of radicalized chains, (e) formation of a submicrocrack by repetition of the steps (c) and (d); ⊕ chain-end radical e.g. $-CH_2-\dot{C}H_2$, x main-chain radical e.g. $-CH_2-\dot{C}H-CH_2-$, ● stable end groups e.g. $-CH_2-CH_3$.

The authors suggest [17, 18] that the 3300 new end groups formed per sub-microcrack in PE-HD (5000 in PP) are created by such a mechanism. Their data indicate also that there is less than a single pair of free radicals available per submicrocrack (0.4 in HDPE, 0.3 in PA 6 and an immeasurable small fraction in PP). These discrepancies, the unfavorable conditions for thermal equilibrium in the splitting reaction, and the absence of a well defined termination of the reaction chain raise considerable concern. They make it difficult to believe that such a reaction is the major reason for the formation of submicrocracks which remain fairly stable during extended and repeated loading periods [220].

Peterlin [58] has based his proposal on the origin of submicrocracks on morphological considerations and analysis of the previously mentioned X-ray data

194

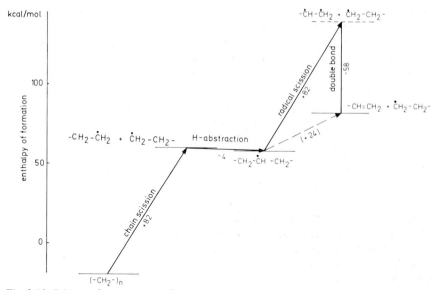

Fig. 8.18. Scheme of energy levels of steps *a–d* in the mechano-chemical reaction chain.

[17–21, 27]. He suggests that the *ends of microfibrils*, preferentially situated on the outer surface of the *fibrils*, retract under stress (Fig. 8.19). In low strength PA 6 they thus open up about 10^{16} oblate spheroids per cm^3 having 6 nm diameter in fiber axis direction and 10 nm in the perpendicular direction. He estimates that the number and size of cracks in high strength polyamide is considerably smaller and is in agreement with the much higher draw ratio employed in producing those fibers. For highly oriented films with microfibrillar substructure similar considerations on the origin of the voids hold.

Fig. 8.19. Fibrillar model of fibrous structure with practically all ends of microfibrils concentrated on the outer boundary of fibrils (after [58]).

Considering the two important regularities of submicrocrack formation [28], the inconsistencies of the Zakrewskii-mechanism [17], and the morphological aspects [58] the retraction of microfibril ends seems to be a logical interpretation of the opening-up of fixed-geometry voids. This means at the same time that the submicrocrack formation is a process inherently independent of chain scission or end-group formation [220]. The voids constitute structural irregularities and contribute as such toward the general non-homogeneity of the distribution of stresses or strains. Their direct influence as potential stress concentrators is weak and ineffective with regard to accelerating chain scission. This conclusion is based on the following facts:

— microfibril ends are free to undergo shear deformation and are free from axial stresses,
— obviously the voids do not prevent the build-up of large elastic strains in the fibers or films — as evidenced by the scissioning of chains and the appearance of deformed-IR bands,
— if chain scission would occur preferably adjacent to microfibril ends further lateral widening of the there existing microcracks would then invariably accelerate chain scission.

E. Energy Release in Chain Scission

A molecular chain segment stressed almost to its load-bearing capacity constitutes an extremely powerful source of elastically stored energy. In thermo-mechanically activated bond scission only a minor fraction of this stored energy is needed to break the chemical bond, namely the mechanical contribution $\beta \psi_b$ to the activation energy U_0. The remaining, major part of the energy is available for mechanical interaction with surrounding chains or is dissipated as heat. The dissipated heat has a twofold effect through the ensuing local temperature rise: it increases the mobility of other chain segments and decreases their rupture strength ψ_b (T). Both effects tend to facilitate the further degradation of the stressed polymer.

The total elastic energy stored in a segment of elastic modulus E_k, length L, and cross-section q amounts to

$$W_k = q L \psi^2 / 2 E_k. \tag{8.7}$$

For a polyamide 6 segment of 5 nm length with $q = 0.189$ nm^2, $\psi_b = 21$ GN/m^2 and $E_k = 200$ GN/m^2 one obtains a value of $1 \cdot 10^{-18}$ J per segment or 600 kJ/mol for the elastically stored energy. To break a C—C bond at room temperature a *mechanical* energy of 110 kJ/mol is needed (cf. Chapter 7 I C 1). In setting up an energy balance the contribution of the elastic forces must not be forgotten which hold the highly stressed *tie-segment ends within the crystal lamellae*. At the rupture stress this energy amounts to about 190 kJ/mol for each segment end (cf. Table 5.5). One thus arrives at an energy of 870 kJ/mol which is liberated at the moment of chain scission. If this energy were to be confined within the volume of the segment and of its ends ($L_{total} \simeq 10$ nm) it would constitute an energy density of $W_{total}/q_{total} = 764$ MJ/m^3.

196

Chain scission events, therefore, have been justly termed *micro-explosion* by Zhurkov and his colleagues.

On the other hand at a (large) concentration of $(0.6 \cdot 10^{+24})$ bond ruptures per m^3 (i.e. of 0.83 mol/m^3) the total energy having been stored in the retracting chain ends amounts to an average of 722 kJ per m^3 which will be dissipated as heat Q_b. The chemical energy U_b of this number of broken bonds amounts to 156 kJ/m^3. These average energy values should also be compared against the density of elastically stored energy, i.e. against $\sigma^2/2E$. This term reaches about 125 MJ/m^3 for very high strength PA 6 fibers, which in turn is only one sixth of the cohesive energy density of that material. The energy dissipation due to ruptured bonds is slightly smaller, therefore, than the pure hysteresis *losses* encountered in loading and unloading a polyamide fiber (for PA-6 at 10 to 30 Hz tan δ is of the order of $5 \cdot 10^{-3}$).

The important problem of how the external work W done on a polymer sample during a loading-unloading cycle serves to increase the internal energy U, to change the entropy S, or to be irreversibly dissipated as heat $Q_{ir} = Td_iS$ was attacked by Müller in 1956 in his classical experiments [59–62]. According to Eqs. 5.2 and 5.3 one can write

$$dW = dU - TdS - Q_{ir}. \tag{8.8}$$

Müller and his co-workers measured several deformation-related thermodynamic quantities: the reversible and irreversible parts of heat production during the plastic deformation of PE, PVC, PETP, PA 6 [59–61], PC [63], PS [64], and of various elastomers [61, 65, 66], the ensuing temperature rises [67], the change of internal energy with the time of storage [68] and its effect on the energy of dissolution [69]. They noted that the entropy of thermoplastics during cold flow is decreased and that the internal energy is increased. They also measured the balance of energy in stressing (s) and during retraction (r) of PIB in a sequence of stretching cycles. The change of internal energy in an i th cycle can be written as:

$$\delta U_i = dW_s - dW_r + d_sS - d_rS + Q_{irs} - Q_{irr} \tag{8.9}$$

of which the entropy terms are equal by definition. The authors observed that δU_i was different from zero only in the first cycle. The increase of internal energy during extension of elastomers and thermoplastics was ascribed by them to a reduction of close range order and to chain scission and formation of radicals as further sinks of external energy.

The latter hypothesis was further tested by Godovskii et al. [31]. PA-6 fibers drawn to $\lambda = 5.5$ at 210 °C were repeatedly stretched at room-temperature. The authors found for PA 6 the same characteristic difference between the first and the following loading cycles as did Müller for PIB: δU was essentially different from zero only in the first loading cycle. The ratio of $\delta W_1/\delta U_1$ was found, independent of the macroscopic stress, to amount to 7.0. This constancy of $\delta W_1/\delta U_1$ is in fact remarkable. It indicates that processes are responsible for the increase of internal energy which occur, independent of σ, if locally a critical excitation is surpassed. Godovskii et al. assume that these processes are chain ruptures. They derive from

δU_1 the number N_1 of chain scission events each of which contributes an increment of internal energy of $1.7 \cdot 10^{-19}$ J (100 kJ/mol). The dissipated energy per chain scission event, $\delta W_1/N_1$, amounted to 700 kJ/mol. These figures are only slightly smaller than the values found above on the basis of elasticity considerations for the mechanical contribution to chain scission (110 kJ/mol) and for the energy dissipated by the retracting segments (870 kJ/mol). This striking coincidence does not prove, however, the above hypothesis that δU_1 might be interpreted solely as an increase in chemical bond energy due to chain scission.

Müller [62] and Kausch and Becht [55] have indicated that slip processes and conformational changes may respond energetically similar to bond ruptures if they are occurring as a consequence of stress induced distortion of the polymer network or as a result of local heating. The increase in internal energy is then related to a decrease of close range order or the decrease of the number of hydrogen bonds or to internal stresses between chains and within crystallites. The existence of inter-molecular forces of considerable magnitude in oriented polymers can be inferred from a number of optical, spectroscopical and mechanical experiments. Particularly noteworthy are the following observations. Vettegren et al. [70] noted that in per-fectly annealed PETP film the maximum of the characteristic backbone vibration band lies at 975 cm^{-1}, whereas in oriented PETP film it lies at 972 cm^{-1}. Lunn and Yannas [71] found that the asymmetric stretch vibration band of methyl groups with a maximum at 2971 cm^{-1} in isotropic PC film splits into two components 1 cm^{-1} apart in a PC film drawn to about 85%. They assigned this frequency shift to the action of *intermolecular* rather than intramolecular forces. With the drawn material they also studied chain-segment orientation by determining the dichroic ratio of the 1364 cm^{-1} band. They obtained an interesting result that chain back-bone motion (with an ensuing change of the degree of orientation) in drawn PC films was clearly observable more than 60 K below the glass transition temperature of 149 °C. The largest deviations from an equilibrium state (highest intermolecular tensions) existed in those films drawn at the lowest temperature (23 °C) were anneal-ing at 81 °C resulted in measurable changes of the orientation distribution. Müller [62] determined that in the cold drawing of non-oriented polymers the fraction $\delta U/\delta W$ of internal energy rise to external work amounted to 0.2 (PE), 0.3 to 0.45 (PA 6, PETP) and 0.3 to 0.5 (PVC).

The question is still open, therefore, as to the extent the internal energy rise is due to the chemical energy of broken bonds or to the local rearrangements trig-gered by chain scission.

F. Fatigue Fracture of Fibers

The preceeding Sections have dealt exclusively with the response of chains and microfibrils to constant or monotonically increasing stress or strain. In service, however, fibers are frequently subjected to intermittent, alternating, or cyclic loads. The behavior of fibers, therefore, to repetitive cyclic loading has been studied for many years (e.g. 72–82). Following an extensive review of Hearle et al. [76] it can be said that in cumulative extension cycling a fiber fails by virtue of having reached its breaking extension. Under such a condition of constantly increasing maximum extension, fatigue can be predicted from an adequate knowledge of the anelastic deformation and the time-dependent breakage conditions of the fiber. As yet no *specific fatigue effects have been identified* in cumulative extension cycling [76].

In one of the earliest reports on fatigue of fibers and fabrics Busse et al. [72] pointed out that the vibration life of nylon, cotton, and rayon cords is inversely proportional to the rate of thermally activated flow steps.

A similar conclusion has been drawn by Regel et al. [74] and Tamush [75] for cyclic stressing. These authors assume the validity of the principle of damage accumulation ("Miner's rule"), i.e. of

$$\int_0^{2\pi N_F/\omega} dt \, / \, \tau(\sigma[t]) = 1 \qquad (8.10)$$

with N_F the number of cycles to failure and ω the loading frequency. If τ, the expectation value of the time-to-break under *constant* stress, is taken to be the common Arrhenius expression one obtains

$$\int_0^{2\pi N_F/\omega} dt \, / \, \tau_0 \exp\left[(U_0 - \gamma \sigma[t])/RT\right] = 1. \qquad (8.11)$$

Regel et al. [74] report that such a damage accumulation law did apply to PAN fibers loaded at 24 Hz up to $1.5 \cdot 10^7$ cycles. For PMM films, viscose fibers, and capron fibers (PA 6) correspondance of experimental data with Eq. (8.11) could be obtained through air cooling of the fatigued samples, after preliminary drawing, or at elevated temperatures. The authors concluded that Eq. (8.11) would describe fatigue fracture within the kinetic concept of fracture if (ambient) temperature T and activation volume γ would be replaced by values T* and γ* which would depend on the parameters of the fatigue experiment (frequency, form of stress or strain pulse).

The fatigue failure of fibers has also been extensively investigated by Prevorsek, Lyons, and co-workers. Their many pertinent papers are referenced in [78]. The authors interpret the fatigue life as the time to void nucleation by rearrangement of molecular segments. They arrive at a kinetic equation relating the number of cycles to failure with various mechanical and molecular parameters [78].

Employing a load-controlled fatigue tester Bunsell and Hearle [77] have more recently found a distinctive fatigue fracture morphology in polyamide 66 fibers. A necessary requisite for this fatigue mechanism to operate was the cycling of loads down to zero stress values. Under such conditions fibers failed at a maximum load only 60 to 70% of the normal breaking load [77, 79]. This observation will be further discussed in the following section (8 I G).

DeVries et al. [49] studied the formation of free radicals under cyclic tensile loading. The radical accumulation curves reflected the stress cycles. The authors explained the kinetics of radical accumulation on the basis of Eq. (7.2), i.e. without assuming a special fatigue mechanism.

Summarizing the above observations one has to conclude that no distinctive fatigue mechanism — other than hysteresis heating — is operating when highly oriented fibers are subjected to cyclic loads which remain always positive. Initiation and propagation of fatigue cracks will be favored, however, if chains and fibrils are given a chance to relax stresses, to degrade possible strain hardening effects, and to reorient. Thus the fatigue mechanism described by Bunsell and Hearle [77, 79] manifests itself as an increase in interfibrillar slip and crack growth almost parallel to the load direction. It will be further discussed in the following section. Characteristic fatigue mechanisms are also well observable in non-oriented polymers. They will be treated in Section II C of this and in the next chapter.

G. Fractography

The visual inspection and the study of broken samples by means of optical or scanning electron microscopy (SEM) are important tools of fracture analysis. They obviously aid
- in the detection of possible causes of crack initiation and
- in the interpretation of the fracture process.

The appearance of a fracture surface undoubtedly is the most convincing evidence that the fracture process has reached the phase of non-homogeneous deformation. But rather frequently the surface and morphology of a failed specimen reveal whether or not a phase of homogeneous material deformation has contributed to the fracture event. It is in this sense that a paragraph on fractography has been introduced into the chapter on *homogeneous* deformation and fracture. Figures 7.8 and 7.9 serve to illustrate this point. These fracture surfaces are more or less an arbitrary result of a large number of homogeneously distributed scission and slip processes of chains and microfibrils. The fracture surface was developed within a very small fraction of the total lifetime and at a location not predictable in advance.

Hendus and Penzel [83] investigated the fracture morphology of polyamide 6 monofilaments. The regularly spun and drawn monofilaments were subsequently subjected to tensile tests at various strain rates. Characteristic fracture surfaces are reproduced in Figures 8.20 and 8.21 [84]. At small strain rates ($\dot{\epsilon} = 0.033$ s^{-1}) frequently v-shaped notches are observed (Fig. 8.20). Such a notch is formed through a crack which initiates at a flaw or material inhomogeneity contained in the filament surface or in a zone close to the surface. While the crack grows slowly the remaining fiber cross-section continues to deform plastically. At a point determined by the sizes of the crack and of the remaining cross-section and by the material properties rapid transverse crack propagation occurs. The measured strength of the monofilament is the higher the smaller the v-shaped notch [83]. Filaments of highest strength contained barely visible small voids.

Figure 8.20 also shows an enlargement of the head of the monofilament. This enlargement corresponds to a shrinkage of the oriented material caused by the warming of the filament during the plastic deformation. If the rate of deformation is increased to 50 s^{-1} then the heat generated through the plastic deformation of the remaining cross-section cannot be dissipated fast enough. Locally the melt temperature is exceeded, the filament heads expand almost to the diameter of the undrawn material (Fig. 8.21).

Fig. 8.20. Corresponding ends of PA 6 monofil broken at a strain rate of 0.033 s⁻¹ (Courtesy H. Hendus, [83]).

Fig. 8.21. Corresponding ends of a PA 6 monofil broken at a strain rate of 50 s⁻¹ (Courtesy H. Hendus, [83]).

The fibrillar nature of drawn monofilaments becomes apparent in Figure 8.22. The filament ends of two different fracture events show a strong axial splitting. As concluded from the formation of small heads at the ends of some microfibrils the fibrillation must have occurred before the catastrophic fracture took place. Strong axial splitting is also known from polyamide 66, Kevlar [148], PETP, acrylic fibers, wool, human hair, and cotton fibers [85].

On the basis of fractographic studies Bunsell and Hearle [77, 79] indirectly identified a fatigue mechanism in polyamide 66. Oriented PA-66 fibers failed in a tensile type manner if they were fatigued uniaxially between an upper stress level equal to the long-time static strength and a lower level of clearly non-zero tensile stresses. These fatigue failures had a morphology similar to that shown in Figure 8.20. The breakage apparently did not involve any specific fatigue effect. If the loading conditions were changed, however, so as to include zero tensile or even compressive

Fig. 8.22. Two different PA-6 fiber fractures under tension showing strong axial splitting (Courtesy H. Hendus, [83]).

stresses a specific fatigue fracture morphology became apparent. Those breaks, occurring after 10^5 cycles, were characterized by the propagation of a crack, almost parallel to the direction of the fiber axis. As shown by Figure 8.23a [86] this leaves a long tail of material at one fiber end. Although this phenomenon is not yet completely understood it may be concluded from Bunsell's and Hearle's observations [77, 79, 85] that compressive yielding and/or buckling of microfibrils or fibrils takes place on a microscopic scale. The repeated lateral rearrangement leads to a gradual loss of interfibrillar cohesion. A crack which has started at some flaw in a direction normal to the tensile stress σ will turn, therefore, into a direction u almost parallel to σ – the angle between u and σ being a function of the lateral stress transfer which still remains.

Fig. 8.23 a and b. Corresponding ends of fatigue failure of polyamide 66 under oscillating load dropping to zero in each cycle (Courtesy J. W. S. Hearle, Manchester).

On the basis of his extensive investigations Hearle [85] has established a classi-fication of the main features of fiber fracture morphologies:

A. Transverse Elastic Crack Propagation

This is the classic form of failure in elastic materials and consists of a relatively smooth "mirror zone" of crack propagation leading to a rougher zone of final failure due to multiple crack-initiation (comparable in morphology to Fig. 1.7).

B. Ductile Transverse Crack Propagation

This is a form of fracture in which a crack propagates stably across a fiber under increasing load and/or strain and is opened out into a V-notch by the continued plastic yielding (the final stages of the drawing process) of the remaining material; the crack leads into a region of final catastrophic failure, occurring when the stress in the remaining reduced cross-section reaches a critical level (Fig. 8.20).

C. Transverse "Fibrous" Break

This type of break runs perpendicularly across the fiber, with a rough texture and no evidence of any propagating cracks; the whole structure appears to be ready to fail at the same time, and the breaks are very similar in appearance to lower-magnification views of the fracture of fiber-reinforced composites (Fig. 7.8 and 7.9).

D. Axial Splitting under Tension

In some fibers failure in axial tension occurs by cracking, or splitting, along planes close to the fiber axis. Due to a nonhomogeneous fibrillar substructure there will be shear stresses in these planes. With reduced interfibrillar cohesion a break is produced characterized by multiple axial splitting over a long length (equal to many fiber diameters) (Fig. 8.22).

E. Splitting Due to Torsion

Under shear stresses generated by torsion, splitting can occur along lines deter-mined by the directions of stress and of material weakness.

F. Axial-fatigue Cracks

Particular localized cracks, deviating slightly from the fiber-axial direction, develop in many fibers as a result of tensile fatigue: they are associated with shear stresses at discontinuities. The resulting breaks show a long tail on one end, a strip off the other end, and a final catastrophic-failure region (Fig. 8.23).

G. Fatigue-cracking along Kink-bands

Kink-bands at angles of about 45° develop in many fibers in compression, for example, or on the inside of bends, and, in repeated flexing, these eventually turn into cracks and lead to fiber failure.

H. Multiple Splitting in Fatigue

In torsional- or flexural-fatigue situations, multiple splitting frequently occurs and can eventually lead to failure.

For a more detailed discussion of *fiber* fracture morphology the reader is referred to the series of publications by Hearle, especially to "An Atlas of Fibre Fracture" [87]. The fracture morphology of *non-oriented polymers* under various loading con-ditions will be treated, where appropriate, in subsequent sections of this and the final chapter.

II. Deformation, Creep, and Fatigue of Unoriented Polymers

A. Impact Loading

Polymeric materials are viscoelastic solids. Their propensity to anelastic and plastic deformation is reduced when they are tested at high load rates and/or at low temperatures. The reduced deformability causes a formerly tough or highly elastic polymer to respond in a brittle manner. The impact fracture of natural rubber at liquid nitrogen temperature is convincing evidence of this fact.

The resistance of polymeric materials to impact loading is of considerable technical importance as illustrated by the following examples:
- aircraft windows have to withstand impinging particles or rain
- protective coverings of high voltage switches must resist the impact of metal chips
- water pipes or plastic floor tiles should not be damaged if hit accidentally
- car bumpers or packaging materials must absorb as much of the energy of a collision as possible
- household items are not expected to break the first time they fall to the ground
- gears should transmit even rapidly increasing forces
- safety belts should yield but not fracture.

In view of the technical relevance the impact resistance of polymers has been intensively investigated. The cited references [88–103] may serve as an introduction into the large body of literature. Vincent [88] and Bucknall et al. [89] give a general survey of the impact testing of polymers. Other references concern the molecular aspects [88–96], instrumentation [97–100], and particle impact [101–103]. The extensive literature on the fracture mechanical analysis of impact bending tests will be referenced in Chapter 9.

The fact that the impact resistance depends strongly on the presence and shape of stress concentrators, on the sample geometry, and on the testing conditions has made it very difficult to define and measure *a unique material property impact resistance.* Since the impact behavior under one condition (e.g. impact bending) cannot very well be predicted from the behavior under different conditions (e.g. falling weight) quite a number of impact tests have been devised. The four best known ones are the three point bending (Charpy unnotched and notched), two point bending (Izod), tensile-impact, and falling-weight tests which have been standardized (DIN 53453, 53448, 53373, 53443E; ASTM D 256, 1822, 2444, 3029).

In the context of this chapter only the three-point bending test (Charpy) will be discussed. In this test the energy loss A_n is measured which a pendulum incurs in striking and breaking a prismatic specimen (of thickness D and width B). The observed energy loss comprises basically four different terms:
- the energy W_e to bend the sample up to the point of crack initiation,
- the energy $G_c BD$ to propagate the crack through the specimen,
- the kinetic energy W_{kin} of the broken sample, and
- vibrational or otherwise dissipated energy.

The simple excess energy measurement of the pendulum does not permit discrimination between these different energy sinks. The separation of the first three terms becomes possible, however, if the bending force P acting between pendulum head and specimen is measured (so-called instrumented impact test).

For the elastic bending of a prismatic beam between supports at a distance L_s the following expressions for energy and strain are obtained.

The total bending energy W_e, as a function of beam deflection δ, is

$$W_e = P\delta/2 = \frac{2E\,\delta^2\,BD^3}{L_s^3}\,, \tag{8.12}$$

the largest density of stored elastic energy in the tensile fiber:

$$w_m = \sigma_m^2/2E = \frac{18E\delta^2\,D^2}{L_s^4}\,, \tag{8.13}$$

and the largest strain can be expressed as

$$\epsilon_m = \sigma_m/E = 6\delta\,D/L_s^2. \tag{8.14}$$

From this relation the strain rate $\dot\epsilon_m$ at a pendulum velocity of 2.9 m/s (DIN 51222, 0.5–4 J) is obtained as 65 s^{-1}. The kinetic energy of a sample of a mass of 1 g ejected by the striker at that speed of 2.9 m/s would amount to $4.2 \cdot 10^{-3}$ J.

The ratio, a_n, of the fracture energy A_n to the sample cross-section $B \cdot D$ is termed impact strength (Schlagzähigkeit). Such a notation creates the allusion that a_n is a fracture surface specific material property. It has frequently been pointed out that this is not the case [88–89]. Neither W_e nor W_{kin} are proportional to the sample cross-section. A comparison of a_n values should only be made, therefore, if all values are obtained from one type of test, preferably even from specimens of identical geometry. Impact strength values a_n of Charpy un-notched specimens (DIN 53453) at 20 °C range from 3.5–12 kJ/m^2 for filled phenol melamine and urea resins, 4–22 kJ/m^2 for various filled epoxy and polyester resins, 12–20 kJ/m^2 for PMMA, PS, and SAN, and 50–90 kJ/m^2 for ethyl cellulose, CA, styrene-butadiene copolymers, and POM. For many thermoplastics (ABS, CAB, PE, PP, PTFE, PC, PVC, PA) no fracture is observed under these test conditions [104].

The above impact strength values designate brittle behavior if $a_n < 40$ kJ/m^2. The materials with a_n between 50 to 90 kJ/m^2 will generally the brittle if bluntly notched. Of those polymers not breaking in the Charpy un-notched test some will respond in a brittle manner if they are sharply notched, whereas others remain tough even then. Vincent [96] and Bucknall et al. [89] propose, therefore, the following qualitative impact resistance rating: *brittle* ($a_n < 40$ kJ/m^2), *brittle if bluntly notched, brittle if sharply notched, tough but crack propagating,* and *very tough and crack arresting.*

The characteristic impact behavior of brittle polymers can conveniently be outlined using polystyrene as an example. Ramsteiner [105] has quite recently performed instrumented impact test on standard prismatic polystyrene bars. An inspection of the broken samples revealed that they had failed through rapid propagation of a crack emanating from the more or less highly crazed tensile zone. The length of the longest crazes was generally identical to the length of the mirror-like zone of the fracture surface. The force-deflection curves obtained in this instrumented impact test show a slightly non-linear increase of the load during 1 ms followed by a rapid drop to zero within less than 50 μs (Fig. 8.24).

The deformation mechanisms taking place during rapid loading can be interpreted as follows: The deformation is homogeneous up to a strain level where crazes are initiated. The craze initiation stress depends on molecular weight, sample treatment, time and temperature (cf. Chapter 9). For the above experiments it would be reasonable to assume that crazes are initiated once the load attains between 60 and 70% of the maximum load. The maximum of the load corresponds to the point where rapid crack propagation sets in.

Fig. 8.24. Load-deflection curve from instrumented impact test of a brittle polymer (after [105]). (A) polystyrene as moulded, (B) annealed 30 min at 130 °C.

The impact toughness of a brittle polymer (area under the load-deformation curve) is essentially determined by the energy W_e to reach the stage of rapid crack propagation (maximum of P). The energy input for separation of the material during the phase of rapid crack propagation is negligible (Fig. 8.24). One of the most important aspects of the impact toughness of brittle polymers concerns, therefore, the limiting conditions under which rapid crack propagation in a material becomes possible. This problem, the fracture mechanics approach, will be treated in Chapter 9. The absolute values of the largest bending force F_m or of the largest stresses in the tensile fiber are not a measure of impact toughness. In fact an annealing treatment (30 min. at 130 °C) of an injection moulded polystyrene bar increased the impact toughness (due to increased deflection δ) from 18 to 21 kJ/m^2 while decreasing F_m from 235 to 215 N [105].

Before characterizing the role of molecular chains in impact loading the fracture of normally tough polymers (e.g. HDPE, PVC, PP, PA) will be studied. Cleavage fracture is difficult to obtain by bending and occurs only in notched specimens of low aspect ratio (L_s/D) or if the rate of loading is high (impact loading). In samples of HDPE with extremely high molecular weight ($M_w > 10^6$ g/mole) cleavage fracture does not occur at all. The fracture surface shown in Figure 8.25 was obtained by Gaube and Kausch [106] in impact loading at 20 °C from a standard-HDPE bar containing a knife cut. The surface clearly reveals three different morphologies:
– a zone of thermally activated crack growth starting from the knife cut (enlarged in Fig. 8.26)
– a zone of unstable crack propagation under a condition of tensile strains

— a zone where the now rapidly propagating crack is deviated through the stresses in the compressive fiber of the bending specimen

Within the zone of thermal crack growth two features seem to be of particular interest, the cellular area II (enlarged in Figs. 8.27 to 8.29) and the void containing area III (enlarged in Figs. 8.30 and 8.31). Figures 8.27 to 8.31 give an impression of the large plastic deformation of the surface within rosettes and ridges where the walls are drawn by up to several hundred per cent. Even the surface of an unstable propagating crack (area IV, enlarged in Figs. 8.32 and 8.33) is not smooth but plasticly deformed throughout. The morphological details of the fracture surface zones

Fig. 8.25. Fracture surface of notched standard HDPE broken under impact loading at 20 °C (taken from [106]).

Fig. 8.26. Enlargement of area I in Fig. 8.25 (taken from [106]).

clearly indicate that crack initiation — even under impact loading — does not simply occur by cutting those molecular chains which happen to traverse the planes of largest tensile stresses. Due to the limited lateral load transfer between unoriented, uncrosslinked chains (cf. Chapter 5) plastic deformation always precedes eventual chain breakage.

207

Fig. 8.27. Enlargement of the cellular area II in Fig. 8.26 (taken from [106].

Fig. 8.28. Higher magnification of a section of area II (Courtesy E. Gaube, Frankfurt-Hoechst).

Fig. 8.29. Apparently smooth area between the rosettes visible in Fig. 8.28 (taken from [106]).

Fig. 8.30. Voids from area III in Fig 8.26 (Courtesy E. Gaube, Frankfurt-Hoechst).

Fig. 8.31. Higher magnification of the indicated section in Fig. 8.30. (Courtesy E. Gaube, Frankfurt-Hoechst).

Fig. 8.32. Enlargement of area IV in Fig. 8.26, the plasticly deformed surface created by an unstably propagating crack (Courtesy E. Gaube, Frankfurt-Hoechst).

Fig. 8.33. Higher magnification of a section of Fig. 8.32. (Courtesy E. Gaube, Frankfurt-Hoechst).

On the basis of the foregoing discussion of the phenomenology of impact fracture it can be said that the impact resistance of a material is increased by all molecular properties or processes which assist the distribution and dissipation of the mechanical energy and permit the attainment of a large deflection before inception of rapid crack propagation. In particular the importance of the relaxation behavior [91, 93–96], of high molecular weight [106], of partial orientation [108], and of the unoccupied volume [92] have been emphasized.

A *maximum fractional unoccupied volume* $(\bar{f})$ was defined by Litt et al. [92] through the densities of the amorphous (ρ_a) and the crystalline state (ρ_c) as:

$$\bar{f} = (\rho_c - \rho_a)/\rho_c. \tag{8.15}$$

The $\bar{f}$ values of common polymers at room temperature vary between zero and 0.12. Litt found a good — although not perfect — correlation between $\bar{f}$ and impact strength. He indicates that polymers with high impact strength all have $\bar{f} > 0.07$.

Considerable qualitative evidence has been cited in the literature (e.g. 88–105) that the intensities of the low-temperature relaxation peaks deriving from main-chain motion (β and γ peaks) are related to impact toughness. A certain correlation has to be expected since an increased molecular mobility
— reduces the level of stored elastic energy at a given sample deformation and load rate,
— prolongs the phase of thermal crack growth, i.e. retards the inception of an unstable crack,
— and increases the energy expenditure during crack propagation.

Although all three effects tend to increase impact toughness a correlation will be limited by the fact that the relaxation behavior is determined in the linear anelastic range of stresses and at small strain rates. These conditions are quite different from those of an impact test. A quantitative discussion also has to be based on a fracture mechanical analysis of crack initiation and propagation (cf. Chapter 9).

The impact resistance of a polymer is reduced by all factors which lead to a general or local increase of stored elastic energy at a given deformation *but are not accompanied by an overproportional increase in strength*. Thus the *stress concentrating effect* of notches, flaws, or inclusions in an otherwise unchanged polymer considerably lowers its impact resistance. Increasing the *degree of cross-linking* beyond the point which insures the loading of all chains only leads to the formation of short, tightly anchored chain segments. Such segments are the first to be overloaded and scissioned upon straining. The small impact resistance of fully cured thermosetting resins underlines this point. Reinforcing thermoplastics by *short* random fibers increases their rigidity more effectively than their strength which leads to a net decrease of impact resistance.

A considerable toughening is to be expected, if strength and deformation can be made to increase simultaneously. Such an effect is realized through the partical orientation of an un-oriented, brittle polymer. For PS drawn to $\lambda = 3.4$ Retting [108] reports increases in tensile strength from 47 to 80 MN/m^2 and of strain at break from 7 to 22%. The IUPAC Working Party "Structure and Properties of Commercial Polymers" studied systematically the effect of orientation of various polystyrene samples (homopolymers and rubber-modified polystyrenes) on optical and mechanical properties [109, 110]. It was observed that the impact strength a_n of unnotched homopolymer specimen increased with draw ratio from about 3 at $\lambda = 1$ to more than 50 kJ/m^2 at $\lambda = 6$ if the molecular orientation was parallel to the main specimens dimension. In a perpendicular direction a_n decreased from 3 to 0.3 kJ/m^2 for the homopolymer and from about 9 to 2 kJ/m^2 for a commercially available rubber-modified polystyrene [109, 110].

The principal effect of chain backbone properties on impact strength has become clear. A chain does not so much contribute to impact strength through the level of stress ψ_b it supports at the moment of chain scission or disentanglement but rather through the energy dissipated before ψ_b is reached. Chains are loaded in shear which necessitates the displacement of chains with respect to each other. A maximum of energy dissipation is obtained, therefore, if the intermolecular shear stresses are *large but insufficient to break a chain* (cf. Eq. 5.41) and if the chains do not easily disentangle so as to insure chain slip over large volume elements (Fig. 8.28).

B. Failure Under Constant Load

Subjecting a specimen to a constant load evidently can have one of three results:
— the specimen breaks upon application of the load
— the specimen fails after some time
— the specimen carries the load for an indefinite period of time.

Generally the third result is technologically desirable. In order to define the region of permissible loads a study of the delayed failure, i.e. of the time-dependent strength, is indispensable.

Owing to its fundamental importance the time-dependent strength of load-bearing polymers has been extensively investigated with temperature as the principal parameter. In Figures 1.4, 1.5, and 3.7 stress-time-temperature diagrams of various thermoplastics are represented. The phenomenon of a delay between load application and final failure has found numerous interpretations. Among these interpretations one group is based on *purely statistical considerations.* There the time-to-failure t_b is inversely related to the probability that a certain detrimental event takes place in an *otherwise unchanged material.* Such events are e.g. the formation of a void of sufficient size or the coincidence of local fracture events. The statistical aspects of fracture have been discussed in Section II of Chapter 3.

A second group of interpretations is based on *reaction rate theory* and on the *gradual exhaustion* of the polymer load carrying capability due to the breakage and/or displacement of elements and/or to the formation of defects (cf. Section IV of Chapter 3). At this point the role of molecular backbone chains in delayed fracture of non-oriented polymers is to be elucidated.

The presence of long and flexible chains causes a polymeric material to show a monotonic, partly anelastic deformation under constant load, viz. a creep deformation. In statistical theories of fracture generally no special consideration is made of the extent of creep deformation. From Section IV of Chapter 3 it may be recalled that the kinetic theory of Zhurkov and Bueche also does not contain creep deformation as a variable. The theory of Hsiao-Kausch, developed for solids not showing large scale anelastic deformation introduces a time-dependent strain which derives, however, as a *consequence* of the gradual network degradation. Bueche and Halpin expressly consider the macroscopic creep function to reflect the creep behavior of molecular filaments which in turn would effect the time-to-rupture. In their theory the retarding response of a rubbery or thermoplastic matrix causes the delay — by t_b — in the growth of a crack nucleus to a critical size.

The mode of creep failure in thermoplastic materials is not unique. As an example the creep fracture morphologies of PVC tubes subjected to different stress levels have been reproduced in Chapter 1. At higher stresses ($\sigma_v = 50$ MN/m^2) PVC tubes fail in a brittle manner after little creep deformation, and one speaks of *delayed brittle fracture* (Fig. 1.1). At medium stresses (42 MN/m^2) and after prolonged times the tubes show large scale plastic deformation, i.e. *delayed yielding* (Fig. 1.2). At lower stresses ($\sigma_v < 20$ MN/m^2) failure does either not occur at all within experimental time scales or by a competing mechanism, the *formation of a creep crack* (Fig. 1.3).

In brittle fracture the time-to-break t_b is determined by the time to initiate a rapidly propagating crack at a flaw or material inhomogeneity. The role of chains in this process will be discussed in Chapter 9.

In ductile failure t_b is equivalent to the time it takes to develop a local yield zone. For this to occur it is necessary that the stress level be insufficient to initiate a running crack at a flaw or material inhomogeneity *before* the homogeneous anelastic deformation of the sample volume deactivates possible stress concentrators. A transition from a ductile to a brittle fracture mode occurs, therefore, if the temperature is lowered and/or the load rate is increased sufficiently. The structural weakening associated with the continued creep deformation sets the stage for the eventual local yielding. The circumferential creep strain of the afore mentioned PVC pipe at $\sigma_v = 42$ MN/m^2 is plotted in Figure 8.34. The curve clearly represents the characteristic parts of a creep curve: the instantaneous (elastic) deformation ϵ_0, a primary phase with a decreasing

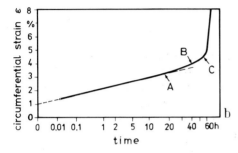

Fig. 8.34. Circumferential creep strain ϵ of a PVC pipe at $\sigma_v = 42$ MN/m^2 and T = 20 °C (after [109]). (a) linear time scale, (b) $\epsilon - \epsilon_0 \simeq (t/h)^{0.23}$. A: limit of primary creep phase, B: inflection point, C: appearance of stress whitening.

strain rate, a secondary phase where the strain rate is constant and a third phase of accelerated creep. During the latter phase creep rates are large, the material approaches the yield condition. Such a condition is generally reached first within the most highly stressed material, i.e. at the smallest sample cross-section.

In order to predict the creep behavior and possibly the ensuing failure a number of approaches have been proposed. These are based respectively on the theory of viscoelasticity — including the concept of free volume — or on empirical representations of $\epsilon(t)$ or of the creep modulus $E(t) = \sigma_0/\epsilon(t)$. The framework of the linear theory of viscoelasticity permits the calculation of viscoelastic moduli from relaxation time spectra and their interconversion. The reduction of stresses and time periods according to the time-temperature superposition principle frequently allows establishment of master-curves and thus the extrapolation to large values of t (cf. Chapter 2). The strain levels presently utilized in load bearing polymers, however, are generally in the non-linear range of viscoelasticity. This restricts the use of otherwise known relaxation time spectra or viscoelastic moduli in the derivation of $\epsilon(t)$ or $E(t)$.

The limits of linearity are discussed in a review article by Yannas [112]. He concludes that for practically all polymers at T-T$_g$ < – 20 K a strain smaller than 1% forms the limit of linear response. For semicrystalline polymers (e.g. PP, PAN, PETP, PA 66) a limit of 0.1 to 0.4% seems to be valid also above the glass transition temperature T$_g$ (even at T – T$_g$ > 100 K).

For amorphous polymers (PC, PIB, NR) the limit increases around T$_g$ from 1 to about 50% of strain [112]. There are some special approaches to a calculation of creep and recovery based on non-linear theories of viscoelasticity [112–114]. At present it is not possible, however, to predict with sufficient generality the onset of accelerated creep and thus of delayed yielding from non-linear viscoelastic theories.

213

Promising results have been obtained, however, using the free volume concept to account for the shift of relaxation times with strain and thermal history and to predict changes of failure stress and failure mechanism [123, 124].

Empirical approaches to represent the creep curves of materials by mathematical functions have been known for more than 60 years. One of the first is the Andrade equation

$$\epsilon(t) - \epsilon_0 = \beta(t/t_0)^{1/3} + Kt \tag{8.16}$$

where $t_0^{-1/3}$ and K are adaptable constants [115]. This equation — originally derived for metals — describes the *uniaxial* creep of different materials such as Pb, Cu, Fe, Cd, PMMA, PA, PVC within certain time periods very well [111, 115, 116]. Various empirical modifications of this equation have been reported [116]. Thus the *circumferential* creep is described by a similar potential law

$$\epsilon(t) - \epsilon_0 = \beta(\sigma_0, T)(t/t_0)^n. \tag{8.17}$$

For PVC tubes the exponent n was found by several investigators to be about 0.23 [111]. For the dependency of β on σ_0 a hyperbolic sine function has been frequently proposed in the literature (e.g. 111, 116). For PVC tubes at $20\,^\circ C\,\beta$ assumes the form [111]:

$$\beta = 0.019 \sinh(\sigma_0/8.85\ \text{MN m}^{-2}). \tag{8.18}$$

Equations (8.17) and (8.18) permit the calculation of $\epsilon(t, \sigma_0)$ at $20\,^\circ C$. They thus also permit one to determine at which stress a specified strain value ϵ_1 is being reached after very long times. One finds that a strain value $\epsilon_1 = 3\%$ — at which the creep rate of moderately stressed PVC tubes ($\sigma_v = 35$ to 48 MN/m^2) changes from phase I to phase II — will be reached at $\sigma_0 = 22$ MN/m^2 only after 50 years. An eventual fracture should take place even later than that.

A still simpler and more accessible approach to predict the creep strain of engineering components is based on a logarithmicly linear approximation of the creep modulus, $E(t)$:

$$\log[E(t)/E(t_0)] = \frac{1}{3}\log(t/t_0)\log[E(1000\ h)/E(1\ h)]. \tag{8.19}$$

Values of the 1-h and 1000-h moduli are generally given in the technical literature (e.g. 104). At $20\,^\circ C$ and at small stresses (5 to 20 MN/m^2) the ratios of $E(1000\ h)$ $E(1\ h)$ of common polymers are of the order of 0.96 (SB), 0.92 to 0.93 (PC, PETP, fiber reinforced thermoplastics), 0.88 to 0.90 (PS, PVC, PMMA, POM), and 0.72 to 0.79 (HDPE, PP, ABS). Whereas extrapolation methods are frequently the only means available they are subject to all the limitations introduced by the gradual development of structural weakening (demonstrated by the change from a decelerating creep rate to an accelerating one). Also the action of competing processes (creep crack development) is generally not taken into account by this extrapolation method.

The development of a creep crack is the dominating failure mechanism of non-oriented thermoplastics at smaller stresses (see diagrams Fig. 1.4 and 1.5). The fracture surface morphology of a typical creep crack is reproduced in Figure 1.7. The final failure of that sample had been caused by crack growth over the whole wall thickness.

The pattern of creep-crack failure shows the three well-known phases: a period t_i of apparently homogeneous deformation up to the point of crack initiation, a period t_a of thermally activated slow crack growth, and a final period t_r of unstable crack growth. An inspection of loaded samples during and after the creep experiment reveals a number of morphological features:

- creep cracks initiate at flaws or material inhomogeneities, more often in the surface zones (Figs. 1.7, 8.35a) but also in the bulk material (Fig. 8.35b) [111, 117, 118];

Fig. 8.35. Fracture surface of creep cracks in LDPE tubes (Courtesy of P. Stockmayer, [118]). (a) crack initiation at inside boundary of tube wall, $\sigma_v = 2$ MN m^{-2}, T = 80 °C, $t_b = 140$ h, (b) crack initiation within tube wall, $\sigma_v = 2$ MN m^{-2}, T = 80 °C, $t_b = 7082$ h.

- the initiated crack grows through thermal activation during a period t_a under apparently constant conditions; this is evidenced by the appearance of the "mirror zone", the frequently ideally circular or semicircular, smooth interior fracture surface (Figs. 1.7, 8.35b); depending on the stress level the mirror zone may extend over almost the complete wall thickness of tubular specimens;
- at higher magnifications of the mirror zones it can be seen that the surfaces are not smooth on a molecular level; in the case of amorphous PVC an irregular, rough, and scaly surface is observed (Fig. 8.36) and in the case of semi-crystalline PE a cellular, locally highly drawn surface morphology becomes apparent (Fig. 1.8);
- inspection of transparent tubular samples under load and of non-transparent samples after an interruption of the creep test showed that in PVC and PE creep crack development started rather late in the experiment, i.e. t_i was larger than 0.90 t_b [111, 117]; in PP tubes the first creep cracks could be detected at t_i being about 0.5 t_b [117].

Before developing a hypothesis as to the origin of creep crack formation it must be discussed which information can be obtained from an evaluation of the stress-time-

Fig. 8.36. Sections of the mirror zone of a creep crack in rigid PVC (Courtesy E. Gaube, [107]). σ_v = 40 MN m^{-2}, T = 20 °C, t_b = 4300 h; (a) Section near the external surface where the crack originated; (b) Section near the internal surface where the crack terminated.

temperature diagrams (Figs. 1.4, 1.5, 3.7, and 3.11). The essential feature certainly is that within a large interval of time the logarithm of the time-to-break t_b decreases linearly with increasing uniaxial or principal stress σ_0. Any one of these straight lines can be expressed by

$$\ln t_b/t_0 = A - \alpha\sigma_0 \tag{8.20}$$

where t_0, A, and α are constant with respect to time and stress. As discussed in Section IV of Chapter 3 various authors have proposed that these "constants" have a more general significance. Most approaches work with an Arrhenius equation:

$$\ln t_b/t_0 = (U_0 - \gamma\sigma_0)/RT. \tag{8.21}$$

In this form U_0, γ, and t_0 are supposedly independent of temperature. The experimental data of Savitskii et al. [119] plotted in Figure 3.7 verify this point for a cellulose nitrate film and the polyamide 6 monofilaments investigated. For each material only one set of constants is necessary to describe all data points. For cellulose nitrate the authors obtained U_0 = 159 kJ/mol, γ = 644 · 10^{-6} m^3/mol, t_0 = 10^{-13} s and for the oriented polyamide 6 (Capron) U_0 = 184 kJ/mol, γ = 113 · 10^{-6} m^3/mol, t_0 = 10^{-13} s.

In the case of PVC tubes one obtains from an evaluation of Figure 1.4 according to Eq. (8.21) values of U_0 = 397 kJ/mol, γ = 1740 · 10^{-6} m^3/mol and (a purely formal value of) t_0 = 1.7 · 10^{-52} s. It should be noted that this set of parameters describes the PVC fracture times notwithstanding the fact that these data belong to three different fracture modes. In the stress-time curves of non-oriented semicrystalline polymers (PE, PP) one observes a drop of strength at larger values of t_b, the well-known bend (Fig. 1.5). The data of the flat portions (relating predominantly to ductile failure) can be represented by

$U_0 = 307$ kJ/mol, $\gamma = 4390 \cdot 10^{-6}$ m^3/mol, and $t_0 = 3 \cdot 10^{-40}$ s

and the data of the steep portions (creep crack failure) by

$U_0 = 181$ kJ/mol, $\gamma = 3610 \cdot 10^{-6}$ m^3/mol, and $t_0 = 8 \cdot 10^{-20}$ s.

For oriented semicrystalline polymers Zhurkov et al. [18] report:

HDPE $U_0 = 113$ kJ/mol, $\gamma = 109 \cdot 10^{-6}$ m^3/mol, $t_0 = 10^{-11}$ s
LDPe $U_0 = 117$ kJ/mol, $\gamma = 247 \cdot 10^{-6}$ m^3/mol, $t_0 = 10^{-12}$ s
PP $\quad U_0 = 125$ kJ/mol, $\gamma = 213 \cdot 10^{-6}$ m^3/mol, $t_0 = 10^{-12}$ s.

These formal evaluations need discussion. The good correspondence between U_0 and the dissociation energy of the weakest main chain bonds for many of the (mostly oriented) polymeric materials investigated by him has led Zhurkov [120, 18] to establish his kinetic theory of fracture. This theory assumes that all important steps contributing to the fracture of a material are not only related to but even controlled by the breakage of primary bonds, hence by U_0 and t_0. As important steps chain scission, microcrack formation, and crack propagation through material separation have been considered [3]. The effect on fracture behavior of structure, orientation, dimensionality of the state of stress, morphology, and homogeneity of chemically "identical" polymers must be expressed, therefore, through the single parameter γ.

The parameter γ has the dimension m^3/mol and is termed activation volume. With reference to Chapter 3, Section IV, it should designate the thermally activated volume if the stress σ_0 indeed acts on this volume directly and without intermediates. This is approximately the case in flow but not in chain scission. The scission of a chain necessitates molecular stresses ψ_b which are clearly much larger than the applied external stresses σ_0. For PA 6 molecular stress concentration factors ψ_b/σ_0 of between 10 and 40 are commonly determined (cf. Chapters 7 and 8 above). Taking $12 \cdot 10^{-6}$ m^3/mol for the activation volume β of the chain scission process a parameter $\gamma = \beta_\psi/\sigma_0$ of 120 to $480 \cdot 10^{-6}$ m^3/mol should be predicted. The activation volume of $12 \cdot 10^{-6}$ m^3/mol amounts to $(0.27$ nm$)^3$ or to a stretching of the rupturing bond by 0.1 nm.

The effect of the structural parameter *chain orientation* on γ becomes apparent from Figure 3.11. There the stress-lifetime data of PAN fibers [74] are plotted with draw ratio λ as a parameter. The highest draw ratio ($\lambda = 17.3$) corresponds to a γ of $248 \cdot 10^{-6}$ m^3/mol, $\lambda = 4$ to 590, $\lambda = 2.62$ to 841 and $\lambda = 1$ to $\gamma = 1200 \cdot 10^{-6}$ m^3/mol. The γ values are the smaller the more homogeneously the macroscopic stress is distributed over the molecular chains.

The foregoing discussion shows that the γ values of PMMA, PVC, and PE can hardly be explained on the basis of chain scission unless extremely large stress concentration values are postulated. In the case of PVC the γ corresponds to $(1.7$ nm$)^3$ which points to the involvement of larger domains in the activation of the creep process. The fact that the PVC fracture data of all modes are described by one equation seems to indicate that all modes depend on the creep deformation. Once a certain structural weakening has resulted from creep the final failure is accomplished by one of the three modes, e.g. by brittle fracture at large stresses, ductile failure, or creep crack development. Since the final period accounts only for a small fraction of t_b no effect on the slope of $\ln t_b/t_0$ is noted.

This is different in the case of semicrystalline polymers which form a "ductile and a brittle branch" of the stress-lifetime curve. There it can be said that two different mechanisms are active, of which the creep crack initiation has the apparently smaller activation energy (181 kJ/mol) and activation volume $(1.8 \text{ nm})^3$. The fact that chain breakages in PE are rarely observed even at high stresses and low temperatures in highly oriented samples makes it very unlikely that the mechanism of creep crack initiation involves chain breakages.

Kagan et al. [121] have studied the effect of supramolecular structure on the time-dependent strength of HDPE. They found a good correlation between the 1-h (plastic) creep strength and the yield strength for materials of different crystallinity, density $(0.945 < \rho < 0.960 \text{ g cm}^{-3})$ crystallite size, and spherulite diameter. These parameters hardly affected the activation volume γ but slightly the activation energy of the plastic deformation process (parallel displacement of the *ductile branch* of the σ-log t/t_0 curve). On the other hand they found a clear tendency to higher *long-term brittle strength* (resistance against creep crack formation) with an increase in density and size of crystallites and with a reduction of spherulite diameter. Gaube et al. [117] also report that an increase in crystallinity (i.e. density) of PE, PP, PETP, POM, PA leads to a higher yield strength — at lower strain values — and to a higher ductile strength. They point out, however, that the decrease of brittle strength in the steep region begins sooner, the higher the crystallinity and the lower the molecular weight. The ultrahigh molecular weight polyethylene GUR does not show creep crack formation at all. Judging from these observations the process of creep crack formation seems to be related to the *gradual disentanglement of chains* and the ease of void opening in the inter-crystalline and/or inter-spherulitic regions. Both mechanisms must be fairly independent of creep deformation. The fact that generally creep cracks are only observable shortly before they cause the final failure indicates that these cracks grow at considerable rates once they are initiated.

Morphological effects on the initiation of creep cracks were also studied by Stockmayer [118]. Using an elaborate peeling technique he transformed whole tube sections into thin films of up to 0.06 mm thickness (Fig. 8.37). The walls of LDPE tubes, transformed into films of 0.9 m^2 area, then were scanned under a microscope for flaws or material inhomogeneities (Fig. 8.38). Inhomogeneities could be detected in every tube and have been related to an imperfect blending of the LDPE with carbon black [118]. Stockmayer also found two interesting correlations:

— creep cracks were initiated preferentially in those zones which also contained (the most) inhomogeneities
— tube lifetimes were greatly reduced if the creep cracks had been initiated in the outside or inside zones of the wall; from an ensemble of tubes tested at 80 °C and having an average lifetime $\bar{t}_b$ of 2340 h tubes with inhomogeneities (and crack initiation) in the boundary zones failed after a $\bar{t}_b$ of 740 h; tubes with crack initiation in the central region of a wall lasted to $\bar{t}_b = 7400$ h, i.e. a factor of 10 longer [118, 149].

Fig. 8.37. Part of a film cut from LDPE tube wall showing inhomogeneous blending in extrusion (Courtesy P. Stockmayer, [118]). Thickness 0.06 mm; width 50 mm.

Comments are necessary concerning the significance of t_0 and σ_0 in Eq. (8.21). Values of 10^{-12} to 10^{-14} s have been frequently determined for t_0; they have been assigned to longitudinal vibrational modes of the backbone chain. For PVC and PE extremely small values are obtained which certainly cannot be interpreted in the same manner. With flow processes it must be recognized that the activation energy is not temperature independent but decreases with temperature:

$$U_0 \, (T + \Delta T) = U_{00} + \frac{\partial U_0}{\partial T} \, \Delta T. \tag{8.22}$$

In this case one has

$$\ln t_b/t_0 = \frac{\partial U_0}{\partial T} \, \frac{\Delta T}{RT} + \frac{(U_{00} - \gamma \sigma_0)}{RT}. \tag{8.23}$$

Fig. 8.38. Section of a film cut from LDPE tube wall showing spots which are free from carbon black. (Courtesy P. Stockmayer, [118]).

219

A value t_0^d formally determined from experimental data according to Eq. (8.21) contains, therefore, the two terms t_0 and the temperature-independent part of

$$\frac{\partial U_0}{\partial T} \frac{\Delta T}{RT} .$$ The latter term seems to be responsible for t_0^d values of 10^{-40} s and less.

The term σ_0 in Eq. (8.21) has been taken to represent the macroscopic uniaxial stress σ_0 in the event of uniaxial stretching — or the largest principal stress component σ_v in the event of multiaxial states of stress. This does not mean that the times to fracture of tubular and uniaxial specimens are always identical if only σ_v is equal σ_0. As discussed above the creep functions in these two cases are described by different potential laws, namely by Eqs. (8.16) and (8.17) respectively.

A systematic study of the effect of the state of stress on the lifetime of PVC tubes has been carried out by Smotrin et al. [151]. They found that at short times to fracture (at stresses of 50 MN m^{-2}) the simple Rankine criterion $\sigma < \sigma^*$ represented their failure data in two-dimensional stress space. With increasing times to fracture, however, the von Mises criterion gave the better representation. Gotham [152] studied the uniaxial creep failure of 15 different polymer materials at 20 °C. Within a time range of up to 10^7 s he observed brittle failure with injection molding grade PMMA, PS, SAN, and glass-fiber reinforced PA 66, and ductile failure for PP, bulk cast PMMA, PC, PSU, PVC, ABS, POM, PA 66, and P4MP.

Concluding this section one may say that there is very little evidence that the loading and breaking of chains controls the creep of unoriented polymers. The softening mechanism — common to all linear polymers [122, 128] — responsible for the change from decelerated to accelerated creep is generally related to segmental motion and changes in intermolecular attraction but not to a reduction of chain length or strength. A search for free radical formation in creep of non-oriented polymers carried out by Jansson et al. [125] is as yet also inconclusive. These authors had included radical scavengers in PVC which led to a considerable reduction in creep rate (by a factor of up to about 10). They state, however, that it has not been possible to separate the stabilisation effect from the effect of mechanical reinforcement [125].

C. Homogeneous Fatigue

1. Phenomenology and Experimental Parameters

Polymeric engineering materials subjected to strong mechanical and environmental excitation show — as any other material — a gradual degradation of their performance including eventual failure. If the property changes are mostly due to *chemical reactions* one speaks of *corrosion,* or of *radiative degradation.* The term *fatigue (Zerrüttung)* is used if a deterioration of material properties is caused by the repeated cyclic or random application of mechanical stresses. The synergistic interaction of mechanical and environmental attack leads to the phenomenon of environmental stress corrosion (ESC). The role of molecular chains in ESC, and in corrosive, thermal, and radiative degradation will be discussed in the final Section of this Chapter. The fatigue of polymeric fibers has already been treated in Sections I F and G of this Chapter. The present Section, therefore, will deal mainly with the principal aspects of fatigue and with isotropic thermoplastics. The fatigue of polymeric engineering materials has received a growing attention in the last thirty years. Review articles on this complex subject have appeared from time to time; the reader is referred to the ones authored by Andrews [126] and more recently by Manson and Hertzberg [127–128], Oberbach [129–130], and Schultz [131]. In addition, a large number of research papers have been published of which an incomplete listing is given in this Section [116, 132–147].

It is the general observation that failure in repeated loading occurs at load levels which are lower than the stresses sustained under the conditions of static loading (creep) or of monotonously increasing deformation (drawing). The lower the stress level to which a material is subjected the larger the number N of load cycles which are sustained. The total time t_f, however, which a fatigued sample spends under load is generally much smaller than the time to failure under static loading conditions. Load reversals or load interruptions have an accelerating effect, therefore, on the loss of load bearing capability; load reversals or interruptions constitute the elements of fatigue. It may already be stated at this point that the accelerating effect of load reversals has to be related to two characteristic material properties:

— the anelastic response which leads within each loading cycle to a dissipation of mechanical energy as heat and
— the occurrence of molecular rearrangements which may lead to the disentanglement and slip of chain segments, to local reorientation, and/or to the formation of voids.

It has to be expected, and must be recognized, therefore, that three principal mechanisms may contribute to fatigue failure: *thermal softening, excessive creep or flow,* and/or the *initiation and propagation of fatigue cracks.*

The experimental parameters of a fatigue test have been aptly defined by Andrews [126]:

— a periodically varying stress system having a characteristic stress amplitude, $\Delta\sigma$;
— a corresponding periodic strain amplitude, $\Delta\epsilon$;
— a mean stress level, σ_m;
— a mean deformation, ϵ_m;
— a frequency, ν; $\omega = 2\pi\nu$;
— a characteristic wave form (sinusoidal, square, etc.) for both the stress and strain;
— ambient and internal temperatures, which in general will not be precisely identical;
— a given specimen geometry, including notches, if any.

More recently the (computer-programmed) random loading has also found growing application in testing structural components [132]. Of course, the above parameters can be combined in different ways and be changed within a test. They partially depend upon each other and on environmental conditions.

The experimental equipment will not be described here, the reader is referred to the cited review articles or pertinent research papers (e.g. 132–138). In practice five different types of stress-strain control are employed which are classified by Andrews [126] and Manson et al. [127] as:

— periodic tensile loading (between fixed stress limits, in tension or compression); [133–135]

221

- periodic loading (between fixed strain limits, in tension or compression); [133–
  –135]
- reversed bending stress (implicit in flexing a sheet in one dimension); [136]
- reversed bending stresses in two dimensions (by rotary deflection of a cylindrical
  specimen); [137]
- reversed shear stresses, e.g. obtained by torsional deformation. [138]

2. Thermal Fatigue Failure

In cyclic loading of a viscoelastic solid an amount of energy ΔW is dissipated per cycle. For sinusoidal loading in the anelastic range, one obtains

$$\sigma(t) = \sigma_0 \sin \omega t \tag{8.24}$$

$$\Delta W = \pi \sigma_0 \epsilon_0 \sin \delta. \tag{8.25}$$

This hysteresis heating leads to a noticeable temperature rise. At low test frequencies and low stress levels the temperature increase of the test specimen generally approaches a finite value. A PA 6 sample for instance fatigued at 50 Hz, at a constant stress amplitude of 8 MN m^{-2}, and at an ambient temperature of 21 °C, assumed a stable temperature of 27 °C after some 10^4 cycles [139–140]. At this temperature mechanical energy input and thermal energy loss were in equilibrium.

At higher frequencies ω, higher ambient temperatures, or higher stress amplitudes, however, the energy input into a sample becomes larger than the heat which can be transferred out of the sample through conduction or radiation. Depending on the experimental setup, failure of the sample then occurs through thermal softening and/ or plastic flow. The stress and frequency range, where this type of thermal failure has to be expected, can be calculated from the energy input, the geometry, and the thermal properties of a sample. Thermal equilibrium in a cylindrical sample of radius R is reached once

$$\frac{\partial T(r, t)}{\partial t} = \frac{\omega \sigma_0^2 \sin \delta}{2 \rho c_p E'} + \frac{\lambda}{\rho c_p} \left(\frac{\partial^2 T}{\partial r^2} + \frac{1}{r} \frac{\partial T}{\partial r} \right) \tag{8.26}$$

approaches zero. The highest temperature, T_e, observable at equilibrium in the center of the cylinder then amounts to

$$T_e = T_a + \left[\frac{R}{2h} + \frac{R^2}{4\lambda} \right] \frac{\omega \sigma_0^2 \sin \delta}{2E'}. \tag{8.27}$$

T_a designates the ambient temperature, h the coefficient of convection at the sample surface; and the constants ρ, c_p, and λ have their usual significance as density, specific heat, and heat conductivity of a material. The term $\lambda/\rho c_p$ is also known as temperature conductivity number. For conventional polymeric materials this number assumes values between $1.0 \cdot 10^{-7}$ m^2/s (PTFE) and about $2.1 \cdot 10^{-7}$ m^2/s (HDPE, POM).

222

It should be mentioned at this point that the hysterises heating ΔW in a fatigue test with *constant-strain amplitude* decreases with increasing temperature, since ΔW is proportional to $\sigma_0 \sin \delta$, i.e. to E''. Under this condition thermal equilibrium can generally be established. The same effect, a decrease of E'', can of course be reached if a sample is plasticized. The plasticization of an engineering component subjected to fatigue at constant-strain amplitude can be, therefore, a suitable way to prolongue the fatigue life of the sample [152]. Machyulis et al. [152] point out that the thermo-stabilization, the fatigue-stabilization, and the plastization effects of various additives on polyamide are not related to each other. For optimal results on fracture energy and cycles to failure Kuchinskas et al. recommend diffusion stabilization from a 5% ethanolic quinhydrone solution at 343 K for 1 h [152].

The conditions of thermal failure are especially discussed by Oberbach [129–130] for a variety of polymers, by Zilvar [139–140] for PA 6, and by Alf [141] for PMMA and unsaturated polyester resins.

It is only through their effect on the above constants and on the complex modulus E^* that the molecular chain structure influences thermal failure. In the context of this monograph this type of failure will, therefore, not be discussed any further.

3. Wöhler Curves

The common observation that the number of cycles to failure, N_F, depends on the stress or strain amplitude is generally represented in the form of Wöhler curves (Wöhlerkurven, $S - N$ curves). Such a representation accounts for the fact that stress (or strain) amplitude and number of load cycles are the most important of the many (and closely interacting) parameters which affect the fatigue life.

The fatigue data represented by Figure 8.39 were obtained from fully reversed, uniaxial tension-compression tests at 0.1 Hz with temperature rises T_e-T_a below 2 K [142]. The three different regions of the $S - N$ curve of polystyrene are fairly typical for the fatigue-life curves of other glassy polymers as well. *In region I* crazes form during the first tensile quarter cycle. The fatigue behavior is essentially that of a pre-crazed sample. Failure occurs through the initial slow growth of a crack through craze material followed by catastrophic crack propagation. The number of cycles to failure N_F is strongly influenced by the stress amplitude, presumably through the strong dependency of the number of crazes on the stress level.

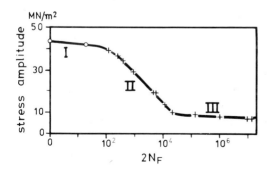

Fig. 8.39. Wöhler curve of polystyrene subjected to fully reversed uniaxial tension-compression at 0.1 Hz at ambient temperature (after [142]). N_F number of cycles to failure.

Region-II fatigue fracture is characterized by the fact that a period of craze nucleation precedes craze growth and the formation, slow growth, and catastrophic propagation of a crack. This type of fatigue fracture is observed at stress levels just below the immediate craze initiation stress σ_i. The dependency of N_F on σ is considerably reduced. This accounts for the delay in craze initiation as well as for the reduced rate of slow crack growth at lower stress levels. The slope ($\simeq 14$ MN m^{-2} per decade of N_F) seems to be characteristic for a variety of polymers [142, 153].

The long-life *region III* essentially forms the endurance limit of the material. The dependency of N_F on σ is extremely small. Crack initiation has a very long incubation period. Analysis of the fracture surfaces of fatigue failures in this region reveals furthermore that the slow crack growth mode has two phases. The first shows little evidence of crazing, the second resembles the region-II failure mode.

The above considerations concern polystyrene fatigued by completely reversed stress ($\sigma_m = 0$, $\Delta\sigma = 2\sigma$). In principle similar observations are also made with other polymers and at other conditions of stress, e.g. under constant maximum stress at various stress amplitudes [127, 153]. At this point is should be emphasized that the "fatigue process" under conditions II and III comprises at least two phases: an initial phase of crack or craze initiation with an apparently *homogeneous* response of the material and a second phase of *heterogeneous* fatigue with localized and growing material discontinuities (fatigue cracks). The second phase in turn comprises several modes of crack propagation (thermal and athermal growth, catastrophic failure). This will be discussed in more detail in Chapter 9.

The discrimination between homogeneous and heterogeneous is certainly useful although not necessarily always significant. Most of the *kinetic theories of failure* treated in Chapter 3 encompass the two phases of crack initiation and slow crack growth. Thus they implicitly assume that the entity of molecular rearrangements between the application of load and the end of the slow crack growth period can be described by one constitutive equation.

4. Molecular Interpretations of Polymer Fatigue

Different experimental observations permit the conclusion that the prolonged periods of craze or crack initiation at lower stresses are not simply due to the reduced probability for crack nucleation in an otherwise unchanged material. The nature of the changes occurring on a molecular level during sample fatigue have been investigated by several authors [e.g. 138, 143–147, 153]. Thus molecular mobility, interaction, and damping have been studied by the torsion pendulum [138, 134–144] and by infra-red technique [138], molecular packing and defect structure by density measurements and X-ray scattering technique [144–146] and by the use of samples of different molecular weight [153], and morphological changes by the recombination kinetics of trapped free radicals [146]. The results obtained from these different experimental methods characterize the molecular rearrangements, but they do not yet permit the establishment of quantitative fatigue limits.

Through simultaneous application of several of these methods Sikka [144] has been able to characterize some molecular rearrangements in homogeneous fatigue. He fatigued in cyclic extension thin films (0.075 mm thick) of polystyrene (Tricite) and polycarbonate. Subsequently he investigated these films by Fourier transform infrared (FTIR) and mechanical spectroscopy and by X-ray diffraction technique. The fatigued PS samples were scanned under an electron microscope to detect any crazes that might have developed due to fatigue. No crazes were observed in the samples fatigued up to 2500 cycles.

The FTIR results obtained on a PS film after 2500 cycles are shown in Figures 8.40 and 8.41. In the *superimposed* FTIR spectra slight distortions of IR bands are noticeable in the frequency range of 900 to 1700 cm^{-1} (Fig. 8.40). To detect minute changes caused by the fatigue process Sikka also recorded *computer-sub-*

Fig. 8.40. Superimposed FTIR spectra I_n of unfatigued (——) and I_f of fatigued (– – –) polystyrene film (from [144]).

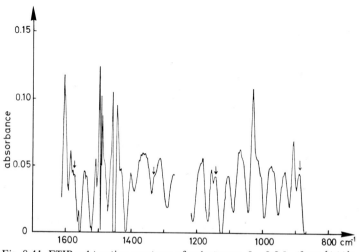

Fig. 8.41. FTIR subtraction spectrum of polystyrene, $I_f - 0.9\, I_n$ (from [144]).

tracted FTIR spectra. One such spectrum (fatigued-90% unfatigued) is plotted in Figure 8.41, the changes observed in this spectrum are listed in Table 8—IV.

From X-ray diffraction data Sikka obtained a mean Bragg distance d_{Bragg} of about 0.48 nm for the unfatigued PS films which showed a decrease of $\simeq 0.01$ nm in the fatigued sample (2500 cycles). This negative shift in d_{Bragg} was believed to be conclusive. It was related to the decrease in interphenyl and intraphenyl distances. The fatigue to 2500 cycles also resulted in a change of dynamic mechanical losses [144]. Figure 8.42 shows a plot of tan δ versus temperature obtained at a frequency of 3.5 Hz.

Table 8.4. Fatigue-affected IR Bands of Polystyrene (from Sikka [144]).

IR Band Frequency (cm^{-1})	Observation
540	Multiple splittings
906	Part of the band shifts to lower frequency at 888 cm^{-1}. Shift amount is -18 cm^{-1}.
1154	Part of the band shifts to lower frequency at 1145 cm^{-1}. Shift amount is -9 cm^{-1}.
1310	Part of the band is contributed by fatigue process.
1376	Part of the band appears at a higher frequency of 1400 cm^{-1}. Shift amounts to 24 cm^{-1}. Rest of the band is resolved into bands at 1352, 1358, 1365, 1371, 1376 cm^{-1}.
1448	Shows strong splitting.
1478	Two additional bands appear at 1489 and 1509 cm^{-1}.
2850	Weak splitting observed.
3062	Medium intensity splitting observed.
3083	Medium intensity splitting observed.

Fig. 8.42. Tensile mechanical loss tangent of polystyrene at 3.5 Hz (from [144]). 1. unfatigued sample; 2. sample fatigued to 2500 cycles.

The FTIR results seem to indicate that the fatigue process has caused changes in the absorption frequencies of various vibrational modes of the *phenyl side groups.* Also the IR bands associated with bending and stretching modes of CH_2 groups and bending modes of CH groups show frequency shifts and band distortions. These results suggest that fatigue has modified the internal and external environment (intrachain and interchain interactions) of PS chains.

On the basis of these assignments and of his X-ray studies Sikka proposes that during fatigue weaker Van der Waal's and possibly some backbone bonds are broken. This has a twofold effect. The loosening of the structure permits on one hand the PS chains to locally rearrange themselves in a more perfect manner and to reduce the average intraphenyl and interchain distances. The local rearrangements are, on the other hand, accompanied by an increase in free volume in interstitial areas between the domains of improved order.

In his experiments with polycarbonate films Sikka [144] used cyclic tensile stresses between 50 and 91% of the yield stress. In the upper stress range the stress-strain curve was clearly non-linear. He employed the same experimental techniques as mentioned before to study molecular fatigue effects. The (occasionally minute) shifts in the IR spectrum are made apparent in the computer-subtracted FTIR spectrum (Fig. 8.43). Sikka's vibrational assignments of the affected bands are compiled in Table 8.5. They indicate a stretching of some backbone chains. X-ray diffraction again revealed change of the mean interchain distance. It amounted to a decrease of 1.25% after 5800 cycles under a tensile stress reaching 83% of the yield stress σ_y. No effects of fatigue on orientation could be detected in such samples by birefringence measurements. The densities decreased by a fraction of up to 0.083% to 1.1966 g/cm^3.

Fig. 8.43. FTIR subtraction spectrum of polycarbonate, $I_f - 0.9 \, I_n$ (from [144]).

Table 8.5. Changes due to Fatigue in IR Bands of Polycarbonate (from Sikka [144])

IR Band (cm^{-1})	Type of Changes Observed	Assignment
560	Multiple splitting	Out of plane bending vibration of the p-disubstituted ring
768	Splitting	Non-coordinated bands
886	Appearance of a shoulder	Non-coordinated bands
1012	Multiple splitting	A$_1$ species modes of the benzene ring
1078	Splitting	Asymmetric C–C stretch of central carbon atom
1150–1268	Absorbs on the negative absorbance side	C–O valence vibration of $$\overset{\backslash\quad\quad\quad\;\;/\!/}{\underset{/\quad\;\;\overset{\|}{O}\quad\;\backslash}{C\!-\!O\!-\!C\!-\!O\!-\!C}}\text{ group}$$
1362	Weak splitting	Aliphatic –CH valence vibration
1408	Very weak splitting	Non-coordinated bands
1502	Multiple splitting	In plane ring vibrations of the phenyl rings
1775	Multiple splitting	C=O stretching vibration
2968	Splitting	conformation around central C atom of the repeat unit has an effect on this band.

Some dynamical mechanical loss data are represented in Figure (8.44). The α and β relaxation peaks in fatigued samples appear at lower temperatures in comparison to the corresponding peaks in unfatigued samples. In all cases the samples fatigued to large numbers of cycles (> 2000 cycles at stress levels $\sigma_{max}/\sigma_y > 0.56$) fractured during the dynamical mechanical tests.

Fig. 8.44. Tensile mechanical loss tangent of polycarbonate at 3.5 Hz (from [144].

Also in this case one has to conclude that under fatigue stresses chains are stretched and locally packed more closely whereas the overall free volume increases.

The observations discussed in detail above seem to be representative for the findings of other authors as well. Bouda et al. [138] report a *decrease of density* with cycle number for PA 6 and PMMA which they ascribe to the *formation of microvoids*. The *decrease of heat capacity* in PMMA in the temperature range 350 to 400 K, the decrease of tan δ at 93 K (1 Hz) and an *increase in shear modulus* G' in the range 100 to 250 K (where no crazes had formed) seemed to indicate a *non-uniform local volume contraction* during the fatigue process.

Schrager [134] investigated epoxy resins of different degrees of cure. A fully cured sample showed a monotonous decrease of G' and an increase in tan δ. Notable is a complete and persistant recovery of G' and tan δ within 20 h. An incompletely cured sample responded to fatigue loading by an initial rise of G' (indicating completion of chemical cure simultaneously to chain scission in other parts of the matrix). After some $20 \cdot 10^3$ cycles chain scission and a decrease of G' dominated.

Wendorff [145] studied fatigue-induced density fluctuations in polyoxymethylene (Hostaform T 1020) by X-ray scattering. He reports several interesting structural details of these defects:

– a constant concentration of about 10^{12} cm^{-3}
– a size growing with stress amplitude and cycle number
– a shape resembling that of a rotation ellipsoid (with axes 2a = 2 to 6 nm, 2b = 9 to 50 nm)
– the defects have a compliance some 3 to 5 times larger than that of the bulk material
– the defects show viscoelastic behavior and a size distribution
– annealing close to the melting temperature leads to partial healing; the defects reappear if cycling is resumed
– form and size of the defects depend on crystallization conditions.

These observations indicate that the defects are microvoids appearing preferentially at interlamellar boundaries oriented perpendicular to the loading direction. Similar defects are observed in homogeneous deformation during *static* loading at those strains where the stress-strain curve begins to deviate from linearity [145].

Whereas Wendorff states that fatigue does not affect the crystalline regions of POM Nagamura et al. [146] report a change of the crystalline mosaic block structure of HDPE. They arrived at this conclusion by analysis of the trapping and decay behavior of γ-ray induced free radicals.

Measurable macroscopic parameters which have been related to the progress of material fatigue are creep strain and strain rate [72, 116, 122, 123, 147]. Mindel et al. [122] studied the creep rate as a function of strain in pure compression of polycarbonate. They found that fatigue loading becomes effective through an increase of strain rate after each interruption of loading. Since the strain level after which accelerated creep began remained constant (at 8.8%), fatigue lives were reduced. Tensile creep has frequently been held responsible for the fatigue failure of polymers. In 1942 Busse et al. [72] suggested this mechanism for polyamide, cotton and rayon. Brüller et al. [147] stated cyclic creep strains were predictable from the Boltzmann superposition principle.

Sauer et al. [153] studied particulary the effect of molecular weight on fatigue of PS and PE under different alternating stresses. They observed that an increase in molecular weight dramatically increases the fatigue life of PS (an increase of the number of cycles by a factor of 10 for an increase of M by a factor of 5). This effect is partly assigned to the retardation in craze initiation because of the fewer number of chain ends and the larger number of entanglements in high molecular weight samples. The major contribution derives, however, from the increased resistance of the crazed material to rupture (cf. Chapter 9.II).

Comparing the fatigue behavior of PE and PS Sauer et al. [153] conclude:
- in both crystalline polymers, like polyethylene, and amorphous polymers, like polystyrene, increases of molecular weight lead to increased fatigue lifetimes, for a given imposed alternating stress, and to increased endurance limits,
- the ratio of the estimated fatigue endurance limit strength to the nominal static fracture strength for specimens prepared from narrow molecular weight polystyrene standards rises appreciably with increasing molecular weight, from a value of 0.19 for a molecular weight of 160,000 to a value of 0.30 for a molecular weight of $2 \cdot 10^6$.
- if fatigue failure due to excessive thermal heating is avoided, the resistance of polyethylene of medium or high molecular weight to alternating loading is considerably greater than that of polystyrene, based on static stress values to produce either yield or fracture. Thus, at comparable molecular weights of about $2 \cdot 10^6$, polyethylene gives an estimated ratio of endurance limit strength to yield stress of about 0.90 as compared to the 0.30 ratio for polystyrene,
- the greater resistance of crystalline PE to fatigue type failure as compared to amorphous PS is attributed primarily to two characteristics. The first concerns the phase of homogeneous fatigue. Because PE has a much lower amorphous-phase T_g and consequent greater molecular and chain mobility of its amorphous chains, it is much less sensitive to surface or volume flaws; and secondly, because of its heterogeneous nature, with crystallites imbedded in an amorphous matrix, it is also much more resistant to fatigue crack *propagation* (cf. Chapter 9).

D. Yielding, Necking, Drawing

Shear yielding, i.e. the onset of large-scale intersegmental displacements within a non-oriented thermoplastic polymer, has a distinct effect on the stress-strain curve. In tensile tests a drop in engineering stress is usually observed and the *yield point* defined as the point of maximum load (Fig. 2.10, curves b and c). In other tests, e.g. in a compression test, there may or may not be a load drop but a sudden decrease of $d\sigma/d\epsilon$ may be noticed. The important phenomenon of yielding has been intensively investigated. Review articles have appeared almost annually in recent years in the general literature [e.g. 114, 154–164].

The classical continuum mechanical criteria for shear yielding have been discussed in Chapter 3, Section III A in terms of the three-dimensional state of stress. At this point molecular interpretations are reviewed in an attempt to assess the role of molecular backbone chains.

230

Experiments at different temperatures, strain rates, and hydrostatic pressures have established that shear yielding is a thermally activated process [154–168, 170–173]. According to Eyring's theory of flow (Chapter 3) strain rate can be expressed as

$$\dot{\epsilon} = A \exp - (U_0 - \gamma\sigma)/RT. \tag{8.28}$$

This gives the following dependence of yield stress σ_y on strain rate and temperature:

$$\frac{\sigma_y}{T} = \frac{U_0}{\gamma T} + \frac{R}{\gamma} \ln \frac{\dot{\epsilon}}{A}. \tag{8.29}$$

This equation was found to be valid for a number of polymers (PVC, PC, PMMA, PS, CA) in more or less extended regions of temperature and strain rate [154, 156, 158]. The (temperature-dependent) activation volumes γ had at room-temperature values between 1.4 (PMMA) and 17 nm^3 (CA). This means that according to this concept polymer deformation at the yield point is due to the thermally activated displacement of molecular domains over volumes which are between 10 (PMMA) and 120 times (PVC) as large as a monomer unit. It has been indicated by several authors [155–158, 160] that the above criterion (Eq. 8.29) corresponds to the Coulomb yield criterion $\tau_0 + \mu p$ = constant. The coefficient of friction μ is inversely proportional to γ. From an analysis of their experimental data on polycarbonate according to Eq. (8.29) Bauwens-Crowet et al. [158] conclude that two flow processes exist. They relate these to an α-process (jumps of segments of the backbone chains) and to the β mechanical relaxation mechanism.

Argon and Bessonov [161, 162] have recently presented a molecular model of shear yielding which involves the rotation of molecular segments against intra- and intermolecular forces and the systematic, pairwise reduction of kinks in a strained polymer. They calculate the free enthalpy of activation of a double kink in a small bundle of collectively acting molecules, ΔG^*, as

$$\Delta G^* = \frac{3\pi G \omega^2 a^3}{16(1-\nu)} \left[1 - 8.5 (1-\nu)^{5/6} \left(\frac{\sigma}{G} \right)^{5/6} \right], \tag{8.30}$$

where G is the shear modulus, ν Poisson's ratio, ω the angle of bundle rotation during the activation, and a the mean bundle radius. Having examined a large number of glassy polymers Argon and Bessonov find that the bundles comprise between about one (PMMA, PS), two (PPO), three to four (PC, PETP) or seven (PI) molecular chains. The distances between double kinks along the molecule range from 0.8 to 1.6 nm [162].

All of these proposed interpretations of yielding state that yielding is brought about by a displacement of neighboring chain segments under change of conformation. In the process of yielding of unoriented thermoplastics no large axial chain stresses are built up and no chain scission has ever been observed in strains *below* the yield strain ϵ_y. Yielding is equivalent to the onset of large scale orientational deformation. It is generally accompanied by a decrease of the material resistance against deformation, by a reduction of sample cross-section in a plane perpendicular to the direction of tensile plastic deformation and by a temperature increase due to the dissipated mechanical work. Material and thermal weakening at constant or increasing true stress give rise to a plastic instability. In uniaxial tensile straining this instability manifests itself as

— shear banding [155—157],
— neck formation [165—175], or
— homogeneous drawing [176—182].

The phenomenon of crazing also can be regarded as a (although localized) deformational instability. It will be discussed in the final chapter.

The large scale "plastic" deformation associated with yielding quite naturally gives rise to shear and extensional deformation of molecular coils. The response of molecular chains in solution to shear or extensional flow has been described in Chapter 5IIE. It has been stated that chain breakages are obtained if strain rates are sufficiently large (generally above 600 s^{-1}). The strain rates in a yielding solid are considerably smaller. The local strain rates across the neck of a sample are typically 1 s^{-1} or less (Fig. 7.12). On the other hand molecular friction coefficients in a solid are so much larger than in a liquid that large forces will be transmitted once a chain molecule slips through the surrounding matrix. As discussed in 5II E axial forces sufficient for chain breakage will be built up if the slipping chains have a highly extended conformation. It must be expected, therefore, that all those chain segments rupture as a consequence of necking or drawing which are properly oriented *and* which initially do have a highly extended conformation or assume such a conformation at the early stages of the yield process. With regard to chain breakage in necking or drawing the total orientational strain λ and the drawing temperature T_d are the most important variables.

There are very few reports on ESR investigations of chain rupture during necking and drawing [21, 169, 174—177]. These reports will be discussed in the following and be supplemented by observations with other methods: the IR spectroscopy of chain-end groups [178] and the determination of molecular weight distributions [179].

Zaks et al. [169] investigated the neck formation in polycarbonate. As inferred from the reduction of sample cross-section in the neck, an orientational strain λ of about 2 had been imparted onto the material while "passing through the neck". At room temperature and at various sample stretching rates corresponding to strain rates in the neck of between 0.02 to 2 s^{-1}, the authors [169] observed a fairly stable but not very well resolved ESR spectrum. The intensity of this spectrum increased with the rate of passage of undrawn PC through the neck from $3 \cdot 10^{15}$ to $1.8 \cdot 10^{16}$ spins/g (Fig. 7.12). The authors also studied the behavior of stable nitroxide radicals and of radicals formed by photolysis during necking of LDPE and PC. The thus observed increased decay of the initially present radicals may be assigned to the reaction with newly formed radicals but, as well, to the strain-induced increased rate of recombination or decay of radicals already present. The existence of the latter phenomenon in highly oriented HDPE, PP, PA 12, and PETP was pointed out by Becht et al. [47].

The ESR technique had also been employed by Lishnevskii [174, 175] who found that during cold drawing of PP, PVC, PA 6, and aromatic polyamide fibers free

radicals were formed in a concentration of between $8 \cdot 10^{14}$ to $3 \cdot 10^{16}$ g^{-1}.
Matthies et al. [176] noted the intense coloring of different polyoxamide and hexa-
methylene isophthalamide fibers during cold drawing. They confirmed by ESR measurement that the formation of free radicals had given rise to the appearance of the color
centers. In the case of the polyoxamide fibers they associated these color centers
with the –NH–CO–CO–NH-group. The largest concentration of spins observed
was $1 \cdot 10^{16}$ g^{-1}.

The above ESR experiments [174–176] were carried out on cold drawn
samples immediately after completion of the drawing procedure. Thus the measured
radical concentrations [Ṙ] are integral values. Nothing can be said about the form of
[Ṙ(λ)], but, in all probability, most of the chain breakages have occurred during the
final stages of the drawing procedure, i.e. within moderately to highly oriented and
even microfibrillar structures. This assumption is supported by the ESR experiments
of Levin et al. [21] and Chiang and Sibilia [177] with fibers partially oriented at elevated temperatures.

The hot drawing of PA 6 at temperatures of e.g. 150 to 210 $^{\circ}$C permits one to apply
orientational strains of up to $\lambda = 6$ without breaking an observable number of chains
[21]. As discussed in detail in Chapters 7 and 8-I, free radicals are formed if such fibers
are stretched at room temperature. For a PA-6 fiber hot stretched to $\lambda = 3.7$ radical
formation began at an "additional cold" strain of 20%, i.e. at a total elongation of
$\lambda = 4.44$ [21]. PETP fibers hot stretched to $\lambda = 3$ showed some radical formation
at "cold" strains of between 5 and 20% but deformed by chain slippage thereafter –
up to a strain of 50%, i.e. to a total elongation of $\lambda = 4.5$ [177]. In HDPE films no
breakage of chains could be observed during cold drawing unless the draw ratio λ
exceeded 5 [178].

An infra-red study of the latter material also revealed [178] that the number N
of coiled rotary isomers (chains containing gg and gtg conformations) depends on
the draw ratio λ. While N (λ) is rather constant if $\lambda < 5$, N decreases linearly at higher
draw ratios to reach about $0.2\ N(1)$ at $\lambda = 12$.

That chain scission occurs during drawing of PA-6 multifilament yarn becomes
apparent from an investigation of the molecular weight. Using a fractional precipi-
tation technique Sengupta et al. [179] determined the molecular weight distribu-
tions of undrawn yarn and of yarn drawn at 35 $^{\circ}$C to $\lambda = 2.92$ and $\lambda = 4.20$ respectively. They observed that for the yarn drawn to $\lambda = 2.92$ the molecular weight
distribution is unimodal as in the undrawn fiber, but the peak is smaller and
slightly shifted toward smaller values of M. At $\lambda = 4.2$ this peak is further diminished and shifted and a second peak at low molecular weights (at less than half
the viscosity number) appears [179].

Indirect evidence of the likely occurrence of chain scission was proposed by
Davis et al. [170–173] who studied the deformation behavior of high molecular
weight low density PE (and of PTFE) at hydrostatic pressure. From the pressure
sensitivity of the rate of plastic deformation they derived a zero-pressure activation
volume of 0.266 nm^3 which they suggested to correspond to the activation volume
γ for the breakage of a PE chain.

From the presented evidence it must be concluded that during the initial stages
of necking and drawing ($\lambda < 3$) of semicrystalline polymers chain breakages do not

occur in large numbers. This conclusion is completely in accord with the Peterlin model of the "plastic deformation" of semicrystalline polymers. As described in 2.II B the early stages of plastic deformation comprise the destruction of an original spherulitic texture and its transition to the new microfibrillar structure. During the transition interspherulitic tie molecules are most highly loaded and some are expected to break. The ties between the crystal blocks, however, are generally not severed at this stage and are instrumental in establishing the microfibrillar structure.

If a semicrystalline polymer is carried from moderate to high extension ratios $(3 < \lambda < 10)$ then chain breakages can occur [21, 169, 174–178]. All of the mentioned breakages must be considered to be of the static-loading type (cf. 5 II B and D). They become possible in large numbers because the *presence of laterally rigid crystalline regions permits:*

— the static transmission of large axial stresses onto chains,
— the balancing of ensuing irregularities of stress distribution, and thus
— the widespread attainment of critical chain stresses before macroscopic break down.

In recent years ultra-high modulus fibers with draw ratios up to 35 have received increased attention [180–182]. Their highly extended chain structure and their extreme mechanical properties (elastic modulus E up to 67 GN m^{-2}, σ_b up to 1.7 GN m^{-2}) have been discussed in 2IC, and the techniques for their processing (drawing, hydrostatic extrusion, solution spinning) are outlined by Bigg [180]. As yet there is no information on the extent of chain breakage during drawing or stressing of these fibers. Capaccio et al. [181] point out that in very highly drawn PE ($\lambda = 30$) much finer fibrils are present and these are characterized by an amazingly high concentration of longitudinal discontinuities in the form of microfibril ruptures. Molecular defects (chain ends, folds) are certainly present in extended chain crystals — but not in quantities indicating major chain breakages during processing [180–182].

With the exception of PC non-oriented amorphous polymers have not given rise to measurable quantities of broken chains during yielding in tension. This behavior is the result of their different morphology. In the absence of crystallites, large axial forces leading to chain scission can only be caused by frictional loading of slipping chain segments. The volume concentration of chain breakages (because of the larger size of the slipping segments) is predictably much smaller than in semicrystalline polymers. In addition (and because of the absence of the equalizing effect of a microfibrillar substructure) macroscopic tensile failure begins before stresses and strains have reached a level sufficient for widespread chain scission.

During necking or homogeneous drawing of many (transparent) polymers an intensive whitening is observed. The origin of this phenomenon are voids which are either formed in a correlated manner within crazes (cf. 9 II) or within shear bands or which appear in an uncorrelated manner within the deforming volume elements. Uncorrelated voids form in semicrystalline polymers (PE, PP) as well as in amorphous ones (PVC) and in elastomers at cryogenic temperatures (BR, PI). Figure 8.45 shows an electron micrograph of a cross-section through the voided area which appeared during necking of a dry PA 6 monofilament [83]. The fact that the voiding starts in the center of the necking region and in dry monofilaments only, indicates that it is related to a triaxial state of stress. The coalescence of such voids may give rise to the

Fig. 8.45. Electron Micrograph of ultrathin cross-section through voided area of drawn 6 polyamide monofilament (Courtesy H. Hendus, [83]). Magnification 4800 x.

fibrillar fracture shown in Figure 8.22 [83]. From the triaxiality of the initiating stresses and from the absence of measurable intensities of free radicals in slightly drawn isotropic polymers ($\lambda < 3$), from the absence of conspicuous molecular weight changes [179, 220], and from the reversibility of the formation of holes it must be concluded that the hole formation depends more on conformational rearrangements of chains and on surface free energy than on the breakage of chains.

The material behavior described above referred mainly to uniaxial tensile deformation. It will be different in techniques of working or forming of polymeric materials, such as rolling, sawing or grinding. These processes involve compressive, tensile, and shear deformation in a generally complex manner. Breakage of chains has been observed in various ways. In 6 IV B and 7 IC 3 a detailed account of the nature and concentration of free radicals formed in comminution has been given. Other mechano-chemical aspects will be discussed in 9 III D.

E. Elastomers

It is evident that the role of highly coiled, loosely interacting but permanently cross-linked chain backbones in the rupture of elastomeric materials must be quite different from the role of chains in thermoplastics. The basic principles and notations of rubber elasticity have been presented in Chapter 2 (Section II A), different fracture models in Chapter 3, and a discussion of the entropy-elastic deformation of a single chain in Chapter 5 (Section I A). If one speaks of a "chain" in relation to a filled and cross-linked polymer network one refers, of course, to the part of a molecule between adjacent points of attachment (filler particle or cross-link). Thus chain rupture refers to the breaking of a chain at or between these points. It has been proposed by Mullins [183] some thirty years ago (and has since been known as Mullins effect) that the breakage of chains in a first loading cycle is responsible for the softening observable in subsequent load cycles. In that respect the phenomenology of the Mullins effect is comparable to that of the breakage of chains in the stretching of fibers.

The values of the entropy-elastic forces acting on highly extended 5-nm chain segments (cf. Table 5.2) indicate that they are by more than one order of magnitude too small to cause main chain bond breakage. For a chain of 125.5 nm length breaking stresses are reached if the chain is extended to an end-to-end distance of 124.7 nm which corresponds to a stretch ratio λ of 18. Rupture of most technically important elastomers takes place, however, at stretch ratios λ smaller than 8 [183–195]. As shown in Figure 3.6 and indicated in the literature the reduced ultimate stresses σ_b plotted versus elongation λ_b form a failure envelope. The elongation to break λ_b is shifted along the failure envelope (towards larger elongations) if the temperature and/or the degree of cross-linking decrease or if the strain rate increases [183–195]. The macroscopic breaking stresses σ_b are of the order of 1 to 30 MN m^{-2} (depending on degree of filling or cross-linking), they decrease with increasing time-to-failure and degree of swelling.

From these observations it has been quite unanimously concluded that an average network chain influences the deformation and breakdown of a stretched rubber in several ways:

— the originally random chain conformation changes to a more extended one at a rate determined by strain, temperature, and internal viscosity,
— in highly cross-linked or filled rubbers numerous (of the shortest) network chains have been completely extended and broken at the fracture elongation (cf. 7 II),
— additional physical cross-links are introduced through crystallization at high elongations,
— voids or cavities can be formed due to non-homogeneous distribution of strains in filled or unfilled elastomers,
— the limits of extensibility are reached once the majority of the network chains has reached a highly extended conformation. In the case of a homogeneously cross-linked network with n random links per network chain the largest breaking elongation $\lambda_{b\,max}$ will approach $\sqrt{n}$. At that strain all chains oriented in stretching direction are fully extended (cf. 5 I A).

Studying the influence of the aforementioned parameters on the load bearing capability of elastomeric networks one may say that forces are transmitted
— through the entropy-elastic deformation of network chains between cross-linking points,
— in the case of highly extended chains through their energy-elastic deformation (cf. 5 I),
— through shearing forces between segments.

The first mode of load transfer prevails in the rubber-elastic time-temperature region, and the third at very low temperatures [190].

For the breakdown of a stressed rubber two basic mechanisms are conceivable: the initiation and growth of a cavity in a moderately strained matrix or the accelerating, cooperative rupture of interconnected, highly loaded network chains. It has generally been concluded that the second mechanism is of dominating importance under experimental conditions, permitting the largest breaking elongation $\lambda_{b\,max}$ to be attained [187, 192, 194, 195]. For a network with chains extended in loading direction a dependence of $\lambda_{b\,max}$ on the inverse square root of the cross-link density ν_e is predicted. In experiment $\lambda_{b\,max}$ was found to be proportional to $(\nu_e)^{-0.5}$ to

$(\nu_e)^{-0.78}$ for a variety of networks (Butyl, BR, EPR, SBR, SI, fluorinated rubber). Smith [187] concludes that the maximum elongation $\lambda_{b\,max}$ is the only feature of large deformation behavior which clearly depends only on network topology.

At ultimate strains which are smaller than $\lambda_{b\,max}$ some chains are already highly extended but their number is insufficient for cooperative accelerating breakdown. It is the general opinion that under such experimental conditions the final breakdown occurs through formation and growth of cavities [184, 185, 190]. It will not be further investigated at this point to what extent chain scission may accompany cacity formation and growth. It may be stated, however, that the chain scission mechanism does not determine the rate of cavity formation.

As a final point it should be emphasized that elastomeric materials are showing a higher strength if they exhibit a multiphase microstructure [183, 184, 186, 188, 195]. This may be achieved by incorporation of suitable filler materials (carbon black, silica); by crystallization under strain, or by blending or copolymerization with an incompatible polymer. The possible role of chain orientation, loading, and scission in these cases has been discussed in 7 II.

III. Environmental Degradation

Within the framework of this book the effect of *thermomechanical* chain scission on the mechanical properties of polymers are to be investigated. Up to this point it has been tried, therefore, to separate and exclude *environmental* effects whenever possible. In many cases it was tacitly assumed that the variables investigated (e.g. strain, stress, temperature, sample morphology, free radical concentration) had been the dominating variables in comparison to e.g. humidity, oxygen content, chemical attack, or effects of irradiation. It goes without saying that these environmental factors are of extreme importance to the service lives of polymer engineering components. A considerable number of comprehensive recent monographs and handbook articles treat the environmental degradation of polymers (e.g. [196—203]). They deal in depth with those aspects of environmental degradation which cannot be discussed any further at this point: thermal degradation, fire and heat stability, chemical degradation, weathering, aging, moisture sensitivity, effects of electromagnetic and particle irradiation, cavitation and rain erosion, and biological degradation. For any detailed information on the above subjects and on methods and agents to protect and stabilize polymeric materials the reader is referred to the cited comprehensive references [196—203, 207—209].

It mainly remains to be discussed in this section the nature of the effect which the *simultaneous* action of mechanical and environmental parameters may have on mechanical properties. One of the most interesting phenomena in this field, the environmental stress crazing (ESC), will be treated in the final chapter.

A characteristic stress-induced degradation process is that of unsaturated rubbers in the presence of ozone. The rates of crack initiation and growth, of formation of free radicals, and of stress relaxation and creep are amplified up to and more than a thousandfold by the action of ozone [196—197, 199, 201, 204—206]. The chemical reaction is not completely understood. It is normally assumed that the first steps of ozone degradation of unsaturated polymers follow the "Criegee mechanism".

Under the effect of molecular ozone the π-bonds of olefines are polarized and ozone joins the double bonds to form primary ozonides [196, 197a]. Primary ozonides,

$$-CH-CH-, \text{ are unstable, they either isomerize into iso-ozonide, } -HC-O-CH-,$$
$$\underset{O_3}{\diagdown\diagup} \qquad\qquad\qquad\qquad\qquad\qquad\qquad \underset{O-O}{\diagdown\diagup}$$

or polymerize, or decompose into chains terminated by a zwitterion, $R_1 C^+OO^-$, and a carbonyl group, $R_2 CO$, respectively. If a bond attacked by ozone is under stress, decomposition can occur during the isomerization reation.

The reaction with ozone obviously is a surface reaction leading to the formation of a surface layer of ozonides and/or further reaction products. This layer grows in thickness in proportion to the square root of the time of exposure [199]. With increasing layer thickness ozone is gradually prevented from reaching undegraded rubber. The mechanical aspects of ozone cracking are extensively reviewed in the cited articles [196–197, 199, 201, 204–206]. It is unanimously reported that the degraded rubber material (natural rubber, SBR, acrylonitrile-butadiene, cis-polybutadiene) has a reduced strength and elasticity. Cracks open up and propagate at strains as low as 5 to 12%; it has been estimated [199] that even at the tip of a slowly propagating crack strains do not exceed 30% (as compared with 700% in an undegraded matrix). As a consequence of the chemical destruction of the material at the tip of a crack the mechanical energy expended in crack propagation (material resistance R) is exceedingly small. Employing the Griffith fracture criterion the various authors have obtained values of 0.05 to 0.12 J/m^2 from observation of macroscopic cracks [197a, 199a, 204, 205] or of 0.4 to 0.5 J/m^2 from micromorphological theories [206].

The formation of free radicals in unsaturated rubbers under simultaneous attack of ozone and stress is of particular interest. The primary steps of the above described reaction of ozone with unsaturated bonds do not lead to free radicals. In cis-polybutadiene (BR), natural rubber (NR), and acrylonitrile-butadiene (ABR), however, large quantities of peroxy radicals have been observed [206, 208]. As one possibility it was pointed out that these radicals may derive from the ozonides or zwitterions through unknown secondary steps possibly involving hydrogen abstraction or proton migration [197, 206, 208]. Another possibility certainly is that radicals are formed through scission of undegraded rubber molecules and that these primary radicals react with molecular oxygen. In BR and ABR the concentration of free radicals has shown the same dependency on strain and ozone concentration as the visual damage, i.e. the ozone cracking of the surface of the degrading rubber specimen. Notably the following observations may be mentioned [206, 208]:

– no (detectable quantities of) free radicals are formed below certain threshold strains (of between 5 and 12%)
– for strains beyond the threshold strain the rate of increase in free radical concentration increases initially linearly with strain and ozone concentration (in the range of 0.08 to 1.6 mg/l)
– increasing strain or ozone concentration lead to a preferential increase in number rather than size of microcracks.

238

The number of free radicals (up to $8 \cdot 10^{14}$ spins per cm^2 rubber surface) corresponds at a given ozone concentration of 2.8 mg/l to the number of ozone molecules in a zone of surrounding atmosphere of 0.02 cm depth. This number is very much smaller than the total number of ozone molecules available in the ambient atmosphere of the degraded specimens. The linear dependency of radical concentration on ozone concentration and strain, therefore, means that potential radical sites are created only after the application of stress and in the presence of ozone molecules. The appearance of free radicals simultaneously with crack opening certainly indicates the breakage of chains during the process of crack opening. In Section I E of this chapter an analysis of the energy content of an unsevered hydrocarbon chain strained elastically up to chain scission had been carried out. This analysis leads to the estimate that the $2 \cdot 10^{14}$ chains per cm^2 of new fracture surface can only be broken elastically if more than 9 J/m^2 were expended in forming that surface. Since the observed surface work parameters are much smaller one has to conclude that under ozone attack the breaking and radical forming chains break at small strains. This interpretation looks identical to that derived from analysis of the behavior of ozone cracks. Its importance lies in the statement, however, that during crack opening chains (of very low strength) are broken and not only broken chains are separated.

The breakage of ozone attacked chains is a conspicuous example of the synergistic effect of the simultaneous action of mechanical and environmental parameters. There are many other environmental parameters (e.g. moisture or oxygen content) which have an accelerating effect in a given situation [196–203]. Of these observations only a few will be discussed here namely the ones which characterize respectively the chemical aging of stressed rubbers [209c–210], the role of moisture in fatigue of PA 66 and PC [211–212], and the accelerating effect of UV irradiation on submicrocrack formation and breakdown of highly oriented polymers [74, 213–214].

The oxidative degradation of polymers has been recognized and studied for quite some time (cf. the extensive review 209a). Most investigations in this field deal with the effect of degradation on the properties of polymers (e.g. 189, 202, 203, 207, 209e) or with an elucidation of the nature and kinetics of the intervening reactions [200, 201, 207, 208, 209a–d]. As one of the first Kim [210] also searched for a possible stress effect on the rate of chain degradation [209d, 210]. She employed a constant-load technique which permitted to derive the relative number of network chains, N(t)/N(O), by comparing the Mooney-Rivlin plot, $\sigma_0/(\lambda-1/\lambda^2)$, of an aging sample with that of a fresh sample. Her interesting results may be summarized as follows:

– The creep rate of polybutadiene rubber films loaded in ambient atmosphere was about a thousandfold larger than the creep rate under dry nitrogen.
– A comparison of the relative potency of the chemical agents present in the ambient atmosphere showed that the few parts per hundred million of naturally occurring ozone had a much stronger effect than the oxygen.
– In ozone-containing air a behavior comparable to that described above for higher ozone concentrations had been found. A critical stress of 0.1 MN m^{-2} had to be exceeded in order to influence the initial creep rate. The (initial) rates of chain cleavage dN(t)/N(O)dt were found to be independent of stress within a stress range

of $0.1 < \sigma_0 < 0.5$ MN m^{-2} and amounted to $2 \cdot 10^{-6}$ s^{-1}. This phenomenon was explained in the same way as for higher ozone concentrations through formation and cracking of a degraded surface layer. Microscopical investigations of the sample surface confirmed such a statement.

– In the absence of ozone the rates of creep and chain scission were greatly reduced. The effect of oxygen *at room temperature* was small and hardly any differences between creep rates in dry nitrogen, dry air without ozone, and ozone-free air of 40% relative humidity had been observed. At *slightly higher temperatures* (49 °C), however, creep rates in dry air accelerated whereas those in inert atmosphere (Argon) remained constant.

– From her experiments Kim concludes that in the initial phase of degradation in ozone-free oxygen stress or strain have no effect on the rate of chain cleavage. After the chemical degradation had reached a stage, however, where cracks opened up under stress, the straining of a sample accelerated the failure process of the specimen through stress concentration and change in cross-linking behavior [210].

The general effect of cross-link density on the elastic modulus of an elastomer is indicated by Eq. 2.3. In their paper Landel and Fedors [189] consider the influence of a time-dependent cross-link density on the shape of the stress-strain curves of silicon, butyl, natural, and fluorinated rubbers. Introducing an additional shift-factor a_x related to the cross-link density, they were able to represent reduced breaking stresses as a function of reduced time in one common master curve.

Thermal and chemical degradation of cross-linked rubbers was also reviewed by Murakami [209c] who especially analyzed whether cleavage occurs along main chains or at cross-links. Depending on the nature of the chain and on that of the cross-links both mechanisms are observed. Natural rubber, radiation cross-linked to various degrees, showed at 20% strain and at temperatures of 80 to 130 °C random scission of network chains. An EPT rubber cross-linked by tetramethyl thiuram disulfide (TMTD) showed degradation at the junction points of network chains. There the S and S$_2$ linkages seemed to be preferred scission points. In TMTD-cured natural rubber both types of degradation occur simultaneously. The dependency of the stress relaxation curves on time, temperature, and initial cross-link density permitted discrimination between thermal and oxidative degradation. For differently cured natural rubber (peroxide-, irradiation-, TMTD-, and sulfur-cured) Murakami [209c] has tabulated the degradation mechanisms.

The accelerating effect of mechanical stresses on chemical stress relaxation clearly became apparent through the comparison of continuous and intermittent stress relaxation. The relative stress relaxation, $1 - \sigma (t, \lambda)/\sigma (O, \lambda)$, under continuous loading was the more pronounced the higher λ and always consistently larger than under intermittent load application. Murakami [209c] also discussed possible increases of relative stresses due to cross-linking reactions and their partial prevention through radical acceptors.

With respect to the general effect of an active liquid environment on a polymer reference to the literature must be made [196–199]. This applies especially to quantitative data concerning the reduction in strength accompanying the swelling or plasticization of a material. The penetration of low molecular weight molecules

240

into a polymer generally reduces the intermolecular attraction and facilitates chain slippage. In the case of hydrogen-bonded materials the possible interaction of the diffusing liquid with the hydrogen bonds has to be considered.

In a series of experiments Hearle et al. [211] studied the effect of air, water, hydrochloric acid, sodium hydroxide, and a laundering detergent on the strength and fatigue of PA-66 fibers. Their results confirm that the breaking strength of a PA-66 fiber immediately after immersion in water is 5% smaller than that in air. Pre-conditioning by immersion for 3 hours causes an additional strength reduction of 3%. Immersion in aqueous solutions of NaOH and HCl with ph-values larger than 2.5 has an effect similar to that of water. In the range of ph-values between 2 and 0, however, strength decreases linearly with decreasing ph up to a strength loss of 20%. Notable is the observation that the pre-conditioning of fibers before making a *fatigue test* (by rotation over a wire) does not have such a great effect on the fatigue life. This may be explained by the fact that the slow growth rates of a fatigue crack starting from the surface permit the environmental agent to always penetrate the crack growth region [211].

The partial penetration of a liquid or vapor into a matrix establishes concentration gradients which do lead to a direct mechanical action through non-uniform swelling or to indirect actions through the nonhomogeneous relaxation or distribution of stresses. These actions are even enhanced in the presence of thermal gradients and may cause rapid crack or craze formation. In the case of a slowly penetrating environment in a well entangled, homogeneous matrix the induced stresses generally can be accommodated elastically or viscoelastically. For example in polycarbonate sheets exposed to artificial weathering no cracking becomes apparent even after severe cycling of temperature and humidity [212]. Within a relatively short period of 30 to 32 months of outdoor weathering, however, a network of surface microcracks developed on the side exposed to the solar radiation. By comparison with artificial UV irradiation the authors [212] were able to show that the photochemical degradation of the surface layer introduces defects and lowers the strength of the polymeric material to such an extent that physically induced non-homogeneous stresses cause microcracking rather than being accommodated. The effect of liquid environment on crack or craze formation will be further discussed in 9 II D.

Photo-chemical degradation is probably the most important factor of environment. The cited monographs [196–203, 207–209] treat in detail the primary processes of photon absorption, electron excitation, energy transfer through excitons, luminescence, phosophorescence, and raditionless transitions, chain scission and free radical formation, secondary reactions, stabilization and protection, and the effect of irradiation on the mechanical properties. Here only one point can be touched: the accelerating effect of UV radiation on microcrack formation in loaded polymers [74, 213–214].

As indicated by Tables 4.4 and 4.5 photon energy in the spectral range of $250 < \lambda < 360$ nm is larger than the dissociation energy of C–C bonds. The high energy tail of solar radiation extends to about $\lambda = 300$ nm. Chain scission in UV and solar light is, therefore, energetically possible. In the near UV in unprotected glassy polymers quantum yields of the order of $1 \cdot 10^{-3}$ to $60 \cdot 10^{-3}$ are observed [209b].

The principle mechanisms of the synergistic action of load and UV radiation may be discussed using the few data available. The simultaneous application of tensile load and UV irradiation to oriented polymers clearly accelerated the formation of free radicals and/or of micro and macro cracks in PA 6 [213–214] and in natural silk, cotton, and "triacetate" fibers [213]. No effect was observed in PMMA [213]. The experiments on cotton and triacetate fibers revealed that at low tensile stresses ($\sigma_0 < 70$ MN m^{-2}) the UV irradiation reduced the lifetimes of the fibers by more than 4 orders of magnitude. Under these conditions the absence or presence of oxygen was of some, but minor, importance since the irradiation in vacuum led to only slightly longer lifetimes than the irradiation in air. In a range of stresses with $70 < \sigma_0 < 220$ MN m^{-2} no oxygen effect on the lifetime of triacetate fibers became apparent. In this stress range the irradiation effect decreased with increasing σ_0. For $\sigma_0 > 220$ MN m^{-2} the lifetime depended only on stress and not on the environmental factors UV irradiation and oxygen content. With cotton fibers a somewhat similar behavior was observed although the upper stress limit was smaller and depended on absence or presence of air [213]. The described behavior points to the existence of three failure mechanisms which are occurring simultaneously and at different rates: oxydation, UV degradation, and creep. An effect of oxydation was observed for the acetate fibers only at lifetimes $t_b > 5 \cdot 10^3$ s and in the presence of UV radiation. At shorter lifetimes, $100 < t_b < 5 \cdot 10^3$ s, failure was essentially due to irradiation. At very short lifetimes (and higher stresses) oxydation and UV degradation were irrelevant if compared to the stress effect.

The determination by X-ray scattering of the number of microcracks in a PA-6 fiber stressed in air at 128 MN m^{-2} gave the notable result [214] that the rate of microcrack accumulation increased almost instantaneously (from $5 \cdot 10^{16}$ m^{-3} s^{-1} to $110 \cdot 10^{16}$ m^{-3} s^{-1}) with the application of UV irradiation. The rate decreased to its initial value in the same abrupt and repeatable manner when UV was switched off after 10^4 s. Irradiation of an unstressed sample did not cause any microcrack formation and did not influence subsequent formation rates. In a previous study [213] it had been shown that UV irradiation of stressed PA-6 and natural silk fibers in a helium atmosphere increased the accumulation of free radicals. In that case the rate of radical accumulation at $200 < \sigma_0 < 600$ MN m^{-2} decreased with the length of the time of irradiation and reached a steady state concentration $N(\dot{R})$ after $5 \cdot 10^3$ s. In PA 6 at a stress of 600 MN m^{-2} the concentration $N(\dot{R})$ was of the order of 10^{24} m^{-3}; this is about the limiting concentration observed in purely stress-induced chain scission also.

From these results the following conclusions can be derived. The photodegradation of *unstressed* PA 6 in air most probably is a random oxidative chain scission process [215]. Based on a study of the formed ESR spectra Heuvel et al. [216] concluded that amide bonds in the amorphous regions are broken leading to R–ĊO and ṄH–R$'$ free radicals. Subramanian et al. [215] propose formation of hydroperoxides and keto-intermediates at the α–CH$_2$ group and a subsequent hydrolytic breakdown at the RCO–NHCOR$'$ bond. It must be assumed that in PA-6 in air under the combined action of UV and stress all chain segments in the amorphous regions are possible fracture sites and not only the most highly stressed ones. The rates of chain degradation and cross-linking may depend on chain stress through the mechanisms discussed below. In the absence of oxygen, UV irradiation accelerated the breakage of chains but did show a saturation at a concentration $N(\dot{R})$. This result can be explained if one assumes that in irradiated *and* stressed PA 6 preferentially those chains are broken and are prevented from recombination which are most highly stressed.

Several mechanisms are conceivable through which the UV degradation of highly extended and stressed chains would be facilitated. In the first place the decrease of bond dissociation energies through electronic excitation and tensile stress has to be mentioned (cf. 4 II B). In the case of ionized hydrocarbon chains, bond dissociation energies are as small as 100 kJ/mol (Table 4.6). As outlined in Chapter 7 bonds of such energy require rather small stresses for breakage (about 1/4 to 1/5 of the strength of a non-ionized chain). In addition to the reduction of the dissociation energy an increase in local thermal excitation — due to energy transfer or dissipation — may have an effect. Chain orientation and segment conformation also influence the effective stability of a chain. Very little is being said in the literature about the relation between molecular order and UV resistance. Reinisch et al. [217] investigated the influence of sample orientation on photo-chemical degradation. They found that highly drawn PA-6 fibers ($\lambda = 3.9$) had an even higher degradation resistance than undrawn material. Krüssmann et al. [218] indicate that the reactivity (in photooxidation) of N-vicinal methylen groups in trans conformation is 60 times that of those in gauche conformation. These data [213—218] were obtained for quite different specimens and experimental conditions. They do not yet permit a quantitative estimate as to the importance and interaction of the above mechanisms.

Electronic excitation, ionization, radical formation, oxidation, and cross-linking are also the principal processes occurring in polymer solids subjected to nuclear radiation (α, β, γ, nucleons). In view of the fact that the molecular mobility influences the kinetics of degradation and cross-linking a synergistic stress effect is conceivable but not yet proven. The current investigations aim at an understanding of the interrelation between irradiation characteristics (dosis and dosis rate), network structure, and macroscopic properties after irradiation [198, 200, 219].

References for Chapter 8

1. S. N. Zhurkov, A. Ya. Savostin, E. E. Tomashevskii: Dokl. Akad. Nauk SSSR *159*, 303 (1964). Soviet Phys. "Doklady" (Engl. Transl.) *9*, 986 (1964).
2. H. H. Kausch, J. Becht: Rheol. Acta *9*, 137 (1970).
3. V. R. Regel, A. I. Slutsker, E. E. Tomashevskii: "The Kinetic Nature of the Strength of Solids" (in Russian), Moscow: Isdat. Nauka, 1974.
4. S. N. Zhurkov, V. I. Vettegren', I. I. Novak, K. N. Kashintseva: Doklady Akad. Nauk SSSR *176*/3, 623—626 (1967). Soviet Phys. "Doklady" (Engl. Transl.) *176*/3, 708—711 (1967).
5. S. N. Zhurkov, V. I. Vettegren', V. E. Korsukov, I. I. Novak: Fizika Tverdogo Tela *11*/2, 290—295 (1969). Soviet Phys. Solid State *11*/2, 233—237 (1969).

6. S. N. Zhurkow, V. I. Vettegren', V. E. Korsukov, I. I. Novak: Fracture 1969, paper IV/47.
7. A. I. Gubanov: Mekh. Polimerov *3*/4, 608–614 (1967).
8. A. I. Gubanov: Mekh. Polimerov *3*/5, 771–776 (1967).
9. A. I. Gubanov: Mekh. Polimerov *4*/4, 586–594 (1968).
10. V. E. Korsukov, V. I. Vettergren', I. I. Novak: Mekh. Polimerov *6*/1, 167–170 (1970). Polym. Mech. (USA) *6*/1, 156–159 (1972).
11. V. A. Kosobukin: Mekh. Polimerov *6*/6, 971–978 (1970) Polym. Mech. (USA) *6*/6, 846–852 (1970).
12. V. A. Kosobukin: Fizika Tverdogo Tela *14*/9, 2595–2602 (1972) Soviet Phys. Solid State *14*/9, 2246–2251 (1973).
13. V. A. Kosobukin: Mekh. Polimerov *8*/1, 3–11 (1972) Polym. Mech. (USA) *8*/1, 1–7 (1972).
14. V. A. Kosobukin: Opt. Spektrosk. *34*/2, 273–277 (1973). Opt. & Spectrosc. (USA) *34*/2, 154–156 (1973).
15. V. I. Vettegren', I. I. Novak: Fizika Tverdogo Tela *15,* 1417–1422 (1973). Sov. Phys. Solid State *15*/5, 957–960 (1973).
16. V. I. Vettegren', I. I. Novak, K. J. Friedland: Int. J. Fracture *11,* 789–801 (1975).
17. S. N. Zhurkov, V. A. Zakrevskii, V. E. Korsukov, V. S. Kuksenko: Fizika Tverdogo Tela *13*/7, 2004–2013 (1971). Soviet Phys. Solid State *13*/7, 1680–1688 (1972).
18. S. N. Zhurkov, V. E. Korsukov: Fizika Tverdogo Tela *15,* 2071–2080 (1973). Soviet Phys. Solid State *15*/7, 1379–1384 (1974)
S. N. Zhurkov, V. E. Korsukov: J. Plym. Sic. (Polymer Phys. Ed.) *12*/385–398 (1974).
19. A. Ya. Savostin, E. E. Tomashevskii: Fizika Tverdogo Tela *12*/10, 2857–2864 (1970). Soviet Phys. Solid State *12*/10, 2307–2311 (1971).
20. V. A. Zakrevskii, V. Ye. Korsukov: Vysokomol. soyed. B*13,* 105 (1971). Vysokomol. soyed. A*14*/4, 955–961 (1972). Polym. Sci. USSR *14,* 1064–1071 (1972).
21. B. Ya. Levin, A. V. Savitskii, A. Ya. Savostin, E. Ye. Tomashevskii: Vysokomol. soyed. A*13*/4, 941–947 (1971). Polymer Sci. USSR *13,* 1061–1068 (1971).
22. S. I. Veliev, V. I. Vettegren', I. I. Novak: Mekh. Polimerov *6*/3, 433–436 (1970). Polym. Mech. (USA) *6*/3, 369–372 (1970).
23. U. G. Gafurov: Mekh. Polimerov *7*/4, 649–653 (1971). Polym. Mech. (USA) *7*/4, 578–581 (1971).
24. V. I. Vettegren', V. Ye. Korsukov, I. I. Novak: Plaste und Kautschuk *19*/2, 86–88 (1972).
25. V. E. Korsukov, V. I. Vettegren', I. I. Novak, A. Chmel: Mekh. Polimerov *4,* 621–625 (1972). Polym. Mech. (USA) *4,* 536–539 (1972).
26. V. Ye. Korsukov, V. I. Vettegren', I. I. Novak, L. P. Zaitseva: Vysokomol. soyed A*16*/7, 1538–1542 (1974). Polymer Sci. USSR *16,* 1781–1785 (1974). E. E. Tomashevskii, V. A. Zakrevskii, I. I. Novak, V. E. Korsukov, V. R. Regel, O. F. Pozdnyakov, A. I. Slutsker, V. S. Kuksenko: Int. J. of Fracture *11*/5, 803–815 (1975).
27. S. N. Zhurkov, V. S. Kuksenko, A. I. Slutsker: Fizika Tverdogo Tela *11*/2, 296–307 (1969). Soviet Phys. Solid State *11*/2, 238–246 (1969).
28. V. S. Kuksenko, A. I. Slutsker: Mekh. Polimerov *6*/1, 43–47 (1970). Polym. Mech. (USA) *6*/1, 36–40 (1970). V. M. Knopov, V. S. Kuksenko, A. I. Slutsker: Mekh. Polimerov *6*/3, 387–392 (1970). Polym. Mech. (USA) *6*/3, 329–333 (1970).
29. V. S. Kuksenko, V. S. Ryskin, V. I. Betekhtin, A. I. Slutsker: Int. J. of Fracture *11*/5, 829–840 (1975).
30. A. V. Amelin, O. F. Pozdnyakov, V. R. Regel: Mekh. Polimerov *4*/3, 467–473 (1968). Polym. Mech. (USA) *4*/3, 376–379 (1968). A. V. Amelin, O. F. Pozdnyakov, V. R. Regel, T. P. Sanfirova: Fizika Tverdogo Tela *12*/9, 2528–2534 (1970). Soviet Phys. Solid State *12*/9, 2034–2038 (1971). A. V. Amelin, Yu. A. Glagoleva, A. O. Podol'skii, O. F. Pozdnyakov, V. R. Regel, T. P. Sanfirova: Fizika Tverdogo Tela *13*/9, 2726–2731 (1971). Soviet Phys. Solid State *13*/9, 2279–2283 (1972).
31. E. E. Tomashevskii: Fizika Tverdogo Tela *12*/11, 3202–3207 (1971). Soviet Phys. Solid State *12*/11, 2588–2592 (1971). Yu. K. Godovskii, V. S. Papkov, A. I. Slutsker, E. E. Tomashevskii, and G. L. Slonimskii: Fizika Tverdogo Tela *13*/8, 2289–2295 (1972). Soviet Phys. Solid State *13*/8, 1918–1923 (1972).

32. E. E. Tomashevskii, E. A. Egorov, A. Ya. Savostin: Int. J. Fracture *11*/5, 817–827 (1975).
33. V. R. Regel, O. F. Pozdnyakov: Plaste u. Kautschuk *19*/2, 99–100 (1972).
34. D. K. Roylance, K. L. DeVries: Polymer Letters *9*, 443–447 (1971).
35. R. P. Wool: PhD Thesis, Univers. of Utah, Dept. of Materials Science and Engng., Salt Lake City, Utah 1974. R. P. Wool: J. Polymer Sci. *12*, 1575–1586 (1974).
36. R. P. Wool: J. Polymer Sci. *13*, 1795–1808 (1975).
37. R. P. Wool: J. Polymer Sci. *14*, 1921–1929 (1976).
38. R. P. Wool, H. H. Kausch: Unpublished results, Salt Lake City 1974.
39. R. P. Wool, W. O. Statton: In press.
40. K. K. R. Mocherla, W. O. Statton: Symposium for High Polymer Physics (1975), 1–13. K. K. R. Mocherla: PhD Thesis, Univers. of Utah, Dept. of Materials Science and Engng., Salt Lake City, Utah 1976.
41. H. Tadokoro et al.: J. chem. Phys. *42*, 4 (1965).
42. T. Miyazawa: J. Polymer Sci. C*2*, 59 (1974).
43. J. P. Luongo: J. Appl. Polymer Sci. *III*/9, 302–309 (1960).
44. J. Becht: Dissertation, Technische Hochschule Darmstadt, 1970.
45. J. Becht, H. Fischer: Kolloid-Z. u. Z. Polymere *240*, 766–774 (1970).
46. F. Szöcs, J. Becht, H. Fischer: Europ. Polymer J. *7*, 173–179 (1971).
47. J. Becht, H. Fischer: Angew. Makromol. Chem. *18*, 81–91 (1971).
48. H. H. Kausch, J. Becht: Deformation and Fracture of High Polymers, 317–333. J. A. Hassell, R. I. Jaffee, Plenum Press 1974.
49. K. L. DeVries, B. A. Lloyd, M. L. Williams: J. Appl. Physics *42*/12, 4644–4653 (1971).
50. B. A. Lloyd, K. L. DeVries, M. L. Williams: J. Appl. Polymer Sci. *10*/A-2, 1415–1445 (1972).
51. D. Klinkenberg: Dissertation, Technische Hochschule, Darmstadt, 1978.
52. T. Nagamura, K. Fukitani, M. Takayanagi: J. Polymer Sci. (Polymer Phys. Ed.) *13*, 1515–1532 (1975).
53. D. C. Prevorsek: J. Polymer Sci. A-2, *4*, 63–88 (1966).
54. H. H. Kausch, K. L. DeVries: Intern. J. of Fracture *11*/5, 727–759 (1975).
55. H. H. Kausch, J. Becht: Kolloid-Z. u. Z. Polymere *250*, 1048–1065 (1972).
56. W. O. Statton: J. Polymer Sci. C*32*, 219 (1971).
57. K. L. DeVries: J. Polymer Sci. C*32*, 325 (1971).
58. A. Peterlin: Internat. J. of Fracture *11*/5, 761–780 (1975).
59. F. H. Müller, A. Engelter: Kolloid-Z. u. Z. Polymere *149*/2-3, 126–127 (1956).
60. F. H. Müller, A. Engelter: Rheolog. Acta *1*, 39–53 (1958).
61. A. Engelter, F. H. Müller: Kolloid-Z. u. Z. Polymere *157*/2, 89–111 (1958).
62. F. H. Müller: Kunststoffe *49*/2, 67–71 (1959).
63. F. H. Müller, N. Weimann: J. Polymer Sci. C*6*, 117–124 (1964).
64. J. Stölting, F. H. Müller: Kolloid-Z. u. Z. Polymere *238*/1–2, 459–470 (1970). J. Stölting, F. H. Müller: Kolloid-Z. u. Z. Polymere *240*, 792–806 (1970).
65. W. Dick, F. H. Müller: Kolloid-Z. u. Z. Polymere *172*/1, 1–18 (1960).
66. W. Dick, A. Engelter, F. H. Müller: Rheol. Acta *1*/4, 6, 506–510 (1961).
67. F. H. Müller: J. Polymer Sci. C*20*, 61–76 (1967).
68. F. H. Müller, A. Engelter: Kolloid-Z. u. Z. Polymere *171*/2, 152–153 (1960).
69. F. H. Müller, J. Stölting: Kolloid-Z. u. Z. Polymere *240*, 790–791 (1970).
70. V. I. Vettegren', A. E. Chmel: Europ. Polymer J. *12*, 853–858 (1976).
71. A. C. Lunn, I. V. Yannas: J. Polymer Sci. (Polymer Phys. Ed.) *10*, 2189–2208 (1972).
72. W. F. Busse, E. T. Lessig, D. L. Loughborough, L. Larrick: J. Appl. Physics *13*, 715–724 (1942).
73. W. J. Lyons: Text. Res. J. *28*, 127 (1958).
74. V. R. Regel, A. M. Leksovskii: Int. J. Fracture Mechanics *3*/2, 99–109 (1967). V. R. Regel, A. M. Leksovskii: Polymer Mechanics *5*, 58–78 (1969) Mekhanika Polimerov *5*, 70–96 (1969). V. R. Regel: Mekhanika Polimerov, *7*, 98–112 (1971).
75. V. P. Tamush: Polymer Mechanics *5*/1, 79–87 (1969). Mekhanika Polimerov *5*/1, 97–107 (1969).
76. J. W. S. Hearle, E. A. Vaughn: Rheologica Acta *9*/1, 76–91 (1970).

77. A. R. Bunsell, J. W. S. Hearle: J. Materials Sci. *6*, 1303–1311 (1971).
78. D. C. Prevorsek, W. J. Lyons: Rubber Chemistry and Technol. *44*/1, 271–293 (1971).
79. A. R. Bunsell, J. W. S. Hearle: J. Appl. Polymer Sci. *18*, 267–291 (1974).
80. V. Regel, V. P. Tamush: Mekhanika Polimerov *3*, 458–478 (1977).
81. J. W. S. Hearle, B. S. Wong: J. Text. Inst. *68*/3, 89–94 (1977).
82. S. F. Calil, J. W. S. Hearle: Fracture 1977, Vol. 2, ICF4, Waterloo, Canada, June 1977.
83. H. Hendus, E. Penzel: Chemiefasern/Textilindustrie (1976/6), 527.
84. Courtesy of H. Hendus, Ludwigshafen 1976.
85. J. W. S. Hearle: Proc. Textile Inst./Inst. Text. de France Conf. Paris (1975), 60–75.
86. Courtesy to J. W. S. Hearle, Manchester 1977.
87. J. W. S. Hearle: "An Atlas of Fibre Fracture". Text. Mfr. (1972), *99*, Jan., Feb., 14; March, 12; May, 20; Aug., 40; Sept., 16; Oct., 40; Nov., 12; Dec., 36; (1973), *100*, Jan., 24; March, 24; April, 34; May, 54; June, 44.
88. P. I. Vincent: Impact tests and service performance of thermoplastics, Plastics Inst., London, 1971.
89. C. B. Bucknall, K. V. Gotham, P. I. Vincent in: Polymer science, A. D. Jenkins (ed.), North-Holland Publ., Chapt. 10, 1972.
90. H. Oberst: Kunststoffe *52*/1, 4–11 (1962).
91. W. Retting: Kolloid-Z. u. Z. Polymere *210*/1, 54–63 (1966).
92. M. H. Litt, A. V. Tobolsky: J. Macromol. Sci.-Phys., B1(3), 433–443 (1976).
93. J. Heijboer: J. Polym. Sci. (C) *16*, 3755 (1968).
94. R. F. Boyer: Polym. Engng. Sci. *8*, 161 (1968).
95. J. A. Sauer: J. Polymer Sci. C*32*, 110–116 (1971).
96. P. I. Vincent: Polymer *15*, 111–116 (1974).
97. W. Retting: Materialprüfung *8*/2, 55–60 (1966).
98. H. Grimminger, G. Koch, J. Penzkofer, E. P. Petermann, W. Retting, J. Steinig: Kunststoffe *59*/6, 375–381 (1969).
99. K. Fujioka: J. Appl. Polymer Sci. *13*, 1421–1434 (1969).
100. H. H. Racké, T. Fett: Kunststoffe *64*/9, 481–487 (1974).
101. G. Hoff, G. Langbein: Kunststoffe *56*, 2–6 (1966).
102. H. Oberst: Kunststoffe *59*/4, 232–240 (1969).
103. W. B. Hillig: Impact Response Characteristics of Polymeric Materials, Techn. Inform. Series, General Electric Co., Report No. 76CRD271, New York 1976.
104. H. Saechtling: Kunststoff-Taschenbuch, 20. Ausg., 910–914, 516, 518, München: Carl Hanser 1977.
105. F. Ramsteiner: Kunststoffe *67*/9, 517–522 (1977).
106. a) E. Gaube, H. H. Kausch: Fracture Theories in Industrial Use of Thermoplastics and Glassfiber Reinforced Plastics, Intern. Conf. Fracture, Munich 1973, Vol. I, Pl VI-311.
106. b) E. Gaube, H. H. Kausch: Kunststoffe *63*/6, 391–397 (1973).
107. Courtesy to E. Gaube, Frankfurt-Hoechst.
108. W. Retting: Angew. Makromol. Chem. *58*/*59*, 133–174 (1977).
109. T. T. Jones: Effect of Molecular Orientation on the Mechanical Properties of Polystyrene, Rep. IUPAC Working Party, IUPAC Internat. Symposium on Macromolecules, 10–14 Sept. 1973, 41–57.
110. W. Retting: The Effect of Molecular Orientation on the Mechanical Properties of Rubber-Modified-Polystyrene, to be published in "Pure and Applied Chemistry".
111. H. Niklas, H. H. Kausch: Kunststoffe *53*/12, 886–891 (1963).
112. I. V. Yannas: J. Polymer Sci. *9*, 163–190 (1974).
113. D. W. Hadley, I. M. Ward: Reports on Progress in Physics *38*/10, 1143–1215 (1975).
114. S. S. Sternstein in: Treatise on Materials Science and Technology, J. M. Schultz, ed., Vol. 10, New York: Academic Press 1977, 567–569.
115. E. N. C. da Andrade: Proc. Royal. Soc. (London) A 84, 1 (1910).
116. Taprogge, R.: Kunststoff-Rundschau *15*/5–12, 3–50 (1968).
117. E. Gaube, G. Diedrich, W. Müller: Kunststoffe *66*/1, 2–8 (1976).
118. P. Stockmayer: Stuttgarter Kunststoff-Kolloquium 1977; Kunststoffe *67*, 470 (1977).

119. A. V. Savitskii, V. A. Mal'chevski, T. P. Sanfirova, L. P. Zosin: Polymer Sci. USSR *16*, 2470–2477 (1974). Vysokomol. soyed. A*16*/9, 2130–2135 (1974).
120. S. N. Zhurkov, B. N. Narzulayev: J. Techn. Phys. *23*, 677 (1953).
121. D. F. Kagan, A. M. Knebel'man, L. A. Kantor: Vysokomol. soyed A*14*/5, 1207–1214 (1972).
122. M. J. Mindel, N. Brown: J. Materials Sci. *9*, 1661–1669 (1974).
123. L. C. E. Struik: Polym. Eng. and Sci. *17*/3, (1977). L. C. E. Struik: Physical Aging in Amorphous Polymers and Other Materials Amsterdam/New York: Elsevier Scientific Publishing Co. 1978.
124. S. Matsuoka, H. E. Bair: J. Appl. Physics *48*/10, 4058–4062 (1977). S. Matsuoka, H. E. Bair, S. S. Bearder, H. E. Kern, J. T. Ryan: Analysis of Nonlinear Stress Relaxation in Polymeric Glasses, Bell Laboratories, Murray Hill, New Jersey 07974.
125. J. F. Jansson, B. Terselius: IUPAC Symposium on Long-Term Properties of Polymers, Stockholm, Sweden 30. 8. – 1. 9. 1976.
126. E. H. Andrews: Fatigue in Polymers, Testing of Polymers, Vol. IV, W. Brown, ed., New York: Interscience 1969, 237.
127. J. A. Manson, R. W. Hertzberg: CRC Critical Reviews in Macromol Sci. 433–500, August 1973.
128. R. W. Hertzberg: Deformation and Fracture Mechanics of Engineering Materials, New York: Wiley 1976.
129. K. Oberbach: Kunststoffe *63*, 35–41 (1973).
130. K. Oberbach: Kunststoff-Kennwerte für Konstrukteure, München: Carl Hanser Verlag 1975, 87–96.
131. J. M. Schultz: Treatise on Materials Science and Technology, Vol. 10, Part B. New York: Academic Press 1973, 80–107.
132. G. Jacoby in: Neuzeitliche Verfahren der Werkstoffprüfung, Düsseldorf: Verlag Stahleisen 1973, 80–107.
133. C. E. Feltner, M. R. Mitchell: ASTM STP 465, Amer. Soc. Test. Mat., 27 (1969).
134. K. Oberbach, G. Heese: Materialprüfung *14*/6, 173–178 (1972).
135. Crawford, R. J., Benham, P. P.: J. Mech. Engng. Sci. (GB) *16*/3, 178–179 (1974).
136. M. E. Graf, M. Ya. Filatov: Probl. Prochn. (USSR) 7/9, 102–106 (1975).
137. R. J. Crawford, P. P. Benham: J. Materials Sci. *9*, 1297–1304 (1974).
138. V. Bouda, A. J. Staverman: J. Polymer Sci. Polymer Sci. (Polymer Phs. Ed.) *14*, 2313–2323 (1976).
139. V. Zilvar: J. Macromol. Sci.-Phys. B5/2, 273–284 (1971).
140. V. Zilvar: Plastics and Polymers, 328–332, October 1971.
141. E. Alf: Untersuchungen zum Verhalten ausgewählter Kunststoffe unter schwingender Beanspruchung (Dissertation), Technische Hochschule Aachen 1972.
142. S. Rabinowitz, A. R. Krause, P. Beardmore: Materials Sci. *8*, 11–22 (1973).
143. M. Schrager: J. of Polymer Sci. Part A-2 *8*, 1999–2014 (1970).
144. S. Sikka: PhD Thesis, Univers. of Utah, Salt Lake City, Utah.
145. J. H. Wendorff: Personal communication, D. Kunststoff-Inst., Darmstadt (1977).
146. T. Nagamura, N. Kusumoto, M. Takayanagi: J. Polymer Sci. (Polymer Phys. Ed.) *11*, 2357–2369 (1973).
147. O. W. Brüller, N. Brand: Kunststoffe *67*/9, 527 (1977).
148. L. Konopasek, J. W. S. Hearle: J. Appl. Polymer Sci. *21*, 2791–2815 (1977).
149. H. H. Kausch: Materialprüfung *20*, 22–26 (1978)
150. K. V. Gotham: Plastics and Polymers *40*, 59–64 (1972).
151. N. T. Smotrin, V. M. Chebanov: Mekhanika Polimerov *6*, 453–467 (1970).
152. A. N. Machyulis, M. I. Pugina, A. A. Zhechyus, V. K. Kuchinskas, A. P. Stasyunas: Mekhanika Polimerov *2*/1, 60–66 (1966). V. K. Kuchinskas, A. N. Machyulis: Polymer Mechanics *4*, 538–543 (1968).
153. J. A. Sauer, A. D. McMaster, D. R. Morrow: J. Macromol. Sci.-Phys. B12(4), 535–562 (1976). J. A. Sauer, E. Foden, D. R. Morrow: Polymer Eng. and Sci. *17*/4, 246–250 (1977).
154. J. C. Bauwens, C. Bauwens-Crowet, G. Homés: J. Appl. Polymer Sci. *7*, 1745–1754 (1969).

155. J. C. Bauwens: J. Appl. Polymer Sci. *8*, 893–901 (1970).
156. I. M. Ward: J. Materials Sci. *6*, 1397–1417 (1971).
157. P. B. Bowden, J. A. Jukes: J. Materials Sci. *7*, 52–63 (1972).
158. C. Bauwens-Crowet, J. C. Bauwens, G. Homès: J. Materials Sci. *7*, 176–183 (1972).
159. R. Raghava, R. M. Caddell, G. S. Y. Yeh: J. Materials Sci. *8*, 225–232 (1973).
160. J. A. Sauer, K. D. Pae: Coll. and Polymer Sci. *252*, 680–695 (1974).
161. A. S. Argon, M. I. Bessonov: Philosophical Mag. *35*/4, 917–933 (1977).
162. A. S. Argon, M. I. Bessonov: Polymer Engng. Sci. *17*/3, (1977).
163. R. P. Kambour, R. E. Robertson: The Mechanical Properties of Plastics, in Polymer Science, Chapter 11, Jenkins Ed., Amsterdam, London: North-Holland, 1972, 687–822.
164. T. E. Brady, G. S. Y. Yeh: J. Appl. Physics *42*/12, 4622–4630 (1971).
165. K. P. Grosskurth: Gummi/Asbest/Kunststoffe *25*/12, 1159–1164 (1972).
166. G. P. Andrianova, V. A. Kargin: Polymer Sci. USSR *12*/1, 1–8 (1970). Vysokomol. Soyed. A*12*/1, 3–9 (1970).
167. G. P. Andrianova, A. S. Kechekyan, V. A. Kargin: J. Polymer Sci. A-2/*9*, 1919–1933 (1971).
168. G. I. Barenblatt in: Deformation and Fracture of High Polymers, Kausch, Hassell, Jaffee Eds., New York: Plenum Press, 1972, 91–111.
169. B. Zaks, M. L. Lebedinskaya, V. N. Chaldize: Vysokomol. Soyed. A12/*12*, 2669–2679 (1970).
170. L. A. Davis, C. A. Pampillo: J. Appl. Physics *42*/12, 4659–4666 (1971).
171. L. A. Davis, C. A. Pampillo: J. Appl. Physics *43*/11, 4285–4293 (1972).
172. J. C. M. Li, C. A. Pampillo, L. A. Davis in: Deformation and Fracture of High Polymers, Kausch, Hassell, Jaffee Eds., New York: Plenum Press, 1972, 239–258.
173. L. A. Davis, R. H. Baughman, C. A. Pampillo: J. Polymer Sci.: Polymer Phys., Ed. *11*, 2441–2451 (1973).
174. V. A. Lishnevskii: Dokl. Akad. Nauk SSSR, *182*/3, 596–599 (1968).
175. V. A. Lishnevskii: Vysokomol. Soedin., Ser. B *11*/1, 44–49 (1969).
176. P. Matthies, J. Schlag, E. Schwartz: Angew. Chem. 77/7, 323–327 (1965).
177. T. C. Chiang, J. P. Sibilia: J. Polymer Sci.: Polymer Phys. Ed. *10*, 2249–2257 (1972).
178. U. G. Gafurov: Vysokomol. Soyed. A14/*4*, 873–880 (1972).
179. A. K. Sengupta, R. K. Singh, A. Majumdar: Textile Res. J., 155–163, March (1973).
180. N. J. Capiati, R. S. Porter: J. Polymer Sci. (Polymer-Phys. Ed.) *13*, 1177–1186 (1975). D. M. Bigg: Polymer Engineering Sci. *16*/11, 725–734 (1976).
181. G. Capaccio, I. M. Ward: Polymer *18*/9, 967–968 (1977).
182. Seminar on ultra-high modulus polymers; La Chimica e l'Industria *59*, Ottobre 1977, 728–735.
183. L. Mullins: Effects of Fillers in Rubber in: The Chemistry and Physics of Rubber-Like Substances, Chapter 11, Bateman Ed., New York: Wiley, 1963, 301–328.
184. T. L. Smith: Pure and Appl. Chem. *23*, 235–253 (1970).
185. N. Sekhar, B. M. E. van der Hoff: J. Appl. Polymer Sci. *15*, 169–182 (1971).
186. F. R. Eirich in Mechanical Behavior of Materials Vol. III, Kyoto: Society of Materials Sci., 1972, 405–418.
187. T. L. Smith, W. H. Chu: J. Polymer Sci. A-2/*10*/1, 133–150 (1972).
188. F. P. Baldwin, G. Ver Strate: Rubber Chemistry and Techn. *45*/3, 709–881 (1972).
189. R. F. Landel, R. F. Fedors in: Deformation and Fracture of High Polymers, Kausch, Hassell, Jaffee Eds., New York: Plenum Press, 1972, 131–148.
190. R. J. Morgan: J. Polymer Sci.: Polymer Phys. Ed. *11*, 1271–1284 (1973).
191. T. L. Smith in: Treatise on Materials Science and Technology, J. M. Schultz, Ed. Vol. 10, Part A New York: Academic Press, 1973, 369.
192. R. F. Fedors, R. F. Landel: J. Polymer Sci. (Polymer-Phys. Ed.) *13*, 419–429 (1975).
193. R. F. Fedors: J. Appl. Polymer Sci. *19*, 787–790 (1975).
194. R. F. Fedors, R. F. Landel: J. Appl. Polymer Sci. *19*, 2709–2715 (1975).
195. R. F. Fedors: The Stereo Rubbers, W. M. Saltman Ed., New York: John Wiley, 1977. 679–804.
196. Encyclopedia of Polymer Science and Technology, Mark Ed., New York: Interscience, 1965. a. W. L. Cox, Vol. 2, 197.   b. N. Grassie, Vol. 4, 702.

197. G. M. Bartenev, Y. S. Zuyev: Strength and Failure of Visco-Elastic Materials, Oxford: Pergamon Press 1968.
    a. pg. 274.
198. D. V. Rosato, R. T. Schwartz: Environmental Effects on Polymeric Materials, Vol. I + II, New York: Interscience 1968.
199. E. H. Andrews: Fracture in Polymers, Edinburgh: Oliver and Boyd 1968.
    a. pg. 168.
200. A. Charlesby, Radiation Effects in Polymers; in: Polymer Science Vol. 1, Chapter 23, Jenkins Ed., Amsterdam London: North-Holland, 1972, 1543–1559.
201. Polymer Stabilization, W. L. Hawkins Ed. New York: Wiley-Interscience, 1971.
202. Natürl. u. künstl. Alterung, Kunststoffe, Fortschrittsberichte, Vol. 3, München 1976: Carl Hanser.
203. F. H. Winslow: Environmental Degradation, in Treatise on Materials Science and Technology, J. M. Schultz, ed., Vol. 10, Part B, New York: Academic Press, 1977, 741–776. Developments in Polymer Degradation, N. Grassie, ed., Vol. 1, London: Appl. Science Publ. 1977.
204. E. H. Andrews: J. Appl. Polymer Sci. 10, 47 (1966).
205. G. Salomon, F. van Bloois: J. Appl. Polymer Sci. 8, 1991 (1964).
206. K. L. DeVries, E. R. Simonson, M. L. Williams: J. Macromol. Sci.-Phys. B4/3, 671 (1970). K. L. DeVries, E. R. Simonson, M. L. Williams: J. Appl. Polymer Sci. 14, 3049 (1970).
207. Degradation and Stabilization of Polyolefins: B. Sedlacek, C. G. Overberger, H. F. Mark, T. G. Fox Eds: Polymer Symposia 57 (1976).
208. B. Ranby, J. F. Rabek: ESR Spectroscopy in Polymer Research, Berlin-Heidelberg-New York: Springer 1977.
209. H. H. G. Jellinek Ed.: Aspects of Degradation and Stabilization of Polymers: Amsterdam-Oxford: Elsevier 1978.
    a) Y. Kamiya, E. Niki: Oxidative Degradation, p. 79–148.
    b) W. Schnabel, J. Kiwi: Photodegradation, p. 149–246.
    c) I. Mita, Effect of Structure on Degradation and Stability of Polymers, p. 247–291.
    d) K. Murakami: Mechanical Degradation, p. 296–392.
    e) H. Kambe: The Effect of Degradation on Mechanical Properties of Polymers, p. 393–430.
210. C. S. Kim: Rubber Chemistry and Technology 42/4, 1095–1121 (1969)
211. J. W. S. Hearle, B. S. Wong: J. of the Textile Institute 63/4, 127–132 (1977).
212. A. Blaga, R. S. Yamasaki: J. Materials Sci. 11, 1513–1520 (1976).
213. G. G. Samoilov, E. E. Tomashevskii: Fizika Tverdogo Tela 10/4, 1094–1097 (1968). Soviet Physics-Solid State 10/4, 866–869 (1968). T. B. Boboev, V. R. Regel': Mekh. Polimerov 5, 929–931 (1969). Polym. Mech. (USA) 5, 824–826 (1972).
214. Kh. Akimbekov, V. S. Kuksenko, S. Nizamidinov, A. I. Slutsker, A. A. Yastrebinskii: Fizika Tverdogo Tela 14/9, 2708–2713 (1972). Soviet Physics-Solid State 14/9, 2339–2343 (1973).
215. R. V. R. Subramanian, T. V. Talele: Textile Research J. 42/4, 207–214 (1972).
216. H. M. Heuvel, K. C. J. B. Lind: J. Polymer Sci., A-2, 8, 401–410 (1970).
217. G. Reinisch, W. Jaeger: Faserforschung und Textiltechnik 19/8, 363–365 (1968).
218. H. Krüssmann, G. Valk, G. Heidemann, S. Dugal: Angew. Chemie 81/6, 226–227 (1969).
219. H. Wilski: Physik-Grundlage der Technik (Physik 1974-Plenarvorträge), Weinheim 1974: Physik-Verlag, p. 183–198.
220. T. M. Stoeckel, J. Blasius, B. Crist: J. Polymer Sci., Polymer Physics Ed. 16, 485–500 (1978).

220: After completion of the manuscript this detailed study of the relation between viscometry, ESR, and IR estimates of bond rupture has appeared. The authors report that in PE and PP fibers and films very little chain scission can be detected through viscometry; the number of broken bonds, N(vis), was $0.4 \cdot 10^{16}$ to $5 \cdot 10^{16} \, cm^{-3}$. N(vis) is by a factor of 1000 smaller than the elsewhere reported IR end-group concentrations. Virtually no change of M was observed in cold drawing of a PA 6 mono-filament whereas the viscosity-average molecular weight of PA 6, PA 66, and PETP fibers decreased measurably during loading to fracture. Evaluating their viscosity and ESR measurements the authors concluded that N(vis) and N(ESR) are of the same order, if nonrandom scission prevails. In the case of random scission a number N(vis) of broken bonds was derived from viscometry which was by a factor of 6 to 20 larger than the value N(ESR) obtained from the intensity of the ESR spectra. Observing opposite influences of an annealing treatment of PA 6 fibers on micro-crack formation and free radical intensity the authors question the close correlation between microcrack formation and bond rupture.

Chapter 9

Molecular Chains in Heterogeneous Fracture

In Chapter 8 principally the contribution of spatially homogeneously distributed molecular rearrangements to fracture initiation was studied. The term *spatially homogeneous* referred to the *absence* of flaws, inclusions, cracks, or notches of a size sufficient to act as a stress concentrator. Under those conditions during an initial phase of external loading damage development or growth is homogeneously distributed on a macroscopic scale. *Heterogeneous fracture* now is defined as the converse of homogeneous fracture or briefly as *fracture through crack propagation*. In this case cracks, notches, inclusions, or accumulated crack nuclei act as macroscopic stress concentrations and essentially confine further damage development to the close proximity of the then existing defect(s). The phenomenon of *crazing* has been included in this chapter because of the well observable structural irregularities and despite the fact that with increasing stress new crazes can be formed at arbitrary nucleation sites.

The development of fracture through cracks will be treated in this chapter following the *fracture mechanics approach*. In view of the extensive, competent, and recent literature on this subject prime attention will be given to the effect of chain properties (length, structure, strength) on crack propagation and on fracture mechanics parameters.

I. Fracture Mechanics

A. Stress Concentration

In Section III B of Chapter 3 Griffith's early considerations [(2)] were outlined. From the balance of *elastically* stored energy (U) and energy to produce new surface area (γ_cA) Griffith derived his well-known fracture criterion for isotropic materials containing an elliptical crack of length 2a (Eq. 3.13). In the last 50 years this *fracture mechanics approach* has been systematically developed to account for partially an-elastic and/or plastic behavior of solids for various crack and sample geometries, and even for material heterogeneities: In all cases it has been — and still is — the aim of the fracture mechanics analysis to derive *generally valid quantitative criteria* for crack stability and crack propagation behavior. Criteria are being sought which are as much independent of the state of external and internal stresses and of crack and sample geometry as possible and which depend principally on material functions such as material resistance R (Rißausbreitungswiderstand), critical stress intensity factor K_c (kritischer Spannungsintensitätsfaktor), and elastic modulus E.

A detailed account of the development and of the present state of the theory of fracture mechanics is given in the series Mechanics of Fracture [2]. From the extensive general literature on this subject only a few works may be cited which deal with the deformation and fracture of engineering materials [3] or specifically polymers [4–6] and standards [7, 8]. For the determination of the material functions R and K_c three experimental methods of controlled crack propagation are predominantly used (Fig. 9.1):
— Mode I, crack opening or tensile mode (einfache Rißöffnung)
— Mode II, sliding or in-plane shear mode
— Mode III, tearing or antiplane shear mode.

For details of these methods, their evaluation, and possible influences of sample geometry on the resulting "material" functions the reader is referred to the general literature cited [2–8]. In this Section the most frequently used crack opening mode, namely mode (I), will be considered (Figs. 9.1 and 9.2).

For a uniaxially stressed isotropic, elastic, *thin* plate containing an infinitely sharp edge crack a state of *plane stress* arises which has a singularity at the crack tip [3–8], for r ≪ a the components of stress are expressed as

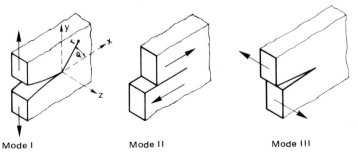

Mode I   Mode II   Mode III

Fig. 9.1. Experimentally used modes of controlled crack propagation: I crack opening or tensile mode, II sliding or in-plane shear mode, III tearing or anti-plane shear mode.

Fig. 9.2. Tensile specimen with single edge crack.

$$\sigma_x = \frac{K_I}{\sqrt{2\,\pi r}}\,\cos\frac{\varphi}{2}\left(1 - \sin\frac{\varphi}{2}\sin\frac{3}{2}\varphi\right) \tag{9.1}$$

$$\sigma_y = \frac{K_I}{\sqrt{2\,\pi r}}\,\cos\frac{\varphi}{2}\left(1 + \sin\frac{\varphi}{2}\sin\frac{3}{2}\varphi\right) \tag{9.2}$$

$$\tau_{xy} = \frac{K_I}{\sqrt{2\,\pi r}}\,\sin\frac{\varphi}{2}\cos\frac{\varphi}{2}\cos\frac{3}{2}\varphi. \tag{9.3}$$

The coordinates r and ϕ are indicated in Figure 9.1. The quantity K_I determines the intensity of the stress field without affecting its shape. It is accordingly termed *stress intensity factor*. The stress intensity factor takes into account both the external (uniaxial) stress σ_0 and the crack length a:

$$K_I = \sigma_0\sqrt{\pi a}\cdot f(a/b) \tag{9.4}$$

where $f(a/b)$ is a correction factor of the order of unity which derives from the finite specimen dimensions and crack configuration [3–8]. The stress intensity factor can be considered a much more general measure of the criticality of the state of a stressed, cracked plate than σ_0 or a individually.

In the case of a *thick* plate the lateral contraction of the material at the crack tip is hindered; a state of *plane strain* prevails and in addition to σ_x and σ_y a normal stress σ_z exists:

$$\sigma_z = \nu(\sigma_x + \sigma_y). \tag{9.5}$$

In plane strain and for $r \ll a$ the components of displacement in the x-direction (u) and the y-direction (v) are obtained [6] as:

$$u = \frac{2\,K_I(1+\nu)}{E}\sqrt{\frac{r}{2\,\pi}}\,\cos\frac{\phi}{2}\left(1 - 2\nu + \sin^2\frac{\phi}{2}\right) \tag{9.6}$$

$$v = \frac{2\,K_I(1+\nu)}{E}\sqrt{\frac{r}{2\,\pi}}\,\sin\frac{\phi}{2}\left(2 - 2\nu - \cos^2\frac{\phi}{2}\right). \tag{9.7}$$

253

For $\phi = \pi$ Eq. (9.7) yields the crack opening displacement

$$2v = \frac{8\,K_I}{E}\sqrt{\frac{r}{2\,\pi}}\,(1 - v^2). \tag{9.8}$$

Whereas the stresses show a singularity for $r \to 0$ the displacements go to zero. Through the introduction of a crack into a (thick) plate under constant uniaxial stress σ_0 the elastically stored energy in the vicinity of the crack tip is *increased* by the finite amount W:

$$W = \frac{\pi a^2\,\sigma_0^2}{2\,E}\,(1 - v^2). \tag{9.9}$$

Concurrently work 2 W is expended by the external system to maintain the uniaxial stress σ_0. The total elastic energy U_e decreases, therefore, by W. This apparent paradox was pointed out by Eshelby [9]. If the crack is extended by Δa then the stress field is shifted by Δa in the x-direction and its energy content *increased* by $\delta W/\delta a$; an energy increment of the same magnitude is set free by the stress release accompanying the failure of the highly stressed crack tip material. It is the latter energy increment which is termed *energy release rate* G_I:

$$G_I = -\frac{\partial U_e}{B \partial a} = \frac{\pi a\,\sigma_0^2}{E(T, t)}\,(1 - v^2)\,f(a/b). \tag{9.10}$$

For a thin plate (state of plane stress) the terms $(1 + v^2)$ and $(1 - v^2)$ in Eqs. (9.6) to (9.10) have to be replaced by unity. In view of the discussion later it is important to determine how the energy G_I is distributed around the crack tip. Following Irwin [10] the elastic strain energy within a cylinder of radius r_1 around the crack tip is (per unit of plate thickness):

$$W(r_1) = \frac{\pi a^2\,\sigma_0^2}{2\,E}\left(\frac{5}{4} - \frac{3}{4}\,v - 2\,v^2\right)\frac{r_1}{a}. \tag{9.11}$$

Whereas the energy content $W(r_1)$ increases proportionally to r_1 the local energy density $w(r)$ decreases as $1/r$; for large distances from the crack tip it attains $w_0 = \sigma_0^2/2\,E$, the strain energy density of an isotropic, elastic solid. From Figure 9.3 it is readily seen that increasing a by $\Delta a = r_1$ sets free an amount of stored energy $G_I\,\Delta a$ which is a factor of $(8 - 8\,v)/(5 - 8\,v)$ larger than the elastic energy stored within a cylinder of radius r_1 around the crack tip. This ratio is independent of the absolute value of r_1. In other words any crack propagation step in an ideal elastic material — even the smallest — releases elastic energy in a material volume several times larger than that immediately surrounding the crack tip (volume $\pi \Delta a^2 B$). Experimentally it is seen that a triangular zone bounded by a plane connecting the crack tip with the lateral face is unloaded. Increasing a requires in any event the breakage of primary and/or secondary bonds and thus energy, say $B\Delta a \gamma_c$. In order to obtain a strict correspondence between $B\Delta a \gamma_c$ and $B\Delta a\,G_I$ one would require that the total of the

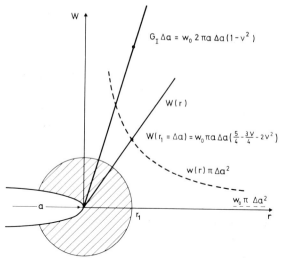

W

$G_I \, \Delta a = w_0 \, 2 \, \pi a \, \Delta a \, (1 - v^2)$

W(r)

$W(r_1 = \Delta a) = w_0 \, \pi a \, \Delta a \left(\frac{5}{4} - \frac{3v}{4} - 2v^2 \right)$

$w(r) \, \pi \, \Delta a^2$

$w_0 \, \pi \, \Delta a^2$

a

r_1

r

Fig. 9.3. Distribution of elastically stored energy in front of a plane crack (mode I). $G_I \, \Delta a$ energy release (per unit specimen thickness) through propagation of crack by Δa, W(r) energy stored within cylinder with radius r about crack tip, w(r) energy density.

elastic energy stored in the contributing volume region be used for the separation of bonds traversing $B\Delta a$. For polymeric solids the smallest conceivable step would be the scission of a single bond, i.e. a step width of $\Delta a = 0.4$ nm. The activation volume for bond scission is of the order of $\beta = 0.008$ nm^3. Such a volume element constitutes only a fraction $0.125/\pi$ of $\pi\Delta a^3$ and contains about 0.015 of $G_I\Delta a^2$. It must be noted, therefore, that on a molecular scale — even in the absence of plastic deformation — fracture propagation requires the release of energy from a volume at least 60 times larger than the activation volume for breakage of primary bonds across the newly formed surface area $B\Delta a$.

Propagation of a crack occurs if the energy release rate G_I is sufficient to account for the material resistance R, i.e. for all modes of energy consumption related to the propagating crack. The condition for crack growth is written, therefore, as

$$G_I(a, t) \geqslant R(a, t). \tag{9.12}$$

At low crack propagation rates kinetic energy terms in R can be neglected. Then the material resistance will include the specific surface energy 2γ (to overcome the cohesion of atoms or molecules across the newly formed surface area), and the energy terms V_{re} of elastic retraction of stressed molecules, V_{pl} of plastic deformation, and V_{ch} of chemical reactions including bond scission. The release of internal stresses (U_i) or chemical reactions with the environment (U_{ch}) may subtract from R:

$$R = 2\gamma + \frac{\partial V_{re}}{B\partial a} + \frac{\partial V_{pl}}{B\partial a} + \frac{\partial V_{ch}}{B\partial a} + \cdots - \left[\frac{\partial U_i}{B\partial a} + \frac{\partial U_{ch}}{B\partial a} \right]. \tag{9.13}$$

In general – and especially with polymeric materials – R will be a function of crack length, specimen geometry, and extent and rate of plastic deformation at the crack tip. A schematic representation of an $R(a)$ of a ductile, strain hardening material is given in Figure 9.4. A crack a_0 grows if $G_I(a_0) > R(a_0)$; if $G_I(a_0) = R(a_0)$ it continues to grow if the slope of the line $G_I(a)$, i.e. $\partial G/\partial a$, is larger than the change of material resistance with crack length, $\partial R/\partial a$. For an arbitrary energy release rate G_I with $R(a_0) < G_I(a_0) < G_c(a_0)$ crack stability will again be achieved at a new value $a_1 > a_0$ where $G(a)$ intersects $R(a)$ (cf. Fig. 9.4). If, however, $G_I(a)$ is equal to

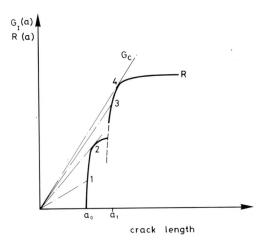

Fig. 9.4. Schematic representation of crack stability in ductile, strain-hardening material:
1. $G < R(a_0)$, 2. $G(a_0) = R(a_0)$, 3. $G(a_1) < G_c$, 4. $G = G_c$.

or larger than $G_c(a)$ – which forms the tangent to $R(a)$ – then the rate of energy release remains always larger than the material resistance and the crack will grow continually. It follows from Figure 9.4 that G_c generally is the larger the larger $R(a \gg a_0)$. Since R depends most strongly on the plastic energy term $\partial V_{pl}/B\partial a$ it also depends on the size of the volume element which deforms plastically during crack propagation. The material resistance R is the smaller, therefore, the more constraint to plastic deformation exists, i.e. it is smaller in plane strain than in plane stress. Under conditions of plane strain G_c approaches a limiting value of G_{Ic} which – through Eqs. (9.4) and (9.10) – defines a critical value of K_I. This critical value, K_{Ic}, is termed *fracture toughness*. For a brittle fracture the numerical value of K_{Ic} can generally be determined from the measurement of crack length and load at the onset of unstable crack propagation [3–8].

Plastic deformation is especially pronounced in polymeric materials. The electron micrographs presented in Chapter 8 provide ample evidence of this fact. It will be necessary, therefore, to study whether and how the fracture mechanics approach derived for an elastic material can be applied to elastic-plastic solids as well. The effect of plastic deformation on the stress distribution at a crack tip is well documented [3–7]. An elastic-plastic material under quasi-elastic conditions, for instance, begins to deform plastically wherever the state of stress meets the yield or flow criterion. Plastic deformation begins in the region of largest stresses, i.e. in the proximity of the crack tip; it limits the components of stress to the yield stress σ_F. In order to maintain the mechanical equilibrium stresses must be raised (up to the yield stress) in more distant regions. Thus the plastic deformation has the effect of increasing the effective crack length [3–7]. Two common approaches to calculate the effective crack enlargement through plastic deformation are based respectively on the von Mises yield criterion [6] and on the consideration of σ_F as an additional compressive stress [7].

From the von Mises yield criterion together with Eqs. (9.1) to (9.3) one derives the shape of the plastic zone ($r < r_F$) for a state of plane stress as [6]:

$$r_{F,Sp} = \frac{K_I^2}{2 \pi \sigma_F^2} \cos^2 \frac{\phi}{2} \left(1 + 3 \sin^2 \frac{\phi}{2} \right) \tag{9.14}$$

and for plane strain as

$$r_{F,V} = \frac{K_I^2}{2 \pi \sigma_F^2} \cos^2 \frac{\phi}{2} \left[\left(1 + 3 \sin^2 \frac{\phi}{2} \right) - 4 \nu(1 - \nu) \right]. \tag{9.15}$$

The plastic zone extends in the plane of the crack by

$$(r_{F,Sp})_{\phi=0} = \frac{K_I^2}{2 \pi \sigma_F^2} \quad \text{for plane stress;} \tag{9.16}$$

and by

$$(r_{F,V})_{\phi=0} = \frac{K_I^2}{2 \pi \sigma_F^2} (1 - 2 \nu)^2 \quad \text{for plane strain.} \tag{9.17}$$

In a moderately thick plate one finds a state of plane strain in the center and of plane stress at the lateral faces. This leads to the well-known dog-bone shape of the plastic zone.

The Dugdate model [7] recognizes the effect of the plastic deformation in front of the crack tip in a different way. It considers a virtual crack which includes the plastic zone – of size r_p – and it treats the yield stress as a superimposed stress acting normally on the boundary between the elastic and the "plastic" zone. The superimposed stress tends to bend the surface of the virtual crack. A consideration of mechanical equilibrium gives [6, 7]

$$r_p = \frac{\pi a \sigma_0^2}{8 \sigma_F^2}, \tag{9.18}$$

$$K_I = \sigma_0 \sqrt{\pi(a + r_p)} \tag{9.19}$$

and a crack opening displacement

$$2 v_a = \frac{K_I^2}{E \sigma_F} \tag{9.20}$$

Equations (9.14) to (9.18) do not constitute a fracture criterion since they do not specify whether and when material separation in the plastically deformed zone occurs. In order to use fracture mechanics to *predict* the stability of an elastic-plastic crack, limits of the plastic deformation have to exist and must be known explicitly or implicitly. Understandably the determination, interpretation, and application of critical stress intensity factors for viscoelastic solids is particularly complicated due to the pronounced time and temperature dependence of the mechanical properties of these materials. The fact that a material shows a ductile behavior does not preclude the use of fracture mechanics but it reduces — or even eliminates — the independence of the fracture mechanics functions (G, K, R) from the geometrical parameters [6].

The occurrence of plastic deformation in a material has a threefold effect:
- it limits the elastic energy density to $\sigma_F^2 / 2E$
- it leads to an increase in R (roughly in proportion to the volume of the plastically deformed zone)
- it may render uncritical any K_I (i.e. any σ_0 and a) unless the extent of plastic deformation is limited.

Following earlier concepts applied to metals the quantitative assessment of the contribution of plastic deformation to material resistance R of polymers has made considerable progress. Recognizing the difference between an R_1 (brittle) observed under plane strain and an R_2 (plastic) resulting from the plastic deformation under plane stress one may rewrite Eq. (9.12) as

$$G_c B = R_1 (B - 2\,r_p) + R_2\,2\,r_p. \tag{9.21}$$

Formally one may equate R_2 with a G_{c_2}, the "plastic component" of the strain energy release rate, and R_1 with G_{c_1} which should correspond to G_{Ic}.

It has been indicated that G_c is the smaller the more pronounced the condition of plane strain, i.e. the larger the specimen thickness B. At the other extreme of comparatively small thicknesses also a dependency of G_c on B is noticed (Fig. 9.5). The

Fig. 9.5. G_c as a function of specimen thickness for polycarbonate (from [21]): (a) tapered slit, crack length taken to be equal to visible crack, (b) tapered slit, plastic zone considered, (c) tapered V groove, crack length taken to be equal to visible crack, (d) tapered V groove, plastic zone considered; + constant thickness slit.

almost linear increase of G_c with B indicates that (in this case of a single groove double cantilever beam) the volume of the plastically deformed zone (shear lips) increases in proportion to B^2. The decrease of G_c due to the beginning mixed mode fracture at larger specimen thicknesses just becomes apparent [21].

At present several different experimental methods of working in the crack opening mode are employed [7–15]. Flat samples with a single edge notch (SEN) are preferentially used under constant load, under slowly increasing strain or under cyclic tensile loading. These conditions permit a subcritical crack growth leading eventually to instability. To determine the energy release rate Eq. (9.10) can be used if the appropriate long-term relaxation modulus E(T, t) is employed. The same is true for plates with a central notch (CN), compact tensile (CT) specimens, Charpy notched samples in slow bending, and wedge open loaded (WOL) plates at small loading rates.

In *shock loading* the short-time moduli are relevant to describe the energy release rate of pre-cracked samples; crack instability is reached primarily because of the increasing stresses and at unchanged crack length. The contribution of thermally activated crack growth to an increase of K_I is, therefore, negligible (the situation is different, though, with uncracked samples and flaws of molecular dimensions as has been shown in Chapter 8).

The WOL and the double-cantilever beam (DCB) specimens are suitable to determine the arrest of an initially unstable crack; the double-torsion (DT) specimen permits the investigation of crack propagation under constant K_I [7–15].

Values of various types of energy release rates G and of fracture toughness values K of polymeric materials have been gathered from the literature [12–75] and are compiled in Table 9.1. Although the experimental method and the principal parameters and observations have also been indicated in Table 9.1, a discussion of some molecular aspects will be attempted in the following.

B. Subcritical Crack Growth

The Griffith concept of an energy balance (G = R) furnishes a criterion of instantaneous *stability* (G < R) or instability (G > R, crack propagation). It does not explicitly contain time as a variable. It has been pointed out above, however, that G depends on t through E(t) and R depends on t predominantly through the plastic energy dissipation term. The larger G, the rate of release of provided elastic energy, the larger $\dot{a}$, the rate of crack propagation. The opposite is not always true; thus material resistances determined in bending, tension, or tearing may decrease with the rate of deformation [17, 20–25, 35–37, 57].

In many polymeric materials three zones of crack propagation behavior can be distinguished:

— no crack growth at $G < G_i$
  (stable crack)
— slow crack growth at $G_i < G < G_{Ic}$
  (subcritical crack growth)
— rapid crack propagation at $G \geqslant G_{Ic}$.

Table 9.1. Energy Release Rates G and Fracture Toughness Values K of Polymeric Materials

Polymer (standard abbrev.)	Method and Specimen	G_{Ic}^* J/m²	K_{Ic}^* MN m⁻³/²	Parameters of Test, Observations	Ref.
Thermoplastic Materials					
PE					
(LDPE)	impact	5000 (35000)		G_{c_1} G_{c_2}	[64]
(medium density)	impact	1300 11900		G_{c_1} G_{c_2}	[64]
(medium density) ($\rho = 0.95$ g cm⁻³, pigmented)	impact SEN	14200	1.7–4	plastic zone correction applied linearly increasing with "time to crack initiation" (inverse of load rate)	[69] [57]
($\rho = 0.96$ g cm⁻³)	mode III tearing	$2 \cdot 10^4$–$20 \cdot 10^4$		G_{III} shows maxima and minima which are related to the α- and β-relaxation maxima	[24]
(PE hard, black)	SEN, CT	3600–20000	3.0–6.5	K_c	[70]
PP					
(ICI, compression moulded)	SEN, SN		6.5–3.6	K_{c_2} decreases with T, ($-180 < T < 20$ °C), change in slope at -60 °C related to β-relaxation peak	[68]
			2.8	K_{c_1} essentially constant	
(BASF Novolen 1120 Lx)	CT		5.2–4.5 3.3–3.75	$v = 0.08$ mm s⁻¹ -40 °C $< T < 20$ °C $v = 8.6$ mm s⁻¹	[183]
PBU 1					
(non-modified)	SEN		2.9–1.2	K_c gradually decreasing with "time to crack initiation"	[57]
(pigmented)	SEN		1.2–3.2	distinct maximum of K_c at "times to crack initiation" around 0.1 to 10 s	[57]

Table 9.1. (Continued)

Polymer (standard abbrev.)	Method and Specimen	G_{Ic}^* J/m²	K_{Ic}^* MN m$^{-3/2}$	Parameters of Test, Observations	Ref.
PS					
–	SEN	1700 ± 600			[34]
–	DCB	300–2500			[17, 35–37]
(Shell Chemical)	DCB	200–7.5		increasing with decreasing á; rapidly decreasing with increasing degree of orientation	[45]
–	SEN		0.6–2.3	$10^{-7} < á < 10^{-2}$ m/s	[51]
	tapered cleavage		3.7–7	$10^{-9} < á < 10^{-2}$ m/s	[51]
(HIPS)	impact	350 (900)		G_{c_1} G_{c_2}	[64]
	impact	1000 15000		G_{c_1}, G_{c_2}	
PVC					
(BP, England)	Charpy notched three point bending		2.3–6	$-200 < T < 50\,°C$ T, rate of bending	[63]
(Cobex)	DCB	3600	2.2–5.6	undrawn sheet	[46]
	DCB	850–50		decrease of G_c with increasing orientation ($1.2 \cdot 10^{-3} < \Delta n < 3.5 \cdot 10^{-3}$)	
–	impact	1230 1440		G_{c_1} G_{c_2}	[64] [64]
ICI Darvic (modified)	DEN		1.94–2.57	K_c	[71]
	impact	10000		G_c	[64]
PMMA					
(ICI)	SEN, DCB, impact, three point bending		1–3.5	$-197 < T < 21\,°C$ $2 \cdot 10^{-8} < á < 0.08$ m/s distinct peak of K_{Ic} related to side chain (γ) relaxation	[114]

Table 9.1. (Continued)

Polymer (standard abbrev.)	Method and Specimen	G_{Ic}^* J/m²	K_{Ic}^* MN m$^{-3/2}$	Parameters of Test, Observations	Ref.
—	SEN	200			[33]
—	SEN		1.7		[52]
(Plexiglas, Röhm)	SEN		1.2–1.8	K_{Ic} $0.1\cdot10^6 < \bar{M}_v < 8\cdot10^6$ increasing with molecular weight	[15, 66]
(Plexiglas, Röhm)	SEN	1000–15000		G_d increasing with $\dot{a}$ and M $200 < \dot{a} < 700$ m/s $0.1\cdot10^6 < M_w < 8\cdot10^6$	[50, 67]
—	cleavage	120–600		various	[17, 21, 35–37, 45, 48]
		237			[48]
(Perspex)	(DCB)	100–35		G_A decreasing with increasing orientation $4\cdot10^{-4} < \Delta n < 14\cdot10^{-4}$	[45]
(Plexiglas G)		0.75–100		log G_{Ic} increases linearly with log $\bar{M}_v$ $(2\cdot10^4 < \bar{M}_v < 10\cdot10^4)$	[65]
		120 ± 20		G_{Ic} essentially independent of molecular weight $(\bar{M}_v > 10\cdot10^4)$	
(Perspex)	CT		1.3–0.75	decreasing with increasing T	[74]
(Perspex)	DT		1.6–1.7		[13]
(Simplex)			1.8–2.1		[13]
—			1.9–2.9	various $\dot{a}$, T	[12]
—	impact	1000		G_{c_1}	[64]
		1300		G_{c_2}	
(Perspex)	impact		1.7–2	notch tip radius	[58]
ABS					
—	impact	12000–16250		G_c increasing with decreasing specimen width	[53]
—	impact	1000–50000		steep increase of G_c between −20 and +10 °C	[64]
—	impact	23000			[69]
(Lustran 244)	impact	47000–49000		G_c	[64]

Table 9.1. (Continued)

Polymer (standard abbrev.)	Method and Specimen	G_{Ic}^* J/m²	K_{Ic}^* MN m$^{-3/2}$	Parameters of Test, Observations	Ref.
PETP					
(medium M_V)	Charpy notched	4300		G_c is calculated from impact resistance of differently notched specimens	[53]
(high M_V)		5200			[53]
(fiber grade)		7200		G_c values for amorphous and crystallized fiber grade PETP are the same	[53]
	SEN, DEN		6.1–5.3–3.8 / 5.1–4.6–3.7	K_{c_2} / K_{c_1} crystalline, 20 °C, −50 °C, −100 °C	[49]
PC					
(Makrolon)	three point bending	5000	3.8–2.5	G_c decreasing with increasing rate of deflection ($10^{-6} < v < 10$ m/s)	[22]
(Makrolon)	three point bending		10–2.5	decreasing with increasing temperature ($-200 < T < -100$ °C), slight maximum at 70 °C	[54]
(Makrolon)	impact		4–2 / 2–15	K_{c_2} decreasing with increasing T (T < 50 °C) / K_{c_1} increasing with increasing T (T > 50 °C)	[54]
(Makrolon)	impact	3500 / 5000 / 4000		G_{c_1} / G_{c_2} / G_c calculated from impact resistance	[64] [64] [53]
(Lexan 101)		1500–2700	3.8–2.3	G_c distinct maximum of G_c at −40 °C; decrease of K_{Ic} between −60 and +20 °C; correlation of G_c with tan δ and fracture surface morphology	[62]
—	cleavage	8000–170		G_{Ic} decreases rapidly with ȧ ($0.25 \cdot 10^{-4} <$ ȧ $< 10^{-2}$ m/s)	[21, 20]
	(DCB)	170–12		G_{Ic} decreases gradually with ȧ ($10^{-2} <$ ȧ < 300 m/s)	[211]

263

Table 9.1. (Continued)

Polymer (standard abbrev.)	Method and Specimen	G_{Ic}* J/m²	K_{Ic}* MN m$^{-3/2}$		Parameters of Test, Observations	Ref.
(Makrolon)	DT		2.6–3.8		various temperatures and crack speeds	[19]
(Makrolon)	SEN		2.2	K_{c_1}	(corresponds to K_c of crack initiation)	[19]
			8–5	K_{c_2}	decreases with increasing T (−140 < T < −30 °C)	
Polyamide 66						
(Maranyl)	three point bending	8800	4.4–2.3		gradually decreasing with higher rates of deflection	[22]
(Maranyl, 0.2% H_2O)	SEN		4–7	K_c	increasing with T (−160 < T < +40 °C)	[68]
	SN		4	K_c	constant in same temperature region,	
			3–4	K_{c_1}	little variation with T	
			8–10	K_{c_2}	maxima corresponding to γ- and β-relaxation peaks are apparent at −140 and −80 °C	
(Nylon 66, dry)	impact	250		G_{c_1}		[64]
		4150		G_{c_2}		
(Nylon)	impact	5000–5300		G_c		[64]
Thermosetting Resins						
PUR						
(unsaturated vinyl-urethane)	impact	1300		G_c		[53]
UP						
(non-modified)	SEN	440	1.2	G_c		[56]
	cleavage	42 ± 6				[59]
EP						
(non-modified)	DT, CT	700	0.55–1.7	G_c	−48 < T < +77 °C, various rates of deformation	[75]
(non-modified)	SEN		1.54			[56]
	cleavage	80 ± 5				[59]

Table 9.1. (Continued)

Polymer (standard abbrev.)	Method and Specimen	G_{Ic}* J/m²	K_{Ic}* MN m$^{-3/2}$	Parameters of Test, Observations	Ref.
(non-modified)	DCB-tapered DT	50–200 210 ± 48		adhesive layer between aluminium substrates	[73] [41]
(rubber-modified) (silica-filled)	cleavage	310 ± 10	2.2 ± 0.1		[59]
Phenolics (phenol-formaldeyhde)	SEN, CN	0.42–50		G_{Ic} calculated from rupture stresses according to Eq. ∴, G_{Ic} decreases with T (20 < T < 205 °C)	[55]
Composite Materials					
Bakelite SR + 40 Vol. % aligned high modulus carbon fiber (Morganite I)	Charpy	20000–36000		G_c values depend on fiber surface treatment (HNO$_3$, silane, brominated, silicone oil treatments resulting in interlaminar shear strengths τ between 12 and 30 MN/m² morganite treatment with τ = 55 MN/m²	[47]
		8800 5000–50000	13.5–19	v_f = 0.4	
Epoxy + aligned RAE carbon fibers	SEN	10000–15000 8000–19000	20–28	v_f = 0.6 v_f = 0.15 } G_c and K_c for various crack geometries and rates of deformation	[56]
UP + randomly oriented E glass fibers	SEN	35000	47	v_f = 0.66	[6, 49, 60]
Epoxy + boron fibers	SEN	8.3	0.8		
Plate Glass	SEN				

Table 9.1. (Continued)

Polymer (standard abbrev.)	Method and Specimen	G_{Ic}^* J/m²	K_{Ic}^* MN m$^{-3/2}$	Parameters of Test, Observations	Ref.
Elastomers					
PBU (mixed cis-1,4,trans- 1,4, and 1,2)	mode III tearing	18–1000		G_{III} increasing with effective rate of tear $a_T\dot{a}$ ($10^{-22} < a_T\dot{a} < 10^{-14}$ m/s)	[28]
		40–78		threshold fracture energy T_0 increasing with M_c ($2400 < M_c < 50000$); $\lambda_s^2 T_0$ independent of test temperature, degree of swelling, type of swelling liquid	
(cis-1,4)	mode III tearing	30–1000		G_{III} for various load rates, degrees of crosslinking, and swelling	[28]
		58 ± 8		threshold fracture energy for $M_c = 3200$	
		81 ± 8		threshold fracture energy for $M_c = 14000$	
(Polysar)	mode III tearing	1000–3000		G_{III} (for trousers and angled test pieces alike) increases with rate of crack growth ($10^{-7} < \dot{a} < 10^{-4}$ m/s)	[26]
(butadiene-styrene and butadiene-acrylonitrile copolymers)	mode III tearing	100–100000		G_{III} increasing with rate of tear ($10^{-7} < \dot{a} < 0.1$ m/s), decreasing with temperature after passing through a maximum	[25]
(polyurethane-solithane 113)	pure shear	2.45		intrinsic fracture energy derived from visco-elastic analysis of crack propagation	[29]

* As indicated in the fifth column, not always critical values G_{Ic} or K_{Ic} are listed.

This behavior is especially well demonstrated by PMMA. In Figure 9.6 the data of Döll [30] are represented.

The slow subcritical crack growth in a polymer is due to thermomechanically activated processes of the same nature as described in the last chapter for the homogeneous fracture, i.e. chain stretching, plastic deformation, void opening, chain disentanglement, and chain scission. The rate with which these processes occur is, however, in the strain field surrounding a crack tip larger than elsewhere in the material.

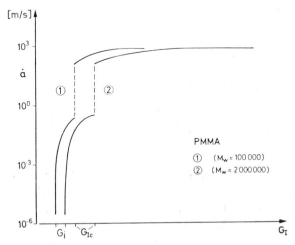

Fig. 9.6. Crack speed as a function of energy release rate G_I (after [30]).

The growth rate $\dot{a}$ in the subcritical region seems to be determined by K_I. (In subcritical crack growth any imposed K_I corresponds *per definitionem* to the K_c for crack growth at the observed speed.) Several relations between K_I and $\dot{a}$ have been established. One of these has been found to describe the subcritical crack growth of a number of polymers [3, 12, 13]:

$$\dot{a} = AK_I^b \qquad (9.22)$$

with A and b being empirical constants, depending on material, environment, temperature, etc. The constant b has been related to $1/\tan \delta$ of the β relaxation [12].

The subcritical crack growth data obtained by several investigators [12–15] for PMMA have been compiled in Figure 9.7. The double-logarithmic plot reveals an extended linear section (growth rates $\dot{a}$ from 10^{-8} to 10^{-2} m/s for stress intensity factors from 0.8 to 1.4 MN m$^{-3/2}$).

For this linear portion the exponent b in Eq. (9.22) can be determined to be 25 ± 2 [13].

From Eqs. (9.22) and (9.4) one can determine the time Δt necessary for the growth of a crack a_1 to length a_2. Neglecting here and in the following $f(a/b)$ one obtains for σ_0 = constant:

$$\Delta t = \int_{a_1}^{a_2} \frac{da}{\dot{a}} = \int_{K_{I1}}^{K_{I2}} \frac{2 K_I dK_I}{\sigma_0^2 \pi A K_I^b} = \frac{2(K_{I1}^{2-b} - K_{I2}^{2-b})}{(b-2)\Delta \pi \sigma_0^2} . \qquad (9.23)$$

Fig. 9.7. Slow crack propagation in PMMA at room temperature (after [11], including data from ref. [12–15]).

If one choses as upper limit K_{Ic} and as lower limit K_{Im}, i.e. that value of K which corresponds to the largest flaw or defect present in the sample, then one obtains as Δt the lifetime t_b of the sample under constant stress. The extremely strong exponential dependence of Δt on K_{Im} makes t_b very sensitive to flaw size. Increasing the size of the largest flaw, a_m, by only 10% decreases the lifetime t_b by a factor of 3.

The above considerations on the dependence of t_b on the size of inherent flaws have been extended by various authors [13, 16–17] to a method of predicting the creep life of glassy polymers. It is assumed that the size a_m of the largest intrinsic flaw can be determined from the breaking stress σ_b in short time loading according to Eq. (9.4):

$$a_m = K_{Ic}^2 / \sigma_b^2 \pi. \tag{9.24}$$

Broutman and McGarry [17] thus obtained intrinsic flaw sizes of 50 μm for PMMA and of 620 μm for PS. Berry [13] reports values of between 62 and 95 μm for PMMA. These values seem to be very high and no direct optical observations of such flaws have been reported. They are also two to three orders of magnitude larger than the diameters of microcracks found in PMMA and listed in Table 8.3. The microcracks are assumed to have dimensions comparable to those of a globular substructure which may be responsible for their appearance under stress. The "intrinsic flaws" must have a different significance, therefore, and may only develop at a stress level close to σ_b.

The higher the crack propagation rate in a sample gets ($\dot{a} > 10^{-3}$ m/s) the more the relation between $\dot{a}$ and K_I — and even the measured value of K_{Ic} — depend on the parameters of the experiment. As shown in Fig. 9.7 K_c increases monotonically up to a K_{Ic} of 1.6 MN/m$^{-3/2}$ if σ_0 is kept constant [12, 13]. In other experiments (Charpy impact tests, crack initiation and arrest in DCB specimens) K-values at higher crack speeds (2 to 300 m/s) are obtained which lie below the K_{Ic} [14, 18].

The different points on the $\dot{a} - K_I$ curve in Figure 9.7 represent quite different energy release rates G_I. According to Eqs. (9.4) and (9.10) G_I increases with the square of K_I. For the material employed one may use an average elastic modulus E of 4.5 GN/m^2 and a ν of 0.36 thus obtaining G_I values of 70 J m^{-2} at the lower range of K_c (0.6 MN m$^{-3/2}$) and 500 J m^{-2} for a K_{Ic} of 1.6 MN m$^{-3/2}$. These G values are equal to the corresponding material resistance R under the given experimental conditions. Their significance and interpretation will be discussed in the following Section.

The slow crack growth behavior of polycarbonate, also a glassy polymer, is even more complex than that of PMMA. At low temperatures (T < -40 °C) K_c values between 2.6 and 3.4 MN m$^{-3/2}$ have been obtained [19] from DT specimen which were independent of crack speed at small crack speeds ($\dot{a} < 10^{-3}$m/s) but dependent on sample thickness and temperature. At higher crack speeds ($\dot{a} \simeq 10^{-1}$ m/s) K_c values slowly increased. Double-side grooved DCB specimen (DG-DCB) however showed a decrease of K_c with increasing $\dot{a}$ in the region of $\dot{a} < 10^{-1}$ m/s [20]. This behavior was confirmed by Kambour et al. [21], who observed a G_c of 8.2 kJ/m^2 at $\dot{a} = 2.5 \cdot 10^{-5}$ m/s and of 12 J/m^2 at $\dot{a} = 300$ m/s. The microphotographical investigation of the slowly fractured sheets (B = 12.7 mm) revealed the $-$ energy-consuming $-$ formation of shear lips 0.4 mm wide (mixed mode crack propagation). At high speeds no shear lips were found. By copolymerisation of PC with silicone blocks the authors [21] were able to increase the fracture toughness in the whole temperature range T > -110 °C. With this material mixed mode crack propagation occurred fairly independently of crack velocity.

The increase of crosshead speed (over 7 decades) in static three point bending of PC resulted [22] only in a slight decrease of the calculated K_c from 3.8 to 3.0 MN m$^{-3/2}$. From the three point bending test a K_{Ic} is derived [14] from:

$$K_I^2 = \frac{EP^2}{2\,B_N} \cdot \frac{\partial C}{\partial a} \tag{9.25}$$

if P is taken as the largest load measured, $\partial C/\partial a$ as the proper change of specimen compliance with growth of a (perpendicular) crack, and B_N as the thickness of the load carrying section. The quantity $\partial C/\partial a$ increases with increasing crack length. At the beginning of crack growth the values of P and $\partial C/\partial a$ (and the corresponding K_{Ii}) will be smaller than the K_{Ic} at catastrophic failure. Arad et al. [22] point out that K_{Ii} and K_{Ic} are the closer together the less material ductility and/or slow crack growth are favored by the experimental conditions.

In relation to studies of the fracture surface energy as a function of loading rate the early and wide application of the tear test (mode III) must be mentioned (e.g. [5, 23−28]). In this fracture mode the material in the crack tip zone is subjected to a complex and highly plastic deformation. Without going into details it may be mentioned that rate effects on the extent of plastic deformation (and thus on fracture surface or tearing energies) have been found [23−29] and correlated with β- and γ-relaxation maxima [5, 23−26]. The tearing energies of thermoplastics and rubbers are generally quite large: e.g. up to 1 kJ/m^2 for PS, 20−200 kJ/m^2 for PE, and 0.1−500 kJ/m^2 for various butadiene copolymers [24−26]. For elastomers

Thomas [27] and Ahagon and Gent [28] report that after adjustment for the changing effective fracture area a common *threshold fracture energy* T_0 of 40 to 80 J/m² could be determined under different experimental conditions. This energy was shown to be independent of temperature and degree of swelling for various types of swelling liquid. The threshold energy decreased slightly with increasing degree of crosslinking (of the polybutadiene samples). In aggressive environment (oxygen, ozone) T_0 is substantially reduced.

An experimental and theoretical analysis of the steady propagation of a crack in a viscoelastic continuum is due to Knauss [29]. Employing a polyurethane elastomer (solithane 113) as material Knauss studied crack propagation under pure shear. From his extensive analysis, Knauss derives a solution to the viscoelastic-boundary-value problem represented by a crack moving in an isotropic, homogeneous, incompressible solid. He finds that the stress singularity at the crack tip vanishes. Under these conditions the stress-intensity factor only describes the far-field loading conditions. Knauss observes that the rate-dependent fracture energy is essentially the *product* of an "intrinsic fracture energy", presumably of molecular origin, and of a non-dimensional function which relates to the rheology of the material surrounding the crack tip. For the polyurethane elastomer this intrinsic fracture energy amounted to 2.45 J/m². As criteria for crack stability are concerned, he found that the criterion of ultimate, constant strain (crack-opening displacement) is as applicable as a generalized energy criterion.

As stated in the introduction to this section the subcritical crack growth in a polymer is due to the thermomechanical activation of different molecular deformation processes such as chain slip and orientation or void opening. The energy dissipated depends on the frequency, nature, kinetics, and interaction of these processes. But there are many and notable attempts to treat the subcritical crack propagation as one thermally activated multi-step process characterized by one enthalpy or energy of activation and one activation volume. Several of these kinetic theories of fracture have been treated in Chapters 3 and 8.

The concept of thermal activation has been extended to the fracture mechanical analysis of crack propagation. Thus Pollet and Burns [179] have discussed earlier approaches and the significance of various variables. Commonly the energy release rate G_I is considered as the mechanical *intensive* parameter. In their treatment, the crack front is thus represented by a line on which a force per unit length, G_I, tends to move it in the forward direction; its motion is, however, restrained by the presence of "thermal" obstacles, i.e., obstacles, or barriers, which can be overcome by thermal activation. Whether the physical origins of those energy barriers are individual obstacles placed on a unique surface along the fracture path or whether they are located somewhere in the surrounding region, does not affect the generality of the treatment if G_I can be considered as an intensive variable.

With the assumption of thermally activated barriers, the average crack velocity can be written according to a general Arrhenius-type equation as:

$$\dot{a} = v_0 \exp\left[-\Delta G(G_I, T)/kT\right] \tag{9.26}$$

where the quantity v_0 can be thought of as the maximum attainable crack velocity and ΔG as the free enthalpy of activation to overcome one "obstacle".

270

As observed by various investigators [12–15, 180] an approximately linear relation holds between the logarithms of crack velocity and crack extension force:

$$\ln \frac{\dot{a}}{v_0} = \frac{C(T)}{kT} \ln \frac{G_I}{G_{Ic}} \; . \qquad (9.27)$$

Pollet and Burns find the data for PMMA taken by Atkins et al. [180] at various temperatures to comply with Eq. (9.27). They determine from $C(T)$ an area of activation $A^* = C(T)/G_I$ of the order of $0.7 \cdot 10^{-4}$ to $2 \cdot 10^{-4}$ nm^2. They interpret this small value as indicating that the thermal obstacles are not lined up on a single surface along the fracture path but distributed in space in a region of material near the apparent fracture path. In view of the consideration in Section IA it should be said that G_I is not an intensive variable. Only a small fraction α of G_I contributes to a reduction of the energy of activation. If slow crack propagation were due solely to the repeated shear displacement of PMMA segments of 0.2 nm length over a distance of 0.8 nm an activation area of 0.16 nm^2 would have been expected. Since $2 \cdot 10^{-4}$ nm^2 have been formally calculated one would have to conclude that α amounts to $2 \cdot 10^{-4}/0.16 = 1/800$.

C. Critical Energy Release Rates

It has been discussed in the previous Section that an increase in stress intensity factor or in G_I, the driving force for crack extension, promotes subcritical crack growth (Figs. 9.6 and 9.7). Since the material resistance R was found to increase with $\dot{a}$ a new equilibrium between G_I and R could be established following any change in G_I. If G_I of K_I are continually increased, however, a point of crack growth instability is reached. The instability may be characterized by the statement that at this point the material resistance $R(\dot{a})$ as given by Eq. (9.13) is not sufficiently rate sensitive to compensate the increase in G_I. Consequently the crack accelerates to a level of velocities where inertia forces and the finite speed v_e of elastic waves have to be considered [67, 181–182]. So far the contribution of the kinetic energy of the receding fracture surfaces to R has been neglected. At the point of incipient crack instability in PMMA with crack velocities of about 0.1 m/s the kinetic energy term amounts to 6 J/m^3. At these velocities this term is, therefore, an insignificant fraction of the average strain energy density of 100 to 500 kJ/m^3. At velocities of between 100 and 1000 m s^{-1}, however, an increasingly larger fraction of the stored elastic energy $G_{I(static)}$ is necessary to account for the kinetic energy of the crack front – reaching 100% at a crack speed equal to the Rayleigh surface wave velocity v_R. The dynamic strain energy release rate G_d is, therefore, only a fraction F of $G_{I(static)}$:

$$G_d = \frac{\pi \sigma_0^2 a (1 - \nu^2)}{E} F(\dot{a}, v_e) \qquad (9.28)$$

Broberg [181] derived the function $F(\dot{a}, v_e)$ for an infinite, isotropic, elastic plate. His expression may be approximated by $(1 - \dot{a}/v_R)$.

Similar considerations apply to the slowdown and eventual arrest of a rapidly propagating crack [18, 182]. An understanding of this phenomenon is of considerable importance in the use of high pressure pipes where the propagation of a crack over long distances must be prevented under any circumstances.

The transition from subcritical to rapid crack propagation is especially pronounced in PMMA but more or less clearly observed in all polymers capable of brittle fracture. The corresponding critical quantities K_{Ic} and G_{Ic} characterize the fracture behavior of a material. In Table 9.1 an (incomplete) listing of measured critical values (G_{Ic} and K_{Ic}) and of energy release rates referring to particular experimental conditions (G_c, G_d, G_{III}) are given. The main experimental parameters and their ranges of variation have been indicated.

Of the various terms of Eq. (9.13) contributing to the material resistance the plastic deformation, $\partial V_{pl}/B\partial a$, is generally by far the largest. It has been attempted in Table 9.2 to identify also surface tension γ and the (hypothetical) energy density of chain breakage U/qN_L. The smallest term is that of the specific surface energy γ (free surface energy, surface tension). The γ values of polymers lie in the range of 0.020 to 0.046 J/m^2 [31]. The energy of elastic retraction is the product of strain energy density and width of the retracting layers. If the layers consist of fully oriented chains stressed to break and thus having the highest possible energy density (of the order of 1000 MJ/m^3) the energy of retraction of chain ends of length L = 5 nm amounts to 8 J/m^2.

The chemical surface energy, $\partial V_{ch}/B\partial a = U_0/qN_L$ also contributes comparatively modestly to R, even if one chain breakage event per molecular cross-section q is considered: 0.4 J/m^2 in PMMA, 2.9 J/m^2 in HDPE. The terms U_i and U_{ch} cannot be evaluated in all generality. The physico-chemical data used for the above and subsequent calculations are collected in Table 9.2 for seven semicrystalline or glassy polymers.

II. Crazing

A. Phenomenology

The phenomenon of crazing has been observed in many glassy polymers, and also in some crystalline polymers, when subjected to tensile stress. Polymer crazes (Fig. 9.8a) in their appearance are similar to the very fine cracks known for a long time to occur on the surfaces of inorganic materials such as ceramics. However, there is a difference between the crazes and the cracks in that crazes have a continuity of material across the craze plane (Figs. 9.8b, 9.9–9.11) whereas cracks do not possess any continuity. Consequently crazed zones are capable of bearing loads as opposed to cracked ones. In the last thirty years a growing number of publications has been devoted to the phenomenon of crazing. Two comprehensive reviews [76–77] on this subject have appeared in 1973. The bibliography on crazing compiled in this monograph [78–178] is mostly to be understood as an extension of the earlier reviews.

Table 9.2. Molecular and Mechanical Data of Selected Polymers (after [11, 31], Average Values for Non-modified, Unoriented Polymers at Room Temperature)

Quantity	Symbol	Unit	HDPE	PP	PA 66	PS	PMMA	PVC	PC
Density	ϱ	g/cm³	0.96	0.91	1.14	1.06	1.19	1.39	1.20
Young's modulus	E	MN/m²	1200	1300	1300	3100	3300	2600	2350
"Chemical surface energy"	U/qN_L	J/m²	2.9	1.3	1.6	0.5	0.4	0.5	0.6
Surface tension	γ_{exp}	J/m²	0.031	0.029	0.046	0.040	0.039	0.039	0.042
	$\gamma_{theor} = \delta^2 \sqrt{q}/4$	J/m²	0.029	0.052	0.084	0.071	0.102	0.058	
Monomer molecular weight	M_m	g/mol	28	42	226	104	100	62.5	254
Activation energy for bond scission	U_0	kJ/mol	335	272	188	230	175	(84)	117
Monomer length	l_m	nm	2.53	2.17	17.3	2.21	2.11	2.55	10.75
Monomer cross-section	$q = M_m/l_m q N_L$	nm²	19.0	35.1	18.9	73.2	65.7	29.1	32.8
Solubility parameter	δ^2	MJ/m³	268	354	(774)	332	502	427	
Undisturbed statistical chain end-to-end distance	$h_0 \, 10^3 M^{-1/2}$	Å (mol/g)^{1/2}	950	765	890	670	640	720	880
Chain rigidity	$s = h_0/h_{0f}$		1.63	1.61	1.63	2.3	2.14	1.83	1.1

273

Fig. 9.8. Crazing of polytrifluoro chloroethylene under constant load (from [130]): (a) appearance of crazed specimen surface, (b) electron micrograph of a craze showing fibrillar craze material which connects rather straight boundaries.

The subject of crazing has been extensively researched for various reasons:
— in brittle fracture of a crazeable material, cracks generally grow and propagate through crazed material,
— the formation of crazes under stress, and/or in presence of an active environment, effect (and generally impairs) mechanical and optical properties, surface quality, and permeability,
— crazes trace and make visible the statistical and geometrical distribution of the variables of their initiation (states of strain or stress, molecular properties, structural defects).

Crazes occur most commonly in amorphous, glassy polymers such as PS, SAN, PMMA, PVC, PSU, PPO, and PC but also in semicrystalline ones (PE, PP, PETP, and POM). Crazes have also been observed in epoxy [131] and phenoxy resins [109]. With reference to the cited literature (e.g. [76—78, 85, 106]) it can be said that crazes are sharply bounded regions filled with "craze matter". The craze matter consists of oriented strands interspersed with voids and has an average density of 40 to 60% of the matrix density (Figs. 9.9 and 9.10). Kambour [76, 85] has described the structure of crazes to be similar to an opencelled foam in which the average diameter of the holes and polymer elements is about 20 nm. There is much evidence that the polymer elements are oriented primarily along the direction of the principal tensile stress. Beahan et al. [106] noted that a fairly uniform fibril structure exists over the whole length of a tapered craze (Fig. 9.9). From the uniformity of this structure, these authors conclude that the increase in the width of the craze with sample extension is mostly due to fibrillation of further matrix material rather than to an increase in fibril extension. More extensive studies [150] have shown, however, that there is a change in the micromorphology of the craze along its length, with coarser, thicker fibrils being found in the tapered section near the craze tip. The morphology of crazes in thin cast films as well as in bulk specimens is rather similar

Fig. 9.9. Newly formed craze in thin slice cut from uncrazed bulk polystyrene; craze growing from left to right in a direction perpendicular to that of the uniaxial tensile stress (Courtesy D. Hull [106]).

Fig. 9.10. Electron micrograph of the central section of a craze grown as that in Figure 9.9 (Courtesy D. Hull [106]).

d   c   b   a

Fig. 9.11. Schematic diagram of the variation in polystyrene craze structure with increasing craze width; the angle at the tip of the craze is exaggerated and the scale in region d is larger than that of a to c. (Courtesy D. Hull [115]).

[115]. In the bulk, fibrils of 25 to 50 nm in diameter have been observed (Fig. 9.11, b and c). At low strain rates, the crazes grow in thickness by drawing further matrix material but also to some extent by an increase in the orientation and elongation of the fibrils leading eventually to the breakage of the fibrils. The first stages have been shown schematically in Figure 9.11. The principal observations with regard to craze morphology are indicated in Table 9.3.

The mechanical response of fibrils in the direction of the applied stress has been determined for polycarbonate [83] and polystyrene [120]. Figure 9.12 shows a stress-strain curve of craze material obtained by Kambour and Kopp [83] for PC. It may be noted that the craze material can sustain tensile stresses only slightly less than the yield stress σ_F of the bulk material. The strains, however, are enormously greater in the craze – between 40 and 140% as compared to the bulk yield strain of about 2%.

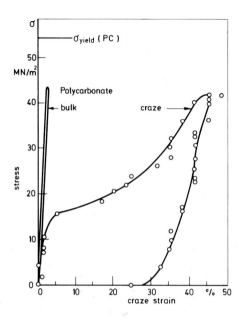

Fig. 9.12. Stress-strain curve of craze material in polycarbonate (after [83]).

It follows from this behavior that the energy of deformation of a thoroughly crazed sample can attain very large values (cf. 8 II A).

B. Craze Initiation

In spite of the fact that a considerable amount of research has been conducted on various aspects of crazing, there has been no exact agreement on the mechanism of craze initiation. So far, there is no theoretical model available which can predict in all generality whether a certain polymer will craze or not under a given set of conditions. And, if it does craze, what effect the temperature and rate of deformation will have on craze formation and propagation. This is certainly due to the fact that the initiation of a craze depends simultaneously on three groups of variables characterizing respectively the *macroscopic state of strain and stress*, the nature of *defects of heterogeneities* in the matrix, and the *molecular behavior* of the polymer in the given thermal and chemical environment. In focussing on particular variables five conceptually different approaches to describe the craze initiation emerge. These are based on stress-bias, critical strain, fracture mechanics, molecular orientation, and molecular mobility respectively. The principle observations and the craze initiation criteria proposed on the base of the above concepts are listed in Table 9.4.

These criteria describe different states of local excitation and deformation of chain segments. The stress bias criterion [86, 139] refers implicitly to two mechanisms: *cavitation* in a dilatational stress field and *stabilization* of cavities through a deviatoric stress component. These mechanisms have been more explicitly considered in the mathematical model of cavity expansion in a rigid plastic by Haward et al. [137] and in the molecular models by Argon [152] and Kausch [11].

Table 9.3. Morphology of Crazes in Glassy Polymers

Observations	Conclusions	Material and References
Optical and mechanical study of the effect of various variables, such as magnitude and duration of stress on initiation and development of crazing.	Crazing is a mechanical separation of polymer chains or groups of chains under tensile stress. Crazing does not occur in materials oriented along the draw direction.	Hsiao and Sauer [78] PS
X-ray scattering and electron microscopy suggests that void content of the craze is dispersed in the form of inter-connected spheroidal holes having a common dimension of 10–20 nm. Stress-strain curves of crazes have been determined showing a yield stress followed by a recoverable flow to roughly 40 to 50% strain at 41 to 55 MN/m². Returned to zero-stress, the craze exhibits creep recovery at a decelerating rate.	Crazes resemble on open-celled foam, the holes and polymer elements of these average about 20 nm in diameter. Craze-tensile behavior is an extension of the craze formation process. Decrease in elastic modulus and yield stress with increasing strain are rationalized in terms of strain-induced decrease in density and resultant increase in stress concentration factor on the microscopic polymer elements of the craze. Polymer surface tension and large internal specific surface areas of the craze are suggested to be important factors in large creep recovery rates of craze.	Kambour [85] PC
Microstructure of crazes in thin films by transmission electron microscope. In thin PS films ϵ (craze initiation) = 1%	Fibrils of 25–50 nm are a common feature of the micro-structure of crazes in cast and bulky films. Major fibrils diameter = 20–30 nm. Minor fibrils connecting major fibrils have a diameter >10 nm and they tend to orient normally to major fibrils. At large total deformation at low $\dot\epsilon$, there is a gradual transition from coarse to fine microstructure of crazes.	Beahan, Bevis & Hull [115] PS
Optical and micromechanical study of PS craze behavior: – average craze thickness = 0.19 μm – spacing between the crazes = 38 μm – crazes start forming at 0.5 σ_F and ϵ = 1.0%	The large number of parallel crazes formed permitted the determination of the stress-strain curve of a craze; it is similar to that found by Kambour and Kopp [83] for solvent-induced crazes in PC (Fig. 9.12)	Hoare and Hull [120] PS

Table 9.4. Criteria of Craze Initiation

Observation	Conclusion	Materials References
Stress-based criteria		
Analysis of craze initiation stresses in tensile samples containing circular holes, in thin-walled biaxial tube samples, and for rods in combined tension-torsion tests	**Stress bias criterion:** $|\sigma_1 - \sigma_2| \geq A(T) + B(T)/I_1$ A, B material constants depending on thermal history and environment, $I_1 = \sigma_1 + \sigma_2 + \sigma_3 > 0$; hydrostatic pressure suppresses crazing	Sternstein et al. [86. 139] PMMA
Analysis of location and direction of crazes formed — in sheets compressed by a die or wedge — during three-point bending of a round-notched bar	Crazes are initiated at the intersection of lines where shear bands are formed. Since crazes can be initiated before or after shear band formation it is assumed that large hydrostatic tensions rather than strains cause the craze initiation. Critical hydrostatic tensions: 87 MN m^{-2} for slowly cooled PC 89 MN m^{-2} for quenched PC	Camwell et al. [141] PS Ishikawa et al. [172] PC
Strain-based criteria		
Analysis of location and direction of crazes formed in tension at the interface between steel balls in PS and rubber balls in PS	Criteria of *principal strain* and of *strain energy* agree best with test data, maximum principle stress, maximum principle shear, and stress bias criterion correspond less well, maximum dilation the least	Wang, Matsuo, Kwei [105] PS
Craze formation in long-term tensile testing	Time-dependent critical principal strain; limiting values ($t \to \infty$), in %: PS 0.3 [110] SAN 0.7 [110] PMMA 0.7–0.9 [104, 110] PVC 0.8–0.9 [104] POM 1.5–2 [104, 121] PP 2 [121] PC 2 [110]	Menges, Pohrt, Schmidt et al. [104, 110, 113, 121]
Craze formation in biaxial stress tests "second quadrant"	$\epsilon_c = A'(T) + B'(T)/I_1$	Oxborough et al. [142] PS

Table 9.4. (Continued)

Observation	Conclusion	Materials References
Fracture mechanics criteria		
Study of craze behavior in sharply notched PMMA immersed in methanol Craze initiation and growth characteristics are controlled by the initial stress intensity factor K_0 not by applied stress: $K_0 < K_m$ no crazing $K_0 < K_n$ craze initiation and arrest $K_m = 0.06$ MN m$^{-3/2}$ $K_n = 0.2–0.3$ MN m$^{-3/2}$ (depending on sample thickness)	Craze growth follows one of two distinct patterns. One leads to eventual craze arrest, the other to a constant speed of craze propagation and final failure. Kinetics of craze growth in methanol explained by liquid flow through the porous crazed material (to which a void spacing of 0.25 μm, a void size of 72 nm and a craze yield stress of 9 MN m^{-2} are assigned).	Marshall et al. [102] PMMA
Study of crack arrest in tensile loading of cast PMMA sheets in various organic liquids	In a plot of σ_t^2 (square of the threshold stresses for crack propagation) versus reciprocal crack length a a distinct lower bound was found which defined a *critical energy release rate* G_I; whereas G_I initially decreases with temperature it becomes independent of temperature at $T > T_c$ where T_c roughly corresponds to the T_g of PMMA in the particular environment.	Andrews et al. [124] PMMA

Photoelastic study of craze initiation and propagation in kerosene and near a Griffith crack. Initial value of K_0 is the governing factor for craze initiation and propagation
$x = At^n$
where x = craze length
A = constant depends on K_0
$n = 0.53$ for PMMA
   $= 0.24$ for PC
   obtained from craze growth histories
	Stress concentration at *crack tip* is sensibly reduced with craze initiation and *not transferred to craze tip*.
	There is a critical stress intensity factor K_m below which no crazing occurs:
	$K_m = 0.13$ MN m$^{-3/2}$ PMMA
	$K_m = 0.29$ MN m$^{-3/2}$ PC
	The difference in sorption kinetics is held responsible for the higher critical stress intensity factor and the lower exponent n in PC, relative to PMMA.
		Narisawa et al. [127]
		PMMA
		PC

279

Table 9.4. (Continued)

Observation	Conclusion	Materials References
Structural and morphological criteria		
Study of craze initiation stress σ_i, of location of crazes within bulk samples and of change of birefringence R across sample thickness x $\sigma_i = 46$ MN m^{-2} for $dR/dx = 0$ $\sigma_i = 38$ MN m^{-2} for $dR/dx = 2 \cdot 10^{-3}$	Craze initiation occurs in regions where chains were oriented transversely to direction of stress (birefringence positive)	Haward et al. [89] PS
Study of birefringence and of tensile and compressive behavior of hot-drawn sheets at various angles to draw direction	The crazing stress increases with amount of orientation when the tensile axis is parallel to the draw direction, and decreases with amount of orientation when the axis is normal to the draw direction. At high draw ratios the crazing stress parallel to the draw direction is higher than the shear yielding stress and the material undergoes a yielding process similar to compression yielding. Molecular orientation has only a small effect on the orientation of the craze plane. The effect of molecular orientation on the crazing behavior of polystyrene is primarily associated with the nucleation stage of the crazing process.	Hull et al. [153] PS
Statistical nature of initiation of stress crazing is explored; a plot of histograms of the time to craze initiation follows approximately an exponential distribution.	The time required for the formation of crazes is an inherent, but statistical, characteristic of the material itself involving a stress depending rate process reflecting to some extent the initial microscopic flaw distribution.	Narisawa et al. [118] PMMA

Table 9.4. (Continued)

Observation	Conclusion	Materials References
Molecular criteria Relation between yield stress σ_F and craze initiation stress σ_i is investigated: $\sigma_F(T) \, \dot{\varepsilon} = $ constant $\sigma_i(T) \, \dot{\varepsilon} = $ constant where σ_i decreases less than σ_F at high temperature. Also $$\left[\frac{\partial \sigma_F}{\partial (\log \dot{\varepsilon})}\right] \text{ and } \left[\frac{\partial \sigma_i}{\partial (\log \dot{\varepsilon})}\right]$$ are represented by the same function $f(T)$	– Both yield stress and crazing stress vary with temperature, but crazing stress varies less than the yield stress. This indicates that crazing has to provide the nucleation surface energy for voids. – Similar response of crazing and yielding data to temperature indicates that in both cases similar molecular conformational changes and backbone motions are involved. – Materials having highest crazing tendency (PMMA, PS, SAN) also have high yield stress and thus a large energy associated with local plastic instability (which is necessary to account for the nucleation and growth of the tiny voids).	Haward et al. [108] PS
Analysis of craze initiation in tubular specimens under combined tension and torsion considering the mechanism of craze growth as one in which craze tufts are produced by repeated break-up of concave air-polymer interfaces at the craze tip (meniscus instability). For a surface energy $\gamma = 0.05$ J/m^2 and a tensile plastic resistance of $\sigma_y = 10^2$ MN m^{-2}, an extension ratio $\lambda_n = 2$ of craze matter gives a craze tuft diameter $= 0.05$ μm.	Crazes initiate at surface or interface stress concentrations where localized plastic flow produces microcavities by intense inhomogeneous plastic shear at a molecular scale. The rate of cavity formation depends primarily on the activation free energy ΔG^* (cf. Eq. 8 – 30) and on the local concentrated deviatoric stress while the subsequent plastic expansion into craze nuclei is in response to global negative pressure.	Argon et al. [152, 166] PS

The strain criteria [104, 110, 113, 121] specify that no crazing will occur if the *largest principal strains* — even at extended times — remain below a limiting value, which was found to be some 2% or less. At short times craze initiation strains may, in some polymers, be of the order of 3 to 5% [104, 113]. In the absence of intensive strain concentration strains of this order are insufficient to cause chain scission but they tend to decrease the intermolecular attraction (cf. 8 II B). This is recognized by Oxborough and Bowden [142] who also use the tensile hydrostatic stress (I_1) in their general critical-strain criterion. In biaxial stress fields the stress bias criterion and their criterion differ only little; as Oxborough and Bowden point out Sternstein's data and their own can be explained by either one of the two criteria.

If one would apply either of the above criteria the craze initiation in a sharply notched plate loaded in tension, one would have to expect instantaneous crazing since both $|\sigma_1 - \sigma_1|$ and ϵ show a singularity at an infinitely sharp crack tip (cf. Eqs. 9.1 to 9.3). Such an expectation would be contrary to the experimental findings. Marshall et al. [102] and Narisawa et al. [127] have established that it is the *initial stress intensity factor* K_0 which controls craze initiation at the crack tip. In the case of PMMA and PC immersed in methanol or kerosene critical values K_m exist below which no craze initiation and growth occurs. This behavior can be understood in view of the discrete sizes of the chain segments and of any voids to be formed, in view of the fact that the density of stored elastic energy is limited (Fig. 9.3), and in view of plastic deformations eliminating the stress singularity. Marshall et al. [102] conclude from their data that crazing occurs when the material at the crack tip reaches a critical strain or crack opening displacement.

The effect of sample structure and morphology on craze initiation has been recognized since the earliest investigations [78]. The structure of the sample surface is for various reasons especially important:

— *defects and/or contaminations* are preferentially found at the surface enhancing craze initiation there [173, 176];
— injection molded or machined samples mostly contain a *surface structure* which differs from that of the bulk material [89, 175, 176];
— molecular chain segments in surfaces have a higher degree of *mobility* [112];
— *environmental attack* (diffusion, plastification, degradation) begins at the surface and penetrates from there.

The investigations of the effect of chain orientation on craze initiation show that a transverse orientation of the chains with respect to the direction of principal stress enhances craze initiation [89, 153]. Because there are fewer chain segments pointing in the direction of principal stress critical local strains are obtained at smaller stresses (cf. 3 IV E). On the other hand the craze initiation stress increases with the degree of alignment of the chains in the stress direction (increasing degree of orientation, small angles θ between draw direction and principal stress). In the case of good alignment craze initiation stresses will be higher than the stress for shear yielding so that no crazing is observed. For PS at 20 °C yielding in tension occurs in samples drawn to $\lambda = 2.6$ or more and at a $\theta(\lambda)$ smaller than 20 to 30° [153]. Particularly noteworthy is the observation of Hull and Hoarse [153] that the molecular orientation has only a small effect on the orientation of the craze plane.

This indicates that the transfer of stresses along the axes of the oriented chain segments does not determine the direction of craze extension.

As an effect of sample structure and morphology the enhancement of craze initiation through surface flaws [118, 152, 173, 176], impurities [155, 161], and heterophase inclusions (reviewed in [114, 168, 190, 191] has also to be mentioned.

The molecular criteria generally take into consideration molecular weight, chain entanglement and the local mobility of chain segments at the given temperature and chemical environment [11, 15, 50, 79, 146, 165–167, 173]. The effect of the chemical environment will be discussed in Section D. The following investigations generally have been conducted in standard atmosphere. An interpretation of craze initiation in terms of local strain softening has been given by Rusch and Beck [95]. They propose that there exists a critical strain for crazing which is dependent on the level of frozen free volume initially distributed in the bulk of the material. Through the imposed dilatant strain sufficient free volume is added to bring the polymer to a state similar to the one reached by it at its T_g. Gent [96], too, suggests that craze initiation results from lowering of the T_g in local high stress fields at surface flaws. However, as shown by various authors [89, 129, 158] surface crazes occur in PS, PMMA, and PC at temperatures of -75 °C and below. This raises a doubt whether the polymer reaches a state characteristic of its T_g at such low temperatures or not.

Lipatov and Fabulyak [112] point to the importance of low temperature relaxation processes due to the motions of side chains. These relaxations shift towards lower temperatures in samples with high surface to volume ratios. This behavior has been interpreted in terms of lower segmental packing on the surface and, therefore, easier molecular motions. It is claimed that this facilitates crazing. Shifts of molecular relaxations towards lower temperature (in PC) have also been observed by Sikka [163] who suggested that this shift may have its origin in the formation of microvoids.

Studies of molecular weight dependence of crazing and of breaking stresses have been carried out by Fellers and Kee [146]. These authors suggest that craze *initiation*

Fig. 9.13. Breaking stress and crazing stress as a function of molecular weight M_n of polystyrene at 25 °C (after [146]). M_e entanglement molecular weight.

stresses are independent of molecular weight when $\overline{M}_n > 2\,M_e$ whereas craze *development* and *breakdown* are clearly molecular weight dependent. This is evident from the plot of stresses versus molecular weight, shown in Figure 9.13. Since the measured elastic moduli (1.5 GN m^{-2}) did not depend on molecular weight the craze initiation strains are also constant (2%). From these results it may be concluded that craze initiation is an event which depends primarily on the interaction between chain segments. Contrary to the apparent independence of stress and strain at craze initiation to molecular weight the T_g of the different PS fractions rises monotonically from 88 °C at $M_n = 70\,000$ to 105 °C at $M_n = 150\,000$. Another observation made by these and other authors is that crazes in high molecular weight samples are very fine, *numerous,* long, and straight as compared to the ones in lower-molecular weight polymers, which are coarser in texture and somewhat shorter in length. This phenomenon seems to be related to the homogeneity of the stress field which is the better the more numerous and the stronger the fibrils. Based on these studies and on the extensive investigations of Döll et al. [15, 30, 50, 60, 61, 67] a model of craze initiation and propagation is presented (Fig. 9.14). The effects of molecular weight

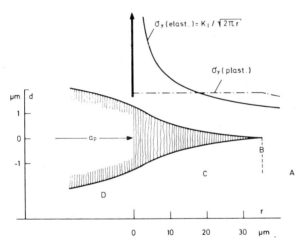

Fig. 9.14. Model of craze initiation and propagation in PMMA (after [11]): A. elastically strained region. B. region of void nucleation and fibril formation, C. craze growth due to fibril extension and continued fibrillation at fairly constant stress σ_y, D. transformation of craze into crack through breakage of stressed fibrils.

$(\overline{M}_n > 2\,M_e)$ on crazing may be characterized using this model:
- $\overline{M}_n$ has very little influence on the elastic straining behavior (region A) and on the stress or strain at which crazes are initiated (B)
- after craze initiation, fibrils are formed which are stronger and which can develop, under increasing stress and time, to greater lengths the higher $\overline{M}_n$; consequently region C is the more extended the higher $\overline{M}_n$
- the force transmitted by region C grows with $\overline{M}_n$ and so does the stress at break.

In an attempt to reconcile the various partially insufficient craze initiation criteria with the experimental data Argon [165–167] and Kausch [11] have proposed models of the craze initiation process which take molecular structure, chain rigidity, conformational changes, and intermolecular interaction into consideration. Argon bases his criterion of the transition from A to B on the break-up of concave air-polymer interfaces (cf. Table 9.4). Kausch has described a three-step craze-nucleation mechanism:
— a primary step of decohesion of molecular coils with low degree of entanglement, i.e. the formation of unstable micro-voids (B_1 in Fig. 9.15).
— a secondary step of anelastic deformation of adjacent molecular coils containing topological entanglements (B_2 in Fig. 9.15),
— a third step of direct lateral transmission of strain onto adjacent coils which leads to a preferred formation of micro-voids in correlation to the first.

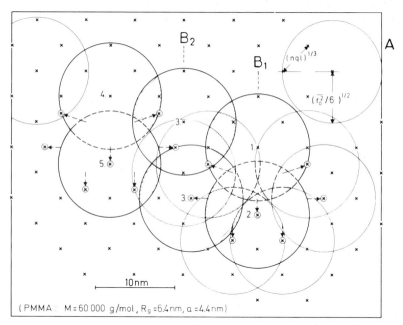

Fig. 9.15. Model of void nucleation and fibril formation in PMMA (after [11]): A. elastically strained region, B_1. possible void nucleation at site of insufficient penetration of molecular coils 1 and 2; shaded areas indicate zones of strong intermolecular interaction and transfer of stresses onto neighboring coils 3, B_2. possible fibril consisting of interpenetrating coils 3; if their response is of sufficient amplitude, void nucleation is furthered in adjacent sites (between coils 4 and 5).

The micro-voids are stabilized through the anelastic deformation of coils in the secondary step. According to a suggestion of Fischer the attainment of a critical macroscopic compliance permitting an anelastic deformation of the coils of sufficient amplitude is the necessary prerequisite for the secondary step to take place [11]. With sustained or increased load the micro-voids coalesce and leave micro-fibrils which connect the opposite craze surfaces. These micro-fibrils are formed, therefore, of moderately elongated, physically entangled molecular coils. The diameter of the micro-voids of 10 to 20 nm and that of the fibrils of 10 to 50 nm indicate that the formation of a micro-void is an event involving only very few molecular coils.

C. Molecular Interpretation of Craze Propagation and Breakdown

The foregoing discussion on the phenomenology of craze initiation has also given a molecular interpretation of the propagation of an individual craze. Taking into consideration the formation of many crazes either simultaneously, or in sequence, a particular craze grows through:
- displacement of the craze tip into uncrazed material by the formation of voids ahead of the craze tip or by coalescence with a separate and smaller craze,
- widening of the craze while extending the craze matter, i.e. the molecular strands; depending on the overall conditions strands may rupture or disentangle giving rise to a crack.

There are three aspects of craze propagation which will be discussed at this point: the kinetics of craze growth, the stress at the craze-matrix interface (region C in Fig. 9.13), and the craze breakdown.

The rate of longitudinal craze growth has been studied by a number of authors. The many problems still existing with regard to the transition of matrix material into craze matter and with the rheological properties of the latter are found to be of double weight in any quantitative description of craze propagation. For this reason no detailed description of the various approaches will be given here – but the basic concepts should be mentioned. The fracture mechanical investigations on PMMA [15, 50, 102, 127, 133] and PC [127, 144] have led to empirical expressions for $d(a + r_p)/dt$, the rate of craze growth, in terms of *stress intensity factors*. Kambour [76] and Marshall et al. [102, 133] stress the importance of *environmental flow* through the porous craze material. Verheulpen-Heymans [155] formulates a model for craze growth based on stress and strain analysis around a craze and on the *rheological properties of the craze matter*. In those cases where the craze length was found proportional to the crack length [15, 144, 177] the *empirical law for crack growth* (e.g. Eq. 9.22) also describes craze growth.

The stresses at the craze-matrix interface may be determined from the craze shape. It was found by Weidmann and Döll [15] for PMMA and by Fraser and Ward [177] for PC that the shape of these interfaces conformed very well to the predicted shape according to the Dugdale-Muskelishvili plastic zone model [7]. In that model it is assumed that a normal stress equal to the matrix yield stress σ_F is acting on the interfaces. This stress σ_F is derived from Eqs. (9.18) and (9.20) as $\pi E v_a/4 r_p$. Using the approximately molecular-weight independent ratio of $r_p/2 v_a \simeq 27$ as given by Weidmann and Döll [15] for PMMA and using an E of 4.5 GN m^{-2} one arrives at an σ_F of 65 MN m^{-2}. This value corresponds to the strength of the isotropic PMMA (Table 1-I). Fraser and Ward [177] obtained in the same way an interfacial stress of 58 MN m^{-2} (at 22 °C) and 130 MN m^{-2} (at −130 °C) for PC.

286

Another method, Knight's analysis of stress distribution along a craze, has been used in a number of investigations (cf. [76, 77]). Verheulpen-Heymans [157] quite recently pointed out, however, that the — mostly unknown — rheological behavior of craze material and of the craze tip zone has such a strong effect on the calculated stress field that at the present time the results of this method cannot be interpreted unambiguously.

Using holographic interferometry Peterson et al. [148] determined the distribution of the incremental stresses caused by a strain increment applied to a craze in PC. At low prestrains (1.3–1.7%) they observed that the craze material near the tip of the craze actually supported a stress increment $\Delta\sigma_y$ well in excess of the average stress increment $\overline{\Delta\sigma_y}$ ($\Delta\sigma_y/\overline{\Delta\sigma_y}$ equal to 1.2 to 1.45). The peaks in the stress increments occurred at a distance of 0.2 to 0.4 of the craze length *behind* the craze tip. In their photoelastic studies of PMMA and PC crack tips Narisawa et al. [127] noted high stress concentrations just before the initiation of a craze and very modest stresses 60 min later. No noticeable new stress concentration appeared at the craze tip.

In a growing craze molecular strands are formed (at B in Fig. 9.14) and extended (zone C in Fig. 9.14). The growth of a craze is slowed down or even halted

— if the craze hits upon an obstacle which prevents further fibrillation of matrix material (e.g. a heterophase inclusion; a zone of higher orientation, larger flexibility or lower stresses; the end of the specimen)
— if the craze has grown to such an extent that the initiating crack or flaw is blunted and rendered uncritical
— if the craze interacts with other crazes or with the environment in such a way as to change the conditions which have led to its initiation and/or growth (stress relaxation of the sample because of craze-related creep deformation, penetration into a lower-stress zone adjacent to another craze).

On the other hand craze growth will continue or be resumed if the molecular strands fail (transition from craze to crack, zone D in Fig. 9.14). It becomes immediately clear from Figure 9.14 and Eq. (9.18) that the energy dissipated in the opening up of a craze is the higher the longer and wider the craze (zone C) and the larger σ_F. By no means all questions are answered concerning the structure and rheology of the molecular strands and of the interface craze-matrix. Nevertheless it can be said that at a void content of 50% a molecular strand is subjected to a tensile stress of about 2 σ_F, a stress which does not vary too much with development of the craze from B to D. The mentioned maximum in *incremental* stress at a distance of 0.2 to 0.4 of the craze length in PC does not contradict the above statement. It indicates, however, that the differential compliance of a freshly developed strand is smaller than that of an extended fibril.

While being subjected to the tensile load the molecular strand obviously degrades and eventually fails and this causes a transition from a craze to a crack. The breakdown occurs by one or more of the following mechanisms:

— slippage and disentanglement of molecular coils leading to fibril rupture (frequently of the central section of the fibril),
— fibril rupture through breakage of entangled chains,
— fibril-matrix separation.

287

The phenomenon of the loading and breakdown of the molecular strands has been studied by various methods. Optical and electron micrographs of crazes are shown in most of the cited references. Examples have been reproduced in Figures 9.8–9.10. Investigations of the craze shape by interference microscopy [15, 155, 177] have also been discussed. At this point some results will be reported which have been obtained by thermal measurements [31, 50, 184–186], from analysis of the influence of molecular weight on crazing [11, 15, 65, 79, 146, 178], through acoustic emission [174, 188], and by the ESR technique [189–190] respectively.

Döll [31, 50] determined the amount of heat Q dissipated in PMMA at the tip of a rapidly propagating crack. To derive Q they measured the temperature rise $\Delta T_1(t)$ at a thermocouple placed on the specimen surface close to the prospective fracture plane. Considering geometry, thermal conductivity, and crack speed $\dot{a}$ they calculated the heat Q dissipated per unit are of crack surface. As expected Q increased with $\dot{a}$ and M_w (from 0.15 to about 5 kJ/m^2) and it was found proportional to the dynamic energy release rate G_d. Extrapolating $Q(G_d)$ down to G_{Ic} (at $\dot{a}$ = 140 m s^{-1}) they observed that the dissipated heat Q ($\dot{a}$ = 140 m s^{-1}) = Q_0 accounted for only 57% of the released mechanical energy G_{Ic}. They ascribe this difference to the fact that the highly stressed molecular strands conserve an amount of energy of the order of 0.43 G_{Ic}, that this energy is released at the moment of fibril rupture and that it is dissipated over wide distances in the form of small shock waves which escape measurement by the employed difference technique [30]. The acoustic emission analysis (AEA) of a crazing PMMA sample by Roeder [188] permits similar conclusions. The formation of crazes (transition A → C in Fig. 9.14) did not give rise to any discernible acoustic signal. Whereas the transition from a craze to a crack (C → D) could be traced by AEA even at an earlier stage than by optical microscopy. In PS, however, craze formation is accompanied by sharp cracking sounds at strains of 0.6% or less [78].

The thermal measurements are not only interesting in view of the energy balance of the crazing process but also because they permit the calculation of the local temperature rise ΔT_0 caused by the opening up and breakdown of the craze in PMMA. Döll [30] assumed that Q_0 was initially confined to the zone of crazed material. Using a density of 0.6 g cm^{-3}, a specific heat of 1.46 J g^{-1} K^{-1}, a craze layer thickness 2 v of 1.65 μm, and a Q_0 of 335 J m^{-2} he arrives at a ΔT_0 = 230 K. This value is in accord with theoretical estimates of Weichert and Schönert [185] and with infrared measurements of Fuller et al. [184] on PMMA. The latter determined within a range of $\dot{a}$ of 200 to 640 m s^{-1} a constant ΔT of 500 K. The simultaneously observed increase in $Q(\dot{a})$ implied that the plastic deformation at the crack tip became more extensive at higher crack speeds. Preliminary experiments on PS showed a ΔT of 400 K and smaller heat values [184]. These temperatures are certainly large although not completely unreasonable. They indicate that not only melting but thermal degradation must occur under these conditions. And they are in proportion with the much higher temperature rises – of several thousand K – which were deduced by Weichert [186] from the spectroscopical analysis of light emitted during fracture of glass.

It was noted at an early stage that sample molecular weight is an important variable in craze propagation and breakdown. Rudd [79] studied the stress relaxation of PS films in contact with the crazing environment butanol. A plot of $\sigma(t)/\sigma(0)$ versus logarithm of time revealed that the time to reach the 40% stress level was three orders of magnitude larger for the higher molecular weight samples. The final decay from 0.4 $\sigma(0)$ to zero stress occurred rather rapidly for all molecular weights [79].

The observation of Fellers and Kee [146] that the breaking stress of PS only gradually increases once $M_n > 2 M_e$ has already been mentioned in the previous section (Fig. 9.13). Their results conform quite well to those of Döll and Weidmann [15, 50]. These authors determined the shape of a craze, the released heat Q and the material resistance R for a series of PMMA samples having well defined molecular weights M_w between $1.1 \cdot 10^5$ and $8 \cdot 10^6$. Measuring the crack opening 2 v, the craze width 2 v_c, and the craze length r_p at a crack speed of 10^{-8} m s^{-1} they noted that these parameters of the craze shape increased with M_w up to an M_w of about $2 \cdot 10^5$. At higher M_w hardly any changes of 2 v and r_p and very small increases of craze width were observed [15]. This means that initially ($M_w < 1.6 \cdot 10^5$) craze widths increase with chain lengths. The craze widths are found to be by a factor of 5.2 larger than the extended chain length. This does not mean, however, that each molecular strand consists of just a few, highly extended chains. One may assume that before craze initiation the molecules have their randomly coiled conformation. Chains of $M_w = 1.1 \cdot 10^5$ for instance have an end-to-end distance of 21 nm. During craze opening this distance will on the average increase by the fibril strain, i.e. to about 30 nm. A total of 2360 of these elongated coils are contained in a fibril of 20 nm diameter and 1200 nm length. At molecular weights comparable to M_e the interpenetration and entanglement of these coils hardly allows the formation of fibrils [11, 146, 187]. At higher molecular weights (up to $M_w = 2 \cdot 10^5$) the coils are larger in size and the absolute number of entanglements per coil increases; the fibrils become stronger and more resistant to failure by slippage or disentanglement. At even higher molecular weights ($M_w > 2 \cdot 10^5$) and under constant load the fibril strength seems to be more determined by the strength than by the flow behavior of the molecules. Under oscillating loads disentanglement may even occur at such molecular weight levels. Recent experiments of Skibo, Hertzberg, and Manson [191] strongly support the contention that the breakdown of molecular strands as a consequence of their disentanglement triggers the stepwise growth of fatigue cracks. This phenomenon will be discussed in more detail in Section 9 III C.

The critical energy release rates G_{Ic} show a behavior similar to that of the craze shape. For small M_w the G_{Ic} are strongly dependent on M_w. Thus they increase from 1.4 J m^{-2} at $M_w = 2 \cdot 10^4$ to 110 J m^{-2} at $M_w = 12$ to $15 \cdot 10^4$ [65]. At higher molecular weights – up to $8 \cdot 10^6$ – only a gradual increase of G_{Ic} to values of between 160 and 600 J m^{-2} is observed [30, 65]. This corresponds to what has been said above on the effect of molecular weight on fibril length and strength.

The occurrence of chain rupture in craze breakdown has been inferred from the above thermal, mechanical and acoustical evidence. Its direct observation by ESR technique is difficult because of the limited number of crazes which can simultaneously be carried to breakdown. The number of chain breakages generally corresponds to that of a single fracture plane. No quantitative data on free radicals formed in crazing have been reported. DeVries et al. [189] have observed a very weak ESR signal in a crazed and cut PMMA fiber. Nielsen et al. [190] did not find any evidence of free radicals in thoroughly crazed – but unbroken – samples of PS and ABS.

The above considerations mostly dealt with an individual craze, with the conditions of its initiation, propagation and breakdown in a brittle polymer. Generally a larger number of crazes is formed in a stressed specimen (cf. Fig. 9.8 a). If those crazes are sufficiently distant from each other then they will grow freely. Opfermann [175] found that in PMMA a lateral distance between crazes of 80 μm assured their unimpeded longitudinal growth. The macroscopic mechanical properties (strain at break, strength at short or long times, energy to break) depend to some extent on the number of crazes per (surface) area, but they are still comparable to those of a brittle solid with a strain at break of some 4 to 5% and a low fracture energy. In order to sensibly increase the macroscopic creep compliance and the energy to break, crazes must be initiated in large numbers throughout the sample volume and they must be prevented from premature breakdown. Both objectives have been achieved with heterophase copolymers or blends.

The obvious technological importance of the toughening of brittle polymers through controlled crazing is reflected by the considerable amount of work done in this area. In a recent monograph by Manson and Sperling [192] a comprehensive elaboration of the physical, chemical, technological, and materials science aspects of polymer mixtures, rubber-toughened plastics, block copolymers, grafted copolymers, interpenetrating networks, and polymer blends has been given. The reader is referred to this monograph for any detailed information on these systems. The conference proceedings [193] from the Bad Nauheim meeting on multiphase polymer systems also treat synthesis, compatibility, and mechanical properties of copolymers and blends. In addition, references are given which focus on individual aspects. Argon [152] and Retting [168] have reviewed the crazing of heterogeneous glassy polymers. A quantitative method to describe creep and impact toughening of HiPS is presented by Bucknall et al. [114]; the effect of crazing on the stress-dependence of creep in PS and SAN is discussed by Moore et al. [94], the surface morphology of stress whitening in ABS by Morbitzer et al. [97], the effect of rubber on the environmental stress-crack resistance of PE by Spenadel [117], and the rapid healing of crazes in ruptured rubber-toughened ABS/MMA sheets by Takahashi [143].

Valuable information on the role of crazing in breakdown of a polymer is obtained, of course, from fractography. Relevant investigations [30, 49, 61, 66, 132, 150, 155, 169, 194—204] on this subject will be discussed in Section III A.

D. Response to Environment

In the previous sections on crazing primarily mechanical and molecular parameters have been discussed while the chemical environment was not considered as a variable. At this point an overview over the physico-chemical response of a crazable material to an active chemical environment will be given. (The *chemical* response to an active *physical* environment such as photodegradation or ozonolysis has been treated or referenced in 8 III).

The physico-chemical actions of a (gaseous or liquid) chemical agent on a polymer involve adsorption and absorption of the agent, the swelling and/or plasticization of the matrix, the reduction of surface energies, and/or chemical reactions such as the hydrolytic depolymerization. These aspects of environmental stress cracking (ESC) have been treated in numerous original articles and some reviews (e.g. [80, 76–77, 171]). From the extremely large body of experimental observations only a few can be discussed in the following.

The physico-chemical actions of a liquid environment may influence the initiation, propagation, or breakdown of a craze in a thermoplastic polymer. There seems to be agreement that a liquid must be able to diffuse into a polymer in order to influence craze initiation. Narisawa [119] determined the critical stresses σ_i for craze initiation in thin films of PS and PC in contact with various alcohols and hydrocrabons. He observed that crazes appear without significant delay and that σ_i decreases with decreasing chain length of the solvent (from 45 to 20 MN m^{-2} for PS, from 70 to 50 MN m^{-2} for PC). From these results he concludes that a slight swelling of a microscopic surface area is necessary and sufficient to effect craze initiation. He derived a criterion for σ_i in the form of Eq. (8.29) with activation volumes of the order of 1.0 to .13 nm^3, activation energies of 109 to 130 kJ/mol, and rate constants of 1 to $10 \cdot 10^{-38}$ s^{-1} (PS) and 2 to $50 \cdot 10^{-45}$ s^{-1} (PC).

The solvent crazing in terms of polymer and liquid solubility parameters has been studied in detail by Andrews et al. [124, 126] and Kambour et al. [125, 128]. Kambour employed a wide range of swelling liquids with solubility parameters δ_S between 5.34 and 19.2 cal$^{1/2}$ cm$^{-3/2}$. He determined the equilibrium solubility S_v, expressed as volume of liquid absorbed per unit volume of polymer, for PS, PPO, and PSU. The swelling of PPO across the whole spectrum of organic liquids was found to be inversely correlated by $|\delta_S - \delta_{PPO}|$ and, consequently, the craze resistance was correlated by $|\delta_S - \delta_{PPO}|$. In PS and PSU craze resistance was less well correlated by the solubility parameters. With all three polymers, however, the equilibrium solubility S_v gave a good measure of the polymer-solvent interaction. Using S_v as independent variable, unique plots of T_g and of craze initiation strain ϵ_i were obtained for two sets of polystyrene data. One set was obtained from samples preplasticized to various degrees by o-dichlorobenzene; the other from "dry" samples in contact with the swelling agent (ϵ_i) or swollen films (T_g). Kambour concludes from these results that the presence or absence of a liquid/polymer interface is immaterial to the crazing effectiveness of a given crazing agent.

The crazing agent thus acts through its presence within the polymer matrix. In increasing the chain mobility (lowering T_g) it facilitates the primary and secondary steps of craze initiation: nucleation and stabilization of a craze. This leads to the lowering of σ_i and ϵ_i in brittle polymers such as PS. Easier nucleation and stabilization even cause the appearance of crazes in otherwise ductile materials such as PPO, PSU, PVC, or PC.

Andrews et al. [124, 126] also studied the equilibrium swelling, of PMMA, in various alcohols and related it to the observed changes in yield stress, σ_F, glass transition temperature, T_g, and material resistance R. They report the interesting phenomenon that R above a certain critical temperature T_c became independent of temperature. Within experimental error T_c corresponded to the T_g of the PMMA under the particular conditions of swelling. They propose a relation for R involving the surface energy of the nucleated cavities and σ_F. They then interpret the constancy of $R(T > T_c)$ with the vanishing of $\sigma_F(T > T_c)$. The absolute values of $R(T > T_c)$

are rather small: about 0.1 J m^{-2} for PMMA in isobutanol, carbon tetrachloride, n-propyl and isopropyl alcohols, 0.3 J m^{-2} for ethanol, 0.45 J m^{-2} for methanol [124].

The swelling, crazing, and cracking behavior of PMMA, PVC, and PSU in contact with some 70 different liquids was analyzed by Vincent and Raha [123]. They considered not only the solubility parameter δ_S but also the value of the hydrogen bonding parameter H_{OD} based on the displacement of the OD infra-red absorption band in CH$_3$OD in presence of the particular liquid and benzene. Better (but still no unique), correlation between the mode of failure and δ_S and H_{OD} was obtained.

An extensive investigation of the role of metal salts in stress cracking of various polyamides was carried out by Dunn and Sansom [90–93]. Metal halides permitted to distinguish two types of action: the formation of complexes between the metal and the carbonyl oxygen and interference with the hydrogen bonding (found for the chlorides of Zn, Co, Cu, Mn) or with solvent cracking, as with Li Cl, Ca Cl$_2$, Mg Cl$_2$, or Li Br [90–91]. The action of metal thiocyanates on polyamide 6 was similar to that of the corresponding metal halides [92]. Of various nitrates Cu(NO$_3$)$_2$ had the strongest effect on the stress cracking of PA6 films [93].

In the investigations of Marshall and Williams et al. [52, 102, 133, 151, 164] Narisawa et al. [127], Kitagawa et al. [144], and Krenz et al. [159–160] craze growth and creep were related to fracture mechanical quantities. Graham et al. [164] model crazes in PMMA as a line plastic zone subjected to a craze stress σ_{cr}. In the presence of active liquids σ_{cr} is reduced from its air value of 100 MN m^{-2} to e.g. 7 (methanol), 5 (ethanol and propanol) or 10 MN m^{-2} (butanol). A clear linear correlation between K_m and σ_{cr} was found to exist. For the craze size at initiation (i.e. just before the beginning of crack growth) the authors found a unique value of 11.5 μm [164]. Similar fracture mechanical studies of solvent crazes in polystyrene were carried out by holographic interferometry by Krenz et al. [159–160].

Studies of the micromechanics of deformation in semicrystalline polyethylene in environmental stress cracking agents [107, 154, 170, 171, 204] have elucidated the role of spherulites and of the mosaic block structure and its disintegration into independent nonuniform fibrils. The morphology of crazes in amorphous PVC in liquid and vapor environment was examined by Driesen [122] and Martin et al. [198].

It has not always been recognized that liquid gases such as N$_2$ and Ar are also acting as environmentally active agents. Parrish and Brown et al. [116, 129] have been the first to report that the fracture stresses of PE and PTFE immersed in helium at 78 K are respectively 21 and 33% larger than those obtained in liquid N$_2$. Later studies of Brown et al. [134, 157–158], Peterlin et al. [135, 137–138, 145] and Kastelic and Baer [140] concern N$_2$, Ar, O$_2$, CO$_2$, and He and a variety of materials (PP, PTFE, PETP, PC, PMMA). It is the common consensus that

- N$_2$ and Ar exert an environmental effect on the tensile stress-strain curves of all polymers at low temperatures
- the environmental effect is the larger the closer a gas is to its condensation point and the greater its thermodynamic activity [145]
- N$_2$, Ar, O$_2$, and CO$_2$ usually cause craze yielding whereas in He or *in vacuo* the polymers undergo brittle fracture without crazing
- the gases have to act at or close to a free surface

— as crazing mechanisms gas absorption and matrix plasticization are proposed as well as gas adsorption and surface energy reduction which facilitates cavity formation.

From the discussion in this section a number of craze *suppression* mechanisms have become apparent which may be summarized here:
— imposition of hydrostatic pressure in addition to tensile stresses [86, 139, 142]
— introduction of compressive stresses into the surface [77, 174, 176]
— orientation of molecules in the surface layer [89, 153]
— plasticization, preferentially of the surface [82, 162, 178]
— coating of the surface by an oligomer of the sample material, e.g. coating of PS by a 600 molecular weight polystyrene oligomer [178] polishing of the surface [173, 178]
— increase of sample molecular weight.

III. Molecular and Morphological Aspects of Crack Propagation

A. Fracture Surfaces

Fractography, the study of the morphology of fracture surfaces, is an obvious tool to elucidate the origin and the mode of propagation of a crack. In this monograph fracture surfaces of fibers (in 8 I G) and of other samples have already been analyzed where appropriate. At this point some additional remarks are necessary concerning particularly the effect of chain length and intermolecular attraction on fracture surface morphology.

The discussion in 9 I C of the various terms contributing to the critical energy release rate G_{Ic} and the data assembled in Tables 9.1 and 9.2 have made it clear that a fracture surface is obviously not simply formed by the breakage of primary and/or secondary bonds across a fracture plane of molecular dimensions. There is always a plastic deformation of the crack tip zone and, consequently, of the ensuing fracture surface. It is to be expected that the extent of plastic deformation is the smaller the smaller the segmental mobility, i.e. the lower the temperature. At liquid nitrogen temperature most high polymers resemble glasses and fracture in a brittle manner.

Fig. 9.16. Central section of a fracture surface from a notched HDPE specimen broken in bending at liquid nitrogen temperature (from [130]).

293

Viewed without magnification the fracture surfaces (e.g. Fig. 9.16) show a macro-scopic roughness but they appear to be locally smooth although not shiny. This indicates that the surfaces contain structural irregularities larger than the wavelength of light. This is the case with e.g. PE, PP, PVC, PS, and also with PMMA which, how-ever, has a very smooth surface.

The macroscopic features of the shown fracture surface (Fig. 9.16) can be ascribed to the propagation of a cleavage crack at high speed normal to the direction of *local* tensile stress. The local stress field is strongly affected by elastic waves generated at earlier stages of crack development and by the initiation of secondary cracks. The fracture surface was obtained by bending a notched polyethylene sample at liquid nitrogen temperature [130]. The surface is locally smooth but otherwise full of steps and ridges. The intersection of wavefronts and crack planes under different (e.g. right) angles leads to the curious patterns as shown in the bottom right of Figure 9.17. The characteristic properties of thermoplastics, molecular anisotropy and strong rate dependency of deformation become apparent to a limited extent only.

Fig. 9.17. Section of the fracture surface in the originally compressive zone from the same specimen shown in Fig. 9.16 (from [130]).

The seemingly smooth parts of the cleavage surface deserve two further com-ments. Figure 9.18 shows that the material disintegration of the "smooth" surface is not restricted to molecular dimensions but has a honeycomb-structure with cells of 100 to 600 nm width. This is in accordance with the earlier statement on the size of surface irregularities. It is also of interest to note that these dimensions are only a few times larger than the radius of gyration of the chain molecules. The ridges be-tween cells are highly deformed. As their width is comparable to that of the cells a plastic deformation of about 100% is indicated. This is in view of the test temperature quite remarkable. It is definitely caused by the fact that the comparatively long and strong molecules are not highly oriented but have a coiled conformation correspond-ing in all probability to the "solidification model" of semicrystalline polymers (cf. 2 I C). Under these circumstances axial forces are only built up if a segment is subjected to a sufficiently large shear displacement. Material separation thus neces-sitates the deformation of several entangled molecular *coils*.

At room temperature molecular mobilities are increased and intermolecular attraction is decreased. The extent of plastic deformation leading to material separa-tion is, therefore, increased rendering some of the formerly brittle plastics tough

294

Fig. 9.18. Successive enlargements of the apparently smooth surface in the center of Fig. 9.16 (from [130]).

(cf. 8 II A). In HDPE at room temperature no cleavage fracture (unstable crack propagation) can be obtained by bending, it occurs only in notched specimens of low aspect ratio or if the rate of loading is high (impact loading). In samples of HDPE with extremely high molecular weight ($M_w > 10^6$) cleavage fracture does not occur at all. Material separation must be enforced through large strains. Figure 8.28 gives an impression of the very large plastic deformation of the fracture surface of a notched bending specimen. The walls of the numerous rosettes and ridges are drawn by several hundred per cent. The energy consumed in forming new surface area is accordingly high. One can state that in short time loading of isotropic bulk polymers chain molecules do not significantly contribute to the fracture surface energy through the energies of bond breakage and elastic retraction. Large values of chain length and chain strength reduce, however, the effect of natural or artificial flaws as stress concentrators due to the larger amount of plastic deformation necessary before material separation occurs. Vice versa, at low molecular weights samples are increasingly more brittle, fracture surfaces are smoother, secondary fracture events are rare or absent, and the fracture surface energy accordingly small.

The above considerations apply to isotropic homogeneous materials, i.e. to amorphous [61, 198, 200] and to semicrystalline polymers having no well defined microstructure [130]. There the direction of the fracture path is more or less determined by the local stress field. In any event no traces of a microstructure became apparent through the fracture surface morphology. This does not preclude the existence of a certain globular microstructure (cf. 2 I C) which can be made visible through ion-etching [132, 208]. Distinct fracture surface features are obtained, however, from polymers with a pronounced microstructure as introduced by extended-chain crystallites or spherulites. There the material resistance depends strongly on the relative orientation of the fracture plane with respect to the structural element.

Extended-chain crystallites have been studied by several authors (cf. 2 I C), most recently by Gogolewski and Pennings et al. [201–203].

295

In Figures 9.19–9.21 electron micrographs of their replicas from the fracture surface of pressure crystallized polyamide 6 are reproduced [202]. The micrographs reveal stacks of lamellae which have a thickness of up to 700 nm. The authors conclude from their extensive IR, WAXS, and electron microscopical studies that the lamellae represent extended chains. As they have proposed in Figure 9.22 a crack may preferentially proceed either along the (010) planes (which contain the chain ends as well as impurities rejected from the crystal growth front) or along the (002) hydrogen bonded sheets of the lamellae. Both processes do not involve the breakage of main chain bonds or of hydrogen bonds.

Fig. 9.19. Electron micrograph of a replica of a fracture surface of 6 polyamide crystallized at 295 °C for 48 h under a pressure of 6.5 kbar. The arrow shows the growth direction of the spherulite (Courtesy S. Gogolewski [202]). (Reproduced by permission of the publishers, IPC Business Press Ltd. ©).

Fig. 9.20. Transmission electron micrograph of a replica of another part of the fracture surface of the sample shown in Fig. 9.19 (Courtesy S. Gogolewski [202]). (Reproduced by permission of the publishers, IPC Business Press Ltd. ©).

Fig. 9.21. Scanning electron micrograph of a fracture surface of 6 polyamide crystallized at 295 °C for 48 h under a pressure of 6.5 kbar. (Courtesy S. Gogolewski [202]). (Reproduced by permission of the publishers, IPC Business Press Ltd. ©).

With respect to extended chain crystallization under high pressure (p > 4 kbar) the authors discuss two mechanisms [201–202]. Based on the observed change in molecular weight distribution and decrease in molecular weight concurrent with chain extension they suggest that the thermal treatment under high pressure makes possible a transamidation reaction between –NH and –CO groups of broken chain folds belonging to adjacent lamellae [201]. On the other hand the appearance of lamellae with a thickness much smaller than the average chain length has led them to the conclusion that fractionation is accompanying the pressure crystallization [202].

The influence of a spherulitic structure of a semicrystalline polymer on its mechanical properties has been implicitly and explicitly recognized for a long time. Only a few references can be given here [183, 204–207]. Clark and Garber

Fig. 9.22. Model proposed for fracture behavior and surface morphology of chain-extended crystals of polyamide (Courtesy S. Gogolewski [202]). (Reproduced by permission of the publishers, IPC Business Press Ltd. ©).

[205] review the effects of industrial processing on the morphology of crystalline polymers. Patel et al. [206] and Andrews [207] are especially concerned with a stress-strain analysis of spherulitic HDPE. They find a maximum of sample elastic modulus of 1.2 GN m^{-2} at a spherulite radius of 13 μm. Studies of the failure morphology of HDPE in an active environment [204] revealed
- that failure occurred almost entirely in a brittle mode
- that failure within an individual spherulite is interlamellar, and
- that the mode of crack propagation depended on the position of the spherulite center with respect to the crack front (leading to interspherulitic fracture in the case of poorly matching interfaces [204].

Extensive morphological studies of polypropylene fracture surfaces have been carried out by Menges et al. [121, 136] and quite recently by Friedrich [183]. Using polypropylenes of different molecular weight and tacticity subjected to different thermal treatment Friedrich [183] obtained samples of different microstructure:

I fine spherulites ($\overline{D}$ = 20 μm)

II coarse spherulites embedded in a matrix of finer spherulites

III fully grown coarse spherulites ($\overline{D}$ up to 500 μm).

The slow crack growth in CT specimens of morphology II is schematically represented in Figure 9.23a. Crazes were seen to develop ahead of the crack preferen-

Fig. 9.23. Slow crack growth following craze formation in bulk polypropylene containing coarse and fine spherulites (Courtesy K. Friedrich, Bochum).
(a) schematic representation, (b) crazes at the interface between coarse spherulites, (c) trans- and interspherulitic crazing, (d) crazing of finely spherulitic matrix.

tially at the interfaces between coarse spherulites (Fig. 9.23b), within larger spherulites (c) and at the boundaries of small ones (d). The behavior of a crack in the zone of rapid crack propagation is illustrated in Figure 9.24. A scanning electron micrograph of the fracture surface reveals the structure-induced deviations of the fracture path from a plane (Fig. 9.24a). The points 1 to 3 identify corresponding fracture paths in Figure 9.24a, b, and c. These paths follow a spherulite boundary (1) or they

Fig. 9.24. Rapid crack propagation in polypropylene. (Courtesy K. Friedrich, Bochum).
(a) scanning electron micrograph of fracture surface showing inter-(1) and transspherulitic fracture (2 and 3), (b) model representation, (c) optical micrographs of sections transverse to the fracture surface showing the same phenomenon.

cut off respectively, a segment (2), or a cone (3). From his extensive morphological studies Friedrich [183] derived an order of the partial material resistance values which decrease from that of the
— separation of finely spherulitic matrix, which is the largest, to
— straight paths through coarse spherulites and their center,
— radial and tangential planes in coarse spherulites, and
— interfaces between coarse spherulites and fine matrix, to that of
— polygonal interfaces between coarse spherulites.

299

The K_c values were found to decrease linearly with $\overline{D}$, increase with decreasing strain rate, and increase with molecular weight, the latter effect being especially pronounced with the coarsely spherulitic morphology. Tentative absolute values of R are determined for the separation of finely spherulitic matrix as 22 kJ m^{-2}, and for interspherulitic rupture as 4 kJ m^{-2} [183]. The first value is of the same order of magnitude as that calculated by Kausch [11] for an oriented PP film (cf. further below).

In the first part of this section, semicrystalline polymers (HDPE, PP, PA) have been discussed. The fracture surface morphology of glassy polymers has received no less attention in the literature. In fact many of the investigations of crazing [76–177] use fractography as a means to elucidate the propagation and breakdown of crazes. Fractographic studies of fracture processes concerning PS are especially reported in refs. [106, 115, 132, 150, 155, 169, 191, 194, 199], concerning PMMA in [61, 66, 197, 200], PVC in [198, 208], and PC in [196].

A feature common to practically all fracture surfaces of glassy polymers are the remnants of craze layers. At low speeds of crack propagation craze breakdown will generally occur in the center of the craze material leaving a more or less homogeneous layer on each fracture surface [15, 50, 150, 194, 199]. In PS at intermediate and higher crack speeds decohesion at the craze/matrix interface becomes possible at room temperature. Beahan et al. [150] have investigated this phenomenon more closely; Figure 9.25 shows their micrograph of (the rather isolated) event of a crack which

Fig. 9.25. Transmission electron micrograph of the craze region ahead of a crack advancing from left to right in polystyrene. (Courtesy D. Hull, Liverpool).

after having propagated at intermediate speeds was stopped in the craze region. The micrograph permits recognition of the decohesion of highly strained craze material at the interface and also the oscillation of this phenomenon between the opposite interfaces. The regular oscillation of the decohesion of craze material creates "mackerel patterns" with typical band repeat distances of between 10 and 70 μm. Hull et al. [150, 194, 199] discuss the dynamic conditions of their formation and the effect of sample temperature. They observe that at temperatures below 235 K no mackerel patterns are formed whereas the initiation of secondary fractures is greatly enhanced.

A more pronounced transition from slow to fast crack propagation is reported for PMMA [61, 66, 197, 200] but the surface features are roughly comparable to those in PS. As observed by Döll and Weidmann [61, 66] at slow crack speeds ($\dot{a} < 0.1$ ms^{-1}), independent of molecular weight, a rather smooth fracture surface is formed which even carries on accidental deviations from the plane in the form of parallel markings (left hand side of Fig. 9.26). After the transition to rapid crack propagation the fracture surface morphology changes completely. A pattern of ribs or lines running approximately parallel to the crack front becomes visible. Although length, direction, and spacing of the ribs are not very regular (Fig. 9.26, right hand side) an average spacing of 90 μm has been measured [61]. At a somewhat higher molecular weight (M_w = 163000) a spacing of about 220 μm is observed (Fig. 9.27, right hand side). At very low molecular weight (M_w = 51000) rib spacing is 20 to

Fig. 9.26. Fracture surface of polymethylmethacrylate (M_w = 115000) showing the transition from slow ($\dot{a} \simeq 0.1$ ms^{-1}) to fast ($\dot{a} > 200$ ms^{-1}) crack propagation; crack propagates from left to right. (Courtesy W. Döll, IFKM Freiburg), Specimen thickness 4.75 mm.

Fig. 9.27. Fracture surface of polymethylmethacrylate (M_w = 163000) showing the transition from a smooth surface (crack speed $\dot{a}$ = 300 ms^{-1}) to a coarse one ($\dot{a} > 400$ ms^{-1}); crack propagation is from left to right. (Courtesy W. Döll, IFKM Freiburg), Specimen thickness 3.75 mm.

24 μm [200]. In view of the symmetry of these markings on the two corresponding fracture surfaces, rupture of the craze material in the center plane seems to be indicated. This phenomenon is somewhat different in origin, therefore, from the mackerel pattern but also related to the formation and breakdown of crazes. The mackerel pattern was caused by the oscillating occurrence of craze/matrix decohesion. The rib markings, however, are better interpreted as *hesitation lines,* also designated as *stick-slip,* i.e. as the alternating development and breakdown of craze material [61, 196, 200]. This is explained by referring to the craze model presented in Figure 9.14. It may be assumed that at $t = t_0$ the crack front just faces a fully developed craze whose width and length (region C) comply with the Dugdale model (Eqs. 9.18 and 9.20). The propagating stress field and the stresses which can be supported by the craze zone are momentarily in equilibrium. At $t = t_0 + \Delta t$, however, the craze is supposed to open further and to support loads which are beyond the load carrying capability of the molecular strands. Consequently the most advanced strands break thus initiating the rupture of neighboring strands as well. The catastrophic breakdown is slowed down once the crack has advanced within region C so far as to reach little-drawn craze material. After the slow-down of crack propagation redevelopment of the craze begins which may be completed at, say, $t = t_0 + t_1$. Considering the average crack speed ($v = 400$ ms^{-1}) and the rib spacing in Figure 9.27 (220 μm) one can calculate that the time interval t_1 for catastrophic breakdown and redevelopment is 0.55 μs. The cyclic crack propagation seems to be influenced, however, by additional mechanisms which are not yet fully understood.

Thus the PMMA samples having the slightly higher molecular weight reveal at crack speeds of 200 to 300 ms^{-1} a rather smooth fracture surface containing no ribs but a large number of parabolic markings (Fig. 9.27, left hand side). These markings are seen in higher magnification in Figure 9.28, they derive from the coalescence of primary and secondary crack fronts travelling in one craze zone but in slightly different planes. The fracture surface of polycarbonate – broken in tension at room temperature – reveals a strikingly similar pattern [196]. In that case the distance between the planes of primary and secondary cracks has been found to be 0.43 μm at room temperature increasing to 0.75 μm at -196 °C.

Döll [61] reports that PMMA samples with a higher molecular weight ($M_w \geqslant 490\,000$) do not show ribs on the high-speed fracture surface (Fig. 9.29). Secondary fractures (this time in different craze zones) may be initiated, however, leading to parabolic markings in that surface region (Fig. 9.30). Kusy et al. [200] find ribs *and* parabolas in the molecular weight range $92\,000 \leqslant M_v < 270\,000$. A certain molecular coil size is (quite obviously) necessary to permit the transfer of stresses sufficient for the initiation and temporary propagation of a secondary fracture plane. So far no generally valid relation between material parameters, experimental conditions, and morphological observations can be given.

In the referenced literature further details are discussed which may be obtained from fractographic analysis, such as the effect of Wallner lines on the disposition of the ribs [61, 196, 200], "crazeless fracture" in low molecular weight PS [155], the retarded breakdown of crazes in fatigue (cf. III C), the ductile fracture of PS at lower loading rates and at temperatures close to T_g through the growth of one or more diamond shaped cavities [169], the globular appearance of ion-etched PS craze matter

Fig. 9.28. Higher magnification of the middle section of Fig. 9.27 showing parabolic markings. (Courtesy W. Döll, IFKM Freiburg).

Fig. 9.29. Fracture surface of PMMA (M_w = 8 000 000) showing the transition from slow ($a \simeq 0.1 \ ms^{-1}$) to fast ($\dot{a} > 100 \ ms^{-1}$) crack propagation; crack propagates from left to right. (Courtesy W. Döll, IFKM Freiburg), Specimen thickness 4.16 mm.

[132] and PVC surfaces [208], and craze-like features on the surface of broken phenol-formaldeyhde samples [195].

In this section the morphology of fracture surfaces has been discussed, a morphology which reflects the local modes of material separation. The *microscopic* size of the structural elements to be broken or separated has been recognized: molecular strands, fibrils or coils, ribs, crystalline lamellae, spherulites. In talking about their ultimate response, however, macroscopic terms had been used: breakage, shear deformation, limits of plastic deformation, material resistance. For good reasons no molecular criteria for material separation had been given. For individual molecules such criteria exist: the temperature of thermal degradation and the stress or strain which causes the chain to scission. For the mentioned structural elements no such simple criteria exist. Perhaps one may mention the critical role of temperature in the transition to rapid crack propagation [30, 50, 184—186, 197] and the constant value of local strain ϵ_y in the extension direction (Fig. 9.31) which was found — independent of crack length — to reach about 60% at the cracktip in biaxially oriented PETP film [209]. One may also mention the critical concentration of chain end groups N_{IR} determined by microscopic infra-red spectroscopical investigation of the near-crack zones of oriented PP film (Fig. 9.32) and of the fracture surface of PE bulk material [210]. Both materials are tough and strong. The given stress distribution in front of the crack in the PP film permits calculation of K_c as $\sigma(r)\sqrt{2\pi r} = 8.3 \pm 2 \ \text{MN m}^{-3/2}$ and G_c as $30 \pm 17 \ \text{kJ m}^{-2}$ [11]. These values, in connection with Table 9.2, indicate quite clearly that chain scission is accompanied by severe plastic deformation. The

Fig. 9.30. Parabolic markings from the surface of a crack travelling at 240 ms^{-1} in PMMA ($M_w = 1.2 \cdot 10^6$); crack propagates from left to right. (Courtesy W. Döll, IFKM Freiburg).

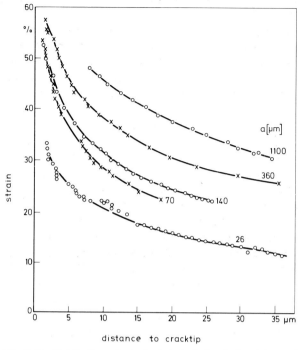

Fig. 9.31. Distribution of principal strain in front of a plane crack in biaxially drawn PETP film (after [202]).

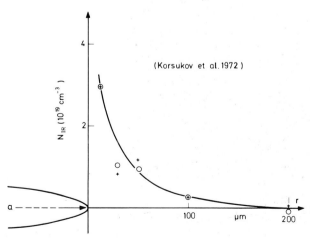

Fig. 9.32. Distribution of chain end concentration in front of a propagating plane crack at $t = 0.2\,\tau$ (+) and $0.75\,\tau$ (○) in polypropylene film (after [11, 210]).

305

possible role of chain scission during or after the application of large orientational strains has been discussed in detail in Chapter 8.

B. Notched Tensile and Impact Fracture

In Chapter 8 tensile and impact behavior of *unnotched* specimens have been discussed. The principal changes in behavior to be expected from the presence of a notch would have to be due to the changed state and intensity of stress in the crack tip zone, to the essential confinement of fracture development to a limited region, and to the increase in rate of local deformation. An analytical description of these effects has been attempted in terms of linear elastic fracture mechanics (9 I). Neglecting the geometrical correction term the uniaxial brittle strength of a cracked thick plate is derived from Eq. (9.10) as:

$$\sigma_b(a) = K_{Ic}/(\pi a)^{1/2} = (E \, G_{Ic}/\pi a \, [1 - \nu^2])^{1/2}. \tag{9.29}$$

In this case the effect of a notch of length a is clearly indicated, therefore. In the general case of rupture involving plastic deformation, however, one would have to use K_c (or G_c) and a corrected crack length $a + f(r_p)$ in Eq. (9.29). All these quantities depend heavily on the degree of plastic deformation at the crack tip, which in turn is influenced by crack length, notch tip radius, and loading conditions. In the case of heavy plastic deformation the rupture stress of a specimen would depend little on the initial crack geometry but strongly on the "strength" of the plastically deforming matrix. Three questions are raised, therefore, which will be discussed in the following: which notch sensitivity is actually observed with different polymers, when are geometrical and when are material parameters of critical importance, and what is the influence of chain length and mobility?

The notch sensitivity of polymeric materials has been comprehensively investigated by Takano and Nielsen [211]. The authors define a notch sensitivity factor k_S of the yield or breaking strength of tensile bars as

$$k_S = \frac{\sigma_F(o) \, A(a)}{\sigma_F(a) \, A(o)} \tag{9.30}$$

with A being the (residual) sample cross-section. In the same way a notch sensitivity factor k_T of the energy to break (area under stress-strain curve) is defined.

The authors tested some 40 different materials at strain rates of 2.5 mm/min for rigid materials and of 25 mm/min for soft elastomers employing six different crack geometries. The "most severe" was a single notch of 3 mm depth (in a 12 mm wide sample) and of 0.25 mm notch tip radius (notch *s*). The "least severe" was a double notch of 3 mm depth and 1.5 mm notch tip radius (notch *d*). The authors measured – where possible – the Izod impact strengths, the yield (σ_F) and breaking strengths (σ_b), and Young's modulus of notched and unnotched specimens.

From their measurements a sequence of $k_S(s)$ values can be established: PMMA (4.71), Kraton ®-1101 (3.68), SAN (2.71), PSU (2.36), EPDM rubber (2.12)

and various particle filled thermoplastics (2.14–1.33). The $k_S(d)$ values exhibit a similar sequence:
PMMA (1.64), Kraton ®-1101 (<1.69), SAN (1.62), and PPO (1.48). There are, however, quite a number of (ductile) materials with $k_S(s)$ values *smaller* than one, notably HDPE (0.88), PTMT (0.96), PTFE (0.94), PA 6 (0.93), and Hytrel ® 4055. Values below unity of $k_S(d)$ are more numerous than those of $k_S(s)$.

Notch sensitivity factors for energy, k_T, show a somewhat different order. With some rare exceptions (PP + 20% glass, PE + 40% glass) k_T values are always larger than unity. The factor $k_T(s)$ reaches values of 625 (semi-transparent PETP), 241 (PSU), 61 (PA 66 + 0.6% H_2O), 48 (PP and PMMA), and 16 to 17 (mineral filled PA 66, PC). For $k_T(d)$ the authors determine 81 (PA 66 + 0.6% H_2O), 34 (PPO), 29 (semi-transparent PETP), 19 (PP), and 14 (annealed HDPE and transparent PETP). All other k_T values measured are smaller than 13.

The data of Pagano and Nielsen [211] confirm the observation that the presence of a crack *not necessarily reduces* the tensile strength of a sample or its energy to break. The authors also note that in some cases (notably in HDPE, PTMT, PTFE, PA 66 + 0.56% H_2O) the "more severe" notch s is the less dangerous and has a smaller k_T value than notch d. The reason for this unexpected reversal is not clear but it may be related to the volume of polymer under elevated stress in the notched region. This volume at the tip of a blunt notch is larger than that at the tip of a sharp notch [211]. Regarding the experimental evidence (e.g. [58, 64, 211, 214]) one may say that in the case of a very sharp notch ($\rho_a < 50\ \mu m$) the onset of rapid crack propagation is determined predominantly by notch length a and material fracture toughness K_c according to Eq. (9.29). In a blunt notch specimen, however, the tangential stress σ_y at the root of the notch has to be considered the critical quantitiy [58, 64]. If the shape of a blunt notch is approximated by an ellipse the well known Neuber-Inglis formula [4] can be used to calculate σ_y:

$$\sigma_y = \sigma_0 (1 + 2 \sqrt{a/\rho_a}). \tag{9.31}$$

The stress concentration increases with decreasing ρ_a whereas the apparent critical energy release rate $(G_c)_{blunt}$ decreases. Considering also a plastic zone of size r_p Plati and Williams [64] obtain:

$$(G_c)_{blunt} = G_c \frac{(1 + \rho_a/2\,r_p)^3}{(1 + \rho_a/r_p)^2}. \tag{9.32}$$

Whenever the measurement of the energy to break a specimen in tension or flexion is used to determine critical energy release rates, it must be recognized, that this energy is the sum of a number of quite different terms. As outlined in 8 II A, the energy loss A_n, which a pendulum incurs in striking and breaking a specimen, is a measure of the elastically stored energy W_e, of the fracture surface energy W_s, of the kinetic energy W_{kin} of the broken pieces, and of otherwise dissipated energy. The elastic energy to bend the specimen to a deflection δ under load P amounts to:

$W_e = \frac{1}{2}\delta P = \frac{1}{2} P^2 C$, with C being the bending compliance. The energy to propagate a crack through the ligament area is taken to be $W_s = G_c B(D-a)$. The problem of how these two energy terms are related to the impact strength a_n has first been examined independently and simultaneously by Brown and Marshall, Williams, and Turner [53]; it has recently been reviewed by Williams and Birch [69].

Crack propagation in a notched bending specimen becomes possible once the release of stored elastic energy is equal to or larger than the material resistance:

$$G_c(a) = \frac{P^2}{2 B} \frac{\partial C}{\partial a} > R(a).$$ (9.33)

The crack will propagate in an unstable manner and without additional external driving forces, i.e. at constant deflection, as long as the release of energy due to the increase in specimen compliance C remains sufficient to supply the energy needed for material separation, i.e. as long as

$$\frac{\partial G_c(a)}{\partial a} > \frac{\partial R(a)}{\partial a}.$$ (9.34)

If one makes use of Eq. (9.33) and of the condition of constant deflection δ, which gives

$$\partial P/P \partial a = -\partial C/C \partial a.$$ (9.35)

one obtains as condition for crack instability

$$\frac{C}{2} \frac{\partial^2 C/\partial a^2}{(\partial C/\partial a)^2} > 1 + \frac{BC}{P^2} \frac{\partial R/\partial a}{(\partial C/\partial a)^2}$$ (9.36)

Introducing the "calibration factor" ϕ,

$$\phi = \frac{C}{\partial C/\partial (a/D)},$$ (9.37)

taking R equal to G_c, and rearranging Eq. (9.36) one arrives at a somewhat simpler form of the instability condition:

$$\frac{\partial \phi}{\partial (a/D)} < -1 - \frac{\partial R/R \partial a}{\partial C/C \partial a}.$$ (9.38)

If R is taken to be constant with a Eq. (9.38) is equal to the condition for crack instability analyzed by Williams et al. [64, 69].

The calibration factor $\phi(a/D)$ has been tabulated by Plati and Williams [64] for the two most widely used impact tests, the three point bending (Charpy) test and the cantilever bending (Izod) test (with a/D ranging from 0.04 to 0.60, 2 L/D from 4 to 12). If a bending specimen fails due to unstable crack propagation then its energy to failure should principally correspond to the elastic energy stored at the point of onset of rapid crack propagation, which can be expressed now as:

$$A_n \simeq W_{el} = RBD\phi \qquad (9.39)$$

A plot of A_n over $BD\phi$ should give, therefore, a straight line with slope $R = G_c$. Testing nine different brittle and ductile polymers (from PS to PE) by the Charpy and the Izod methods, Plati and Williams [64] were able to arrive at fairly unique values of G_c (cf. Table 9.2). They point out, however, that in order to obtain linear $BD\phi$ plots they had to consider an effective crack length consisting of the initial crack length and the size of a suitably chosen plastic zone.

If the condition for unstable crack propagation is not fulfilled or maintained throughout the available range of crack lengths then a continued supply of mechanical energy is necessary to propagate the crack through the material. In that case the failure energy A_n should be mainly proportional to the ligament area:

$$A_n \simeq W_s = RB(D - a_0). \qquad (9.40)$$

Corresponding plots of A_n over $B(D\text{-}a_0)$ for HiPS and ABS [53, 64, 69] revealed this linear relationship.

The fact that even the rapid deformation of a glassy polymer under concentrated stresses entails considerable local plastic deformation immediately suggests that the molecular properties, which influence yielding and flow, are also effecting G_c and thus the impact strength. The data compiled in Table 9-II reveal this dependency of G_c on temperature, rate of deformation, and molecular properties. A possible relation between molecular relaxation processes and fracture surface energy of polymers has been pointed out in many of the cited references (e.g. [14, 19, 22, 24, 25, 54, 63, 64, 212–214]).

In 8 II A an account of the role of mechanical relaxation mechanisms in the impact loading of *unnotched* specimens has been given. The reasons for an expected positive correlation (and for the also observed deviations) between impact strength and magnitude of mechanical losses have been indicated there. At this point the molecular aspects of *notched* impact strength are to be investigated. Sauer [213] and Vincent [214] have reviewed the impact and stress-relaxation data from a large number of publications including their own extensive works and those referenced in 8 II A. They arrive at various general conclusions which may be listed in the following:

— The *impact resistance rating* of a polymer is generally determined by the storage component E' of its dynamic modulus. For some twenty different polymers tested, 65% of the data complied with the following correlation [214]:

Impact resistance rating	Modulus
— brittle	$E' > 4.49$ GN m^{-2}
— brittle if bluntly notched	$E' \simeq 3$ GN m^{-2}
— brittle if sharply notched	$E' \simeq 2.2$ GN m^{-2}
— tough but crack propagating	$E' < 1.5$ GN m^{-2}

Deviations from this scheme were either related to molecular structure (bulky side groups of PMMA), to mechanical loss peaks (PTFE, HDPE, PC, PPO), to morphology (injection molded PP), or to heterogeneous reinforcements (short glass fibers, particulate fillers).

— Polymers with high impact strength at room temperature also have a significant low temperature loss peak (tan $\delta \simeq 5 \cdot 10^{-2}$ or higher). Polymers in this class include, PE, PC, PB, PTFE, PCTFE, POM, and the polyamides.

— Polymers with low impact strength at room temperature usually have no significant low temperature loss peaks in the 100 K to 300 K range (PS, PMMA).

— Impact strengths have been measured as a function of temperature in several polymers and it has been shown that in many cases one gets a rise in brittle impact strength in the vicinity of the low temperature γ-relaxation (PTFE, PE, POM). For example, in PTFE at about 210 K, in PE at 150 K. Good correlation has also been observed between the magnitude of the low temperature γ-loss peak in polysulfone and impact resistance for a series of specimens containing various amounts of an antiplasticizer.

— An increase in molecular weight will generally affect the impact strength through the increase in G_c; this is most notably observed in HDPE (cf. 8 II A).

— In high impact polystyrene (HiPS) the normally brittle PS matrix has been blended with a suitable rubber-type component such as a styrene-butadiene copolymer which has a glass transition well below room temperature. The impact strength increases in proportion to the amount of the second component added. In this case of a two-phase polymer, however, impact resistance is increased through the initiation of a myriad of fine crazes and not through the generally increased extensibility of a homogeneously deforming matrix.

C. Fatigue Cracks

The characteristics of fatigue loading, i.e. of the repeated application of varying stress or strain amplitudes have been introduced in detail in Section 8 II C. In that discussion one point was left open: that of the mechanism of fatigue crack propagation. A comprehensive elaboration of this subject has quite recently been given by Hertzberg in his book on "Deformation and Fracture Mechanics of Engineering Materials" [3]. The reader is referred to this work or to the review articles by Plumbridge [217] and Manson and Hertzberg [218] for a detailed discussion of the different stages of fatigue crack growth, of the distinct characteristics of fatigue fracture surfaces, of the various theoretical approaches to derive crack growth-rate equations, and of the S—N curves of a wide variety of homogeneous and fiber reinforced polymers — and of metals, for that matter [3, 217, 218]. At this point only some recent observations will be discussed which seem to be directly related to the chain-like nature of macromolecules [173, 178, 191, 215—220].

As to be expected, the growth of a crack under constant or increasing load and fatigue crack propagation have quite a few features in common. Thus both the static and the fatigue fracture surfaces reveal a slow growth and a rapid crack propagation zone. The stress intensity factors reached in both cases at the transition from stable to unstable crack propagation correlate very well [218]. The rate of stable fatigue crack growth is empirically expressed by most authors in form of

310

$$\frac{da}{dN_F} = A(\Delta K)^n,\tag{9.41}$$

with A and n material parameters and ΔK the applied stress intensity factor range. This is in complete analogy to the expression for the rate of crack propagation under static loading (Eq. 9.22). Marshall et al. [133] and Radon et al. [219] substitute ΔK by the difference of K^2 and K^3 terms. Andrews et al. [215] employ in their analysis of the fatigue of PE the energy release rate G_c (instead of ΔK) as independent parameter. Such approaches are not significantly different because of the relation between K and G and of the form of Eq. (9.41). In their detailed reports [3, 218] Hertzberg and Manson analyse the various theoretical approaches, furnish fatigue crack growth data for some 20 materials at different temperatures and frequencies, and discuss the effect of environmental and material parameters.

Skibo, Hertzberg, and Manson [191] studied fatigue crack growth characteristics in polystyrene as a function of stress intensity factor range and cyclic frequency. Precracked single edge notched and compact-tension type specimens made from commercially available polystyrene sheet (mol. wt. $= 2.7 \cdot 10^5$) were cycled under constant load at frequencies of 0.1, 1, 10 and 100 Hz, producing growth rates ranging from $4 \cdot 10^{-7}$ to $4 \cdot 10^{-3}$ cm/cycle. For a given stress intensity level, fatigue crack growth rates were found to decrease with increasing frequency, the effect being strongest at high stress intensity values. The variable frequency sensitivity of this polymer over the test range studied was explained in terms of a variable creep component. The macroscopic appearance of the fracture surface showed two distinct regions. At low stress intensity values, a highly reflective, mirror-like surface was observed which transformed to a rougher, cloudy surface structure with increasing stress intensity level. Raising the test frequency shifted the transition between these areas to higher values of stress intensity. The microscopic appearance of the mirror region revealed evidence of crack propagation through a single craze while the appearance of the rough region indicated crack growth through many crazes, all nominally normal to the applied stress axis. Electron fractographic examination of the mirror region revealed many parallel bands perpendicular to the direction of crack growth, each formed by a discontinuous crack growth process as a result of many fatigue cycles. The size of these bands was found to be consistent with the dimension of the crack tip plastic zone as computed by the Dugdale model. At high stress intensity levels a new set of parallel markings was found in the cloudy region which corresponded to the incremental crack extension for an individual loading cycle [191].

In these experiments two chain-related phenomena have been observed: the discontinuous growth of a fatigue crack and the frequency sensitivity of the growth rate. The discontinuous growth of a fatigue crack had been noted earlier for other polymers as well: PVC [216], PMMA, PC, and PSU [191]. Elinck et al. [216] report that, depending on frequency, between 130 and 370 load cycles were necessary to form one "arrest line" in PVC; Skibo et al. [191] found 600 to 1400 cycles for PS. The width of these growth bands corresponded well to the plastic zone size r_p calculated from the Dugdale model. Based on these findings Skibo et al. [191] propose as a possible crack growth mechanism the gradual development of a craze of width r_p and its subsequent and rapid breakdown. Such a mechanism would also be supported by the model of craze formation given in G II C: once the critical and frequency-dependent fibril extension is reached at the crack tip cooperative breakdown of the fibrils occurs and the crack propagates from D/C almost to B (Fig. 9.14).

The frequency-dependent rate of disentanglement of the molecular coils in the fatigued fibrils seems to account for part of the frequency effect on growth rate. In addition hysteresis heating occurs in the strained craze material. Both effects combine to give rise to a distinct frequency sensitivity of A for a variety of materials such as PC and PMMA [219, 220], and PPO, PVC, PA 66, PC, PVDF, and PSU [220]. As has been noted by Skibo et al. [220] the frequency sensitivity varies with temperature. It reaches a maximum at that temperature where the external (fatigue) frequency corresponds to the internal segmental jump frequency (of the β-relaxation process).

The effects of strain amplitude, sample molecular weight, environment, and surface coating on the fatigue properties of PE and PS have been studied by Sauer et al. [173, 178]. Their results have been discussed in 8 II C. It is especially noteworthy that the initiation of a fatigue crack can be retarded by a decade or more through the application of a compatible, viscous coating. A 600 molecular weight PS oligomer served this purpose for both polished and unpolished surfaces of cylindrical PS specimens [178].

The relation between microstructure of LDPE and fatigue crack growth was investigated by Andrews et al. [215]. They found growth rates according to Eq. 9.41 in two regions: that of the initial brittle transpherulitic crack propagation and that of the ductile crack propagation at higher K values. Between these two regions a distinct transition region was found where da/dN_F depended little on ΔK.

D. Mechano-Chemistry

In a broader sense mechano-chemistry comprises all aspects of the stress-induced scission of chain molecules. In a narrower sense, however, one speaks of mechanochemical methods if one refers to the *intentional mechanical degradation* of (solid) polymers. The objectives in the latter cases are the comminution or softening of the material or the production of large and highly reactive surfaces as a means to initiate permanent chemical bonding between different polymers. In Table 9.5 an overview on methods and processes is given which may result in a mechanical degradation of chain molecules. The objectives of these processes are indicated with respect to the deformation mechanism. It should be pointed out that in the mechano-chemical methods, in the narrower sense, the degrading solids are subjected to ill defined, complex states of stress leading to a deformation which always includes simultaneously the yielding and flowing of material and the breakage of chains. Table 9.5 lists the deformation mechanisms which are the most important in view of the corresponding objective. Reference is made to those chapters and sections of this book where the particular deformation mechanisms are treated.

In recent years a number of comprehensive review articles on the mechanochemistry of polymers have appeared [221–226]. A competent two-volume handbook by Porter and Casale on this subject [227] has just become available. The

Table 9.5. Mechano-chemistry of Solid Polymers: Methods and Objectives

Method, Process	Objective	Deformation Mechanism	Reference to Chapter Section
Grinding, sawing, cutting	comminution	compressive yielding,	8.II. D
Ball or vibromilling Impact loading	comminution shock absorption	elastic compressive and/or tensile deformation	5.II. E
Mastication (of elastomers)	softening	stretching and breakage of chain molecules	5, 8.II.E
Extrusion, injection, drawing	forming	shear flow	8.II. D
Mixing, stirring	dispersion of additives	frictional interaction, compressive yielding	
Synthesis through mechanical blending	grafting, copolymerization of incompatible or mechanically different polymers	chain scission and free radical reactions	6.IV. A, 7.III.

reader is referred to these references [222–227] for any extended information on the effect of experimental variables (type of equipment, temperature, rate of mechanical working, atmosphere) or polymer characteristics (chemical structure, initial molecular weight) on the resulting material properties.

Within this monograph special consideration is given to the loading and breakage of molecular chains. With regard to mechano-chemical processes such as comminution or mechanical synthesis one has to study, therefore,

— the response of the polymer matrix to mechanical working
— the consequences of chain scission: reduction of molecular weight and radical formation
— the results of radical reactions, e.g. oxidative degradation, cross-linking, copolymerization.

The property changes resulting from a mechanical treatment of polymeric materials have been known and utilized for a long time. Thus knowledge of the fact that mechanical working of natural rubber leads to its softening already dates back 120 years [225]. But it was only after the introduction of the concept of macromolecules some fifty years ago that this effect of mechanical treatment was related to the rupture of molecular chains. The fact that the intensive milling of a heterophase mixture (of rubber and maleic acid anhydride) can induce a chemical reaction between the components was first observed in 1941 [224]. Systematic studies of the nature of these chemical reactions and particularly of the role of the formed free radicals began about ten years later [224–225].

The response of polymer matrices and of chain molecules to mechanical working have been intensively discussed in the previous chapters. The ESR investigations of ground polymers (6.IV A and 7.I C 3) have shown that in a comminution process (e.g. milling) chain molecules are broken in large numbers and as a *consequence* of the formation of new surface during the breaking-up of the polymer particles. The chain ruptures have an immediate and notable effect on the sample molecular weight M [221–338]. In Figure 9.33 M is plotted as a function of particle diameter D. The linear sections may be represented by a potential law

$$M(D) = A(D/D_0)^n \tag{9.42}$$

where A and n are material-related constants. Curves *a* and *b* in Figure 9.33 reveal that chain scission is more effective in polystyrene (n = 0.76) than in PMMA (n = 0.50). In both cases comminution was carried out [228] in the presence of radical scavengers (NO_2) to prevent chain recombination or cross-linking. Comminution in air seems to have permitted cross-linking reactions to such an extent that the average molecular weight of a particle size fraction did not change with D (curve *c*). The opposite effect, decrease of M without increase of specific surface was noted by Komissarov et al. [222] in the dispersion of PA and PETP fibers *without cooling*.

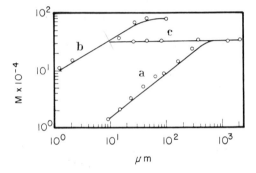

Fig. 9.33. Molecular weight of comminution products as a function of particle size (after [200] from [201]): (a) polystyrene in NO_2, (b) polymethylmethacrylate block copolymer in NO_2, (c) polystyrene in air.

It is a general observation that in continuous mechanical working D (and M) do not decrease indefinitely but reach certain limiting values. These limiting average M values are of the order of the entanglement molecular weight M_e. This does not preclude the possibility that in mechanical degradation fractions of low molecular weight and even volatile components can be produced [221–225].

The effect of environmental temperature on the rate of degradation has also been widely studied [221–227]. In view of the composite nature of the degradation process no simple rate equations can be expected. It becomes clear from Eqs. (5.41) and (7.3) that the softening of the matrix (decrease of ζ_0) and the reduced effective bond strength U(T) partly balance each other. According to an extensive review by Casale [226] the temperature effect on matrix rigidity seems to be the dominant one. The slower relaxation times at lower temperatures lead to an increase in mechanical degradation with decreasing temperature (negative temperature coefficient of the overall mechano-chemical reaction).

References for Chapter 9

1. A. A. Griffith: Phil. Trans. Roy. Soc. (London) A 221, 163–198 (1920).
2. Mechanics of Fracture, G. C. Sih, ed., Vol. I-V, Leyden: Noordhoff International Publishing. 1975–1977.
3. R. W. Hertzberg: Deformation and Fracture Mechanics of Engineering Materials, New York: John Wiley 1976.
4. J. G. Williams: Stress analysis of polymers, London: Longman 1973.
5. E. H. Andrews: Fracture in Polymers, Edinburgh: Oliver + Boyd, 1968.
6. F. Kerkhof: Kolloid-Z. u. Z. Polymere 251, 545 (1973).
7. D. S. J. Dugdale: Mech. Phys. Solids 8, 100 (1960).
8. ASTM Book of Standards, Part 31: Physical and Mechanical Testing of Metals (1969).
   H. J. G. Blauel, J. F. Kalthoff, E. Sommer: Materialprüfung 12, 69 (1970).
9. J. D. Eshelby: Proc. Royal Soc. A241, 376–396 (1957).
10. G. R. Irwin: Handbuch der Physik 6, Berlin: Springer Verlag 1958, 559.
11. H. H. Kausch: Kunststoffe 66/9, 538–544 (1976).
12. G. P. Marshall, L. H. Coutts, J. G. Williams: Materials Sci. 9, 1409 (1974).
13. P. W. Beamont, R. J. Young: J. Materials Sci. 10, 1334–1342 (1975).
14. F. A. Johnson, J. C. Radon: J. Polymer Sci., Polymer Chem. Ed. 11, 1995–2020 (1973).
15. G. W. Weidmann, W. Döll: Colloid and Polymer Sci. 254, 205–214 (1976).
16. J. P. Berry, in: Fracture Processes in Polymeric Solids, B. Rosen, ed., New York: Interscience 1964, Chap. II.
17. L. J. Broutman, F. J. McGarry: J. appl. Polymer Sci. 9, 589–626 (1965).
18. M. F. Kanninen, A. R. Rosenfield, R. G. Hoagland, in: Deformation and Fracture of High Polymers, Kausch, Hassell, Jaffee Eds., New York: Plenum Press 1974, 471–484.
19. M. Parvin, J. G. Williams: J. Materials Sci. 10, 1883–1888 (1975).
20. Y. W. Mai: Int. J. Fracture 9, 349 (1973).
21. R. P. Kambour, S. Miller: General Electric Report No 77CRD009, March 1977.
    R. P. Kambour, A. S. Holik, S. Miller: General Electric Report No 77CRD125, June 1977.
22. S. Arad, J. C. Radon, L. E. Culver: J. appl. Polymer Sci. 17, 1467–1478 (1973).
23. W. Retting: Materialprüfung 5, 101 (1963).
24. W. Retting: Kolloid-Z. Z. Polymere 210, 54 (1966).
25. H. W. Greensmith, L. Mullins, A. G. Thomas: Trans. Soc. Rheol. 4, 179 (1960).
26. G. J. Lake, P. B. Lindley, A. G. Thomas: Paper 43 (Session IV) Int. Conf. on Fracture 1969, Brighton.
27. A. G. Thomas, in: Deformation and Fracture of High Polymers, New York: Plenum Press 1973, 467–470.
28. A. Ahagon, A. N. Gent: J. Polymer Sci., Polymer Physics Ed. 13, 1903–1911 (1975).
29. H. K. Mueller: Ph. D. Thesis, California Inst. of Technology, Pasadena, California, June 1968.
30. W. Döll: Energieumsetzung in Wärme beim kritischen Bruchbeginn in PMMA. Report of the Institut für Festkörpermechanik, Freiburg, 1978.
31. D. W. van Krevelen: Properties of Polymers, Amsterdam: Elsevier Publ. Co. 1972, Table 6.6.
    Polymer Handbook, Brandrup, J., Immergut, E. H., J. Wiley + Sons, New York 1975.
32. J. J. Benbow, F. C. Roesler: Proc. Phys. Soc. B 70, 201–211 (1957).
33. J. P. Berry: J. Polymer Sci. 50, 107–115 (1961).
34. J. P. Berry: J. Polymer Sci. 50, 313–321 (1961).
35. N. L. Svensson: Proc. Phys. Soc. 77, 876–884 (1961).
36. J. J. Benbow: Proc. Phys. Soc. 78, 970 (1961).
37. J. P. Berry: J. appl. Physics 34, 62 (1963).
38. A. van den Boogaert: The Physical Basis of Yield and Fracture, Oxford, 1966, pg. 167.
39. P. I. Vincent, K. V. Gotham: Nature 210, 1254 (1966).
40. J. P. Berry: J. appl. Physics 34, 62–68 (1968).
41. J. O. Outwater, D. J. Gerry: J. Adhesion 1, 290 (1969).
42. S. Mostovoy, P. B. Crosely, E. J. Ripling: J. Materials, 2, 661 (1967).

43. G. P. Marshall, L. E. Culver, J. G. Williams: Polymer *37*, 75–81 (1969).
44. R. Griffith, D. G. Holloway: J. Materials Sci. *5*, 302–307 (1970).
45. J. W. Curtis: J. of Physics *3*, 1413–1422 (1970).
46. L. E. Miller, K. E. Puttick, J. G. Rider: J. Polymer Sci. *33*/C, 13–22 (1971).
47. B. Harris, P. W. R. Beaumont, E. Moncunill de Ferran: J. Materials Sci. *6*, 238–251 (1971).
48. A. R. Rosenfield, P. N. Mincer: Polymer Sci. Symposium *32*, New York: John Wiley (1971), 283–296.
49. J. S. Foot, I. M. Ward: J. Materials Sci. *7*, 367–387 (1972).
50. W. Döll: Kolloid-Z. u. Z. Polymere *250*, 1066–1073 (1972).
51. G. P. Marshall, L. E. Culver, J. G. Willliams: Int. J. Fracture *9*, 295 (1973).
52. J. G. Williams, G. P. Marshall, in: Deformation and Fracture of High Polymers, H. H. Kausch, J. A. Hassell, R. I. Jaffee Eds., New York: Plenum Press 1973, 557.
53. H. R. Brown: J. Materials Sci. *8*, 941–948 (1973); G. P. Marshall, J. G. Williams, C. E. Turner, ibid., 949–956 (1973).
54. F. A. Johnson, A. P. Glover, J. C. Radon: Proc. 1973 Symposium on Mechanical Behavior of Materials, 141–148, The Society of Materials Sci., Japan (1974).
55. B. E. Nelson: J. Colloid and Interface Sci. *47*/3, 595–599 (1974).
56. P. W. R. Beaumont: J. Adhesion *6*, 107–137 (1974).
57. G. Goldbach: Kunststoffe *64*/9, 475–481 (1974).
58. R. A. W. Fraser, I. M. Ward: J. Materials Sci. *9*, 1624–1630 (1974).
59. A. D. S. Diggwa: Polymer *15*, 101 (1974).
60. W. Döll: Int. J. Fracture *11*, 184–186 (1975).
61. W. Döll: J. Materials Sci. *10*, 935–942 (1975).
62. R. Ravetti, W. W. Gerberich, T. E. Hutchinson: J. Materials Sci. *10*, 1441–1448 (1975).
63. F. A. Johnson, J. C. Radon: J. Polymer Sci., Polymer Chemistry Ed. *13*, 495–516 (1975).
64. E. Plati, J. G. Williams: Polymer *16*, 915–920 (1975). E. Plati, J. G. Williams: Polymer Engng. and Sci. *15*, 470–477 (1975).
65. R. P. Kusy, M. J. Katz: J. Materials Sci. *11*, 1475–1486 (1976). R. P. Kusy, D. T. Turner: Polymer *17*, 161–166 (1976).
66. W. Döll, G. W. Weidmann: J. Materials Sci., Letters *11*, 2348–2350 (1976).
67. W. Döll: Int. J. Fracture *12*/4, 595–605 (1976).
68. Y. W. Mai, J. G. Williams: J. Materials Sci. *12*, 1376–1382 (1977).
69. J. G. Williams, M. W. Birch: Fracture 1977, Vol. 1, ICF4, Waterloo Canada, June 19–24. 1977, pg. 501–528.
70. W. Döll, E. Gaube: unpublished results.
71. N. J. Mills, N. Walker: Polymer *17*, 335–344 (1976).
72. P. S. Leevers, J. C. Radon, L. E. Culver: Polymer *17*, 627–632 (1976).
73. R. A. Gledhill, A. J. Kinloch: Polymer *17*, 727–731 (1976).
74. G. P. Morgan, I. M. Ward: Polymer *18*, 87–91 (1977).
75. S. Yamini, R. J. Young: Polymer *18*, 1075–1084 (1977).
76. R. P. Kambour: J. Polymer Sci. D (Reviews) *7*, 1–154 (1973).
77. S. Rabinowitz, P. Beardmore: CRC Critical Reviews *1*, 1 (1973).
78. C. C. Hsiao, J. A. Sauer: J. of Appl. Physics *21*/11, 1071–1083 (1950). J. A. Sauer, C. C. Hsiao: Trans. ASME *75*, 895 (1953).
79. J. F. Rudd: J. Polymer Sci. *B1*, 1 (1963).
80. H. A. Stuart, D. Jeschke, G. Markowski: Materialprüfung *6*, 77 (1964).
81. A. C. Knight: J. Polymer Sci. A, *3*, 1845 (1965).
82. G. Rehage, G. Goldbach: Die Angewandte Makromolekulare Chemie *1*, 125–149 (1967).
83. R. P. Kambour, R. W. Kopp: General Electric, Report No. 67-C-374 (1967).
84. L. C. Cessna, Jr., S. S. Sternstein: Fundamental Phenomena in the Materials Sci. *4*, (Plenum Press), 45–79 (1967).
85. R. P. Kambour: Applied Polymer Symposia *7*, 215–235 (1968).
86. S. S. Sternstein, L. Ongchin, A. Silverman: Applied Polymer Symposia *7*, 175–199 (1968).
87. R. P. Kambour, A. S. Holik: General Electric, Report No. 69-C-022 (1968).
88. G. M. Bartenev: Vysokomol. soyed. A11/*10*, 2341–2347 (1969). Polymer Sci. USSR 11/*10*, 2663–2671 (1970).

89. R. N. Haward, B. M. Murphy, E. F. T. White: Paper 45 (Session IV), Fracture 1969, Brighton, 45/1–45/13.
90. P. Dunn, G. F. Sansom: J. appl. Polymer Sci. *13*, 1641–1655 (1969).
91. P. Dunn, G. F. Sansom: J. appl. Polymer Sci. *13*, 1657–1672 (1969).
92. P. Dunn, G. F. Sansom: J. appl. Polymer Sci. *13*, 1673–1688 (1969).
93. P. Dunn, G. F. Sansom: J. appl. Polymer Sci. *14*, 1799–1806 (1970).
94. R. S. Moore, C. Gieniewski: J. appl. Polymer Sci. *14*, 2889–2904 (1970).
95. K. C. Rusch, R. H. Beck, Jr.: J. Macromol. Sci.-Phys. B 3(3) 365 (1969), B 4(3), 261 (1970).
96. A. N. Gent: J. Materials Sci. *5*, 925 (1970).
97. L. Morbitzer, R. Holm, U. Brenneisen, K. Heidenreich, H. Röhr: Kunststoffe *60*/11, 861–866 (1970).
98. J. Murray, D. Hull: Polymer Letters *8*, 159–163 (1970), J. Murray, D. Hull: J. Polymer Sci. A-2, *8*, 1521–1543 (1970).
99. Yu. M. Malinskii, V. V. Prokopenko, N. A. Ivanova, V. A. Kargin: Mekh. Polimerov (USSR) *6*/2, 271–275 (1970), Polymer Mech. (USA) *6*/2, 240–244 (1973).
100. Yu. M. Malinskii, V. V. Prokopenko, N. A. Ivanova: Mekh. Polimerov (USSR) *6*/3, 445–448 (1970), Polymer Mech. (USA) *6*/3, 382–384 (1970).
101. F. Fischer: Z. f. Werkstofftechnik *1*/2, 74–83 (1970).
102. G. P. Marshall, L. E. Culver, J. G. Williams: Proc. Roy. Soc. Lond. A*319*, 165–187 (1970).
103. G. Menges, H. Schmidt, H. Berg: Kunststoffe *60*/11, 868–872 (1970).
104. G. Menges, H. Schmidt: Plastics & Polymers *38*, 13–19 (1970).
105. T. T. Wang, M. Matsuo, T. K. Kwei: J. appl. Physics *42*/11, 4188–4196 (1971).
106. P. Beahan, M. Bevis, D. Hull: Phil. Magazine *24*/12, 1267–1279 (1971).
107. N. Fujimoto, M. Takayanagi, Y. Yamaguchi: Proceeding of Int. Conf. Mech. Behavior of Materials, Kyoto, 1971, 572.
108. R. N. Haward, B. M. Murphy, E. F. T. White: J. Polymer Sci. A-2/*9*, 801–814 (1971).
109. B. J. Macnulty: J. Materials Sci. *6*, 1070–1075 (1971).
110. J. Pohrt: J. Macromol. Sci.-Phys. B5(2), 299–316 (1971).
111. J. Pohrt: Gummi Asbest Kunststoffe *24*, 594–606, and 700–713 (1971). K. V. Gotham: Plastics and Polymers *40*, 277–282 (1972).
112. Yu. S. Lipatov, F. B. Fabulyak: J. Appl. Polymer Sci. *16*, 2131 (1972).
113. H. Schmidt: Kunststoff-Rundschau *19*/1, 1–7 (1972). H. Schmidt: Kunststoff-Rundschau *19*/2,3, 56–65 (1972).
114. C. B. Bucknall, D. Clayton: J. Materials Sci. *7*, 202–210 (1972).
115. P. Beahan, M. Bevis, D. Hull: J. Materials Sci. *8*, 162–168 (1972).
116. N. Brown, M. F. Parrish: Polymer Letters Ed. *10*, 777–779 (1972).
117. L. Spenadel: J. Appl. Polymer Sci. *16*, 2375–2386 (1972).
118. I. Narisawa, T. Kondo: Int. J. of Fracture Mechanics *8*/4, 435–440 (1972).
119. I. Narisawa: J. Polymer Sci. Part A-2/*10*, 1789–1797 (1972).
120. J. Hoare, D. Hull: Phil. Magazine *26*/2, 443–455 (1972).
121. G. Menges, E. Alf: Kunststoffe *62*/4, 259–267 (1972).
122. H. E. Driesen: gwf-gas/erdgas *113*/6, 265–270 (1972).
123. P. I. Vincent, S. Raha: Polymer *13*/6, 283–287 (1972).
124. E. H. Andrews, L. Bevan: Polymer *13*/7, 337–346 (1972).
125. R. P. Kambour, E. E. Romagosa, C. L. Gruner: Macromolecules *5*/4, 335–340 (1972).
126. E. H. Andrews, G. M. Levy, J. Willis: J. Materials Sci. *8*, 1000–1008 (1973).
127. I. Narisawa, T. Kondo: J. Polymer Sci. – Poly. Phys. Ed. – *11*, 223–232 (1973).
128. R. P. Kambour, C. L. Gruner, E. E. Romagosa: J. Polymer Sci. – Poly. Phys. Ed. – *11*, 1879–1890 (1973).
129. N. Brown: J. Polymer Sci. – Poly. Phys. Ed. – *11*, 2099–2111 (1973).
130. E. Gaube, H. H. Kausch: Kunststoffe *63*/6, 391–397 (1973).
131. J. Lilley, D. G. Holloway: Phil. Mag. (8) *28*/1, 215–220 (1973).
132. K. P. Grosskurth: Kautschuk u. Gummi-Kunststoffe *26*/2, 43–45 (1973).
133. G. P. Marshall, J. G. Williams: J. Appl. Polymer Sci. *17*, 987–1005 (1973).
134. S. Fischer, N. Brown: J. Appl. Phys. *44*/10, 4322–4327 (1973).

135. H. G. Olf, A. Peterlin: Polymer *14*, 78–79 (1973).
136. G. Menges: Kunststoffe *63*, 95–100 a. 173–177 (1973).
137. R. N. Haward, D. R. J. Owen: J. Materials Sci. *8*, 1135–1144 (1973).
138. H. G. Olf, A. Peterlin: Macromolecules *6*, 470–472 (1973). H. G. Olf, A. Peterlin: J. Colloid and Interface Sci. *47*/3, 628–634 (1973).
139. S. S. Sternstein, F. A. Myers: J. Macromol. Sci.-Phys. *8*/3–4, 539–571 (1973).
140. J. R. Kastelic, E. Baer: J. Macromol. Sci.-Phys. *7*/4, 679–703 (1973).
141. L. Camwell, D. Hull: Phil. Magazine *27*/5, 1135–1150 (1973).
142. R. J. Oxborough, P. B. Bowden: Phil. Magazine *28*/3, 547–559 (1973).
143. K. J. Takahashi: Polymer Sci. – Poly. Phys. Ed. – *12*, 1697–1705 (1974).
144. M. Kitagawa, K. Motomura: J. Polymer Sci. – Poly. Phys. Ed. – *12*, 1979–1991 (1974).
145. H. G. Olf, A. Peterlin: J. Polymer Sci. – Poly. Phys. Ed. – *12*, 2209–2251 (1974).
146. J. F. Fellers, B. F. Kee: J. Appl. Polymer Sci. *18*, 2355–2365 (1974).
147. E. H. Andrews, G. M. Levy: Polymer *15*/9, 599–607 (1974).
148. T. L. Peterson, D. G. Ast, E. J. Kramer: J. Appl. Phys. *45*/10, 4220–4228 (1974).
149. K. P. Grosskurth: Kautschuk u. Gummi – Kunststoffe *27*/8, 324–328 (1974).
150. P. Beahan, M. Bevis, D. Hull: Proc. R. Soc. Lond. A *343*, 525–535 (1975).
151. G. W. Weidmann, J. G. Williams: Polymer *16*/12, 921–924 (1975).
152. A. S. Argon: Pure and Appl. Chemistry 43/1–2, 247–272 (1975).
153. D. Hull, L. Hoarse: Plastics and Rubber: Materials and Appl. *5*, 65–73 (1976).
154. P. Filzek, R. Süselbeck, W. Wicke: Kunststoffe *66*/1, 38 (1976).
155. M. J. Doyle: J. Polymer Sci., Polymer Phys. Ed., *13*, 127–135 (1975).
156. N. Verheulpen-Heymans, J. C. Bauwens: J. Materials Sci. *11*, 1–6 (1976). N. Verheulpen-Heymans, J. C. Bauwens: J. Materials Sci. *11*, 7–16 (1976).
157. N. Verheulpen-Heymans: J. Polymer Sci.: Polymer Phys. Ed., *14*, 93–99 (1976).
158. Y. Imai, N. Brown: J. Materials Sci. *11*, 417–424 (1976). Y. Imai, N. Brown: J. Materials Sci. *11*, 425–433 (1976).
159. H. G. Krenz, D. G. Ast, E. J. Kramer: J. Materials Sci. *11*, 2198–2210 (1976).
160. H. G. Krenz, E. J. Kramer, D. G. Ast: J. Materials Sci. *11*, 2211–2221 (1976).
161. D. L. G. Lainchbury, M. Bevis: J. Materials Sci. *11*, 2222–2234 (1976).
162. D. L. G. Lainchbury, M. Bevis: J. Materials Sci. *11*, 2235–2241 (1976).
163. S. Sikka: PhD Dissertation, University of Utah, Salt Lake City, Utah (1976).
164. I. D. Graham, J. G. Williams, E. L. Zichy: Polymer *17*/5, 439–442 (1976).
165. A. S. Argon, M. I. Bessonov: Reports of Research in Mech. Process. of Polymers *12*, MIT (1976).
166. A. S. Argon, J. G. Hannoosh, M. M. Salama: Reports of Research in Mech. Process. of Polymer *17*, MIT (1977); Fracture 1977, Vol. 1, Waterloo, Can., pg. 445. A. S. Argon, J. G. Hannoosh: Reports of Research in Mech. Process. of Polymers *21*, MIT (1977), Phil. Mag. *36*/5, 1195–1216 (1977).
167. A. S. Argon, M. M. Salama: Reports of Research in Mech. Process. of Polymers *22*, MIT (1977), Phil. Mag. *36*/5, 1217–1234 (1977).
168. W. Retting: Die Angew. Makrom. Chemie *58/59*, 133–174 (1977).
169. K. Smith, R. N. Haward: Polymer *18*/7, 745–746 (1977).
170. C. J. Singleton, E. Roche, P. H. Geil: J. Appl. Polymer Sci. *21*, 2319–2340 (1977).
171. R. P. Kambour: General Electric, Report No. 77CRD169, August 1977.
172. M. Ishikawa, I. Narisawa, H. Ogawa: J. Polymer Sci. (Polymer Phys. Ed.) *15*, 1791–1804 (1977).
173. J. W. Sauer, E. Foden, D. R. Morrow: Polymer Eng. and Sci. *17*/4, 246–250 (1977).
174. R. Bardenheier: DVM, Vorträge der 9. Sitzung des Arbeitskreises Bruchvorgänge, 29–38, 11. 10. 1977.
175. J. Opfermann: DVM, Vorträge der 9. Sitzung des Arbeitskreises Bruchvorgänge, 39–46, 11. 10. 1977.
176. K. P. Grosskurth: DVM, Vorträge der 9. Sitzung des Arbeitskreises Bruchvorgänge, 47–60, 11. 10. 1977.
177. R. A. W. Fraser, I. M. Ward: Polymer *19*/2, 220 (1978).

178. J. A. Sauer: Polymer *19*, NN (1978). S. Warty, D. R. Morrow, J. A. Sauer: Polymer, to be published.
179. J.-C. Pollet, S. J. Burns: Int. J. of Fracture *13*, 667–679 (1977). J.-C. Pollet, S. J. Burns: Int. J. Fracture *13*/6, 775–786 (1977).
180. A. G. Atkins, C. S. Lee, R. M. Caddell: J. Mat. Sci. *10*, 1381 (1975).
181. K. B. Broberg: Arkiv för Fysik *18*, 159–192 (1960).
182. G. T. Hahn, M. F. Kanninen: Fracture 1977, Vol. 1, ICF4, Waterloo, Canada, June 1977, 193.
183. K. Friedrich: Fracture 1977, Vol. 3, ICF4, Waterloo, Canada, June 1977, pg. 1119. K. Friedrich: Dissertation, Ruhr-Universität Bochum, 1978.
184. K. N. G. Fuller, P. G. Fox, J. E. Field: Proc. R. Soc. A(GB), Vol. 341, No. 1627, 537–557 (1974).
185. R. Weichert, K. Schönert: J. Phys. Mech. Sol. 22, 127–133 (1974).
186. R. Weichert: Dissertation, Universität Karlsruhe, 1976.
187. P. C. Moon, R. E. Barker, Jr.: J. Polymer Sci. (Polymer Phys. Ed.) *11*, 909–917 (1973).
188. E. Roeder, H.-A. Crostack: Kunststoffe *67*/8, 454–456 (1977).
189. K. L. DeVries, D. K. Roylance, M. L. Williams: Report UTEC DO 68-056, Univ. of Utah, Salt Lake City, 1968.
190. L. E. Nielsen, D. J. Dahm, P. A. Berger, V. S. Murty, J. L. Kardos: Polymer Sci. (Polymer Phys. Ed.) *12*, 1239 (1974).
191. M. D. Skibo, R. W. Hertzberg, J. A. Manson: J. Mat. Sci. *11*, 479–490 (1976).
192. J. A. Manson, L. H. Sperling: Polymer Blends and Composites, New York/London 1976: Plenum Press.
193. Mehrphasige Polymersysteme, Angew. Makromol. Chem. 58/59, 60/61 (1977).
194. J. Murray, D. Hull: J, Polymer Sci., Part A-2 *8*, 583–594 (1970).
195. B. E. Nelson, D. T. Turner: Polymer Letters *9*, 677–680 (1970).
196. D. Hull, T. W. Owen: J. Polymer Sci. (Polymer Phys. Ed.) *11*, 2039–2055 (1973).
197. W. Döll: Colloid & Polymer Sci. *252*, 880–885 (1974).
198. J. R. Martin, J. F. Johnson: J. Polymer Sci. (Polymer Phys. Ed.) *12*, 1081 1088 (1974).
199. J. Hoare, D. Hull: J. Materials Sci. *10*, 1861–1870 (1975).
200. R. P. Kusy, D. T. Turner: Polymer *18*, 391–399 (1977).
201. S. Gogolewski: Polymer *18*, 63–68 (1977).
202. S. Gogolewski, A. J. Pennings: Polymer *18*, 647–660 (1977).
203. J. E. Stamhuis, A. J. Pennings: Polymer *18*, 667–674 (1977).
204. S. Bandyopadhyay, H. R. Brown: Polymer *19*, 589–592 (1978).
205. E. S. Clark, C. A. Garber: Int. J. Polymeric Mater. *1*, 31–46 (1971).
206. J. Patel, P. J. Philips: Polymer Letters Ed. *11*, 771–776 (1973).
207. E. Andrews: Pure and Appl. Chem. *39*/1–2, 179–194 (1974).
208. G. Menges, N. Berndtsen: Kunststoffe *66*/11, 735–740 (1976).
209. P. I. Vincent, S. Picknell, G. F. Harding: An Investigation of Fracture Criteria for Anisotropic Polyethylene Terephthalate Film Using Mechanical and Optical Techniques, Div. Pol. Sci. Case Western Reserve Univ., Cleveland, Ohio 44106 (1967).
210. V. E. Korsukov, V. I. Vettegren', I. I. Novak, A. Chmel': Mekh. Polimerov 4, 621–625 (1972). Polymer Mechanics 4, 536–539 (1972).
211. M. Takano, L. E. Nielsen: J. Appl. Polymer Sci. *20*, 2193–2207 (1976).
212. H. Oberst: Kunststoffe *53*, 4 (1963).
213. J. A. Sauer: J. Polymer Sci. C (Polymer Symposia) *32*, 69–122 (1971).
214. P. I. Vincent: Polymer *15*, 111–116 (1974).
215. E. H. Andrews, B. J. Walker: Proc. R. Soc. Lond. A. *325*, 57–79 (1971).
216. J. P. Elinck, J. C. Bauwens, G. Homès: Int. J. Fracture Mech. *7*, 277–287 (1971).
217. W. J. Plumbridge: J. Materials Sci. *7*, 939–962 (1972).
218. J. A. Manson, R. W. Hertzberg: CRC Crit. Reviews in Macromol. Sci. *1* (4), 433–499, August 1973.
219. J. C. Radon, L. E. Culver: Polymer *16*/7, 539–544 (1975).
220. M. D. Skibo, R. W. Hertzberg, J. A. Manson: Fracture 1977, Vol. 4, ICF4, Waterloo, Canada, June 1977, 1127–1133.

221. N. K. Baramboim: Mechanochemistry of Polymers (translated from the Russian), W. T. Watson, ed., MacLaren 1964.
222. S. A. Komissarov, N. K. Baramboim: Vysokomol. soyed. A11/5, 1050–1058 (1969). Polymer Sci. USSR 11/5, 1189–1198 (1969).
223. N. K. Baramboim, W. G. Protasow: die Technik 30/2, 73–81 (1975).
224. W. Lauer: Kautschuk + Gummi, Kunststoffe 28/9, 536–613 (1975).
225. A. Casale, R. S. Porter, J. F. Johnson: Rubber Chem. and Techn. 44/2, 534–577 (1971).
226. A. Casale: J. of Appl. Polymer Sci. 19, 1461–1473 (1975).
227. A. Casale, R. S. Porter: Polymer Stress Reactions, New York: Academic Press 1978.
228. T. Pazonyi, F. Tüdös, M. Dimitrov: Plaste und Kautschuk 16/8, 577–581 (1969).

List of Tables

Table A. 1. List of Abbreviations of the most important Polymers

(according to DIN 7723, DIN 7728, ASTM D1600-64 T, ASTM 1418-67, ISO/R 1043-1969)

ABS	Acrylonitrile-butadiene-styrene copolymer	NBR	Acrylonitrile-butadiene rubber
ABR	Acrylonitrile-butadiene rubber	NCR	Acrylonitrile-chloroprene rubber
AMMA	Acrylonitrile-methyl methacrylate copolymer	NR	Natural rubber
		PA	Polyamide
		PA 6	Polycaprolactam
ASA	Acrylester-styrene-acrylonitrile copolymer	PA 66	Poly(hexamethylene adipamide)
BR	Polybutadiene rubber	PA 6.10	Poly(hexamethylene sebacamide)
CA	Cellulose acetate		
CAB	Cellulose acetobutyrate	PA 11	Polyamide of 11-amino-undecanoic acid
CAP	Cellulose acetopropionate		
CN	Cellulose nitrate	PA 12	Polylaurolactam
CP	Cellulose propionate	PAA	Poly (acrylic acid)
CPE	Chlorinated polyethylene	PAN	Polyacrylonitrile
CPVC	Chlorinated poly(vinyl chloride)	PB	Polybutylene
		(PBA)	Polybutyl acrylate
EC	Ethyl cellulose	PBTP	Poly(butylene glycol terephthalate)
EP	Epoxide resin		
EPDM	Ethylene propylene terpolymer rubber	PC	Polycarbonate
		PCTFE	Poly(chlorotrifluoroethylene)
EPR	Ethylene-propylene rubber	PDAP	Poly(diethyl phthalate)
EPTR	Ethylene-propylene terpolymer rubber	PE	Polyethylene
		PEC	Chlorinated polyethylene
ETFE	Ethylene-tetrafluorethylene copolymer	PED	Deuterated polyethylene
		PEH	PE (as opposed to PED)
EU	Polyether-urethane	PEO	Polyethylene oxide
EVA	Ethylene-vinyl-acetate copolymer	PEOB	(Poly[p-(2-hydroxyethoxy) benzoic acid])
HDPE	High-density polyethylene	(PES)	Polyethersulfone
HiPS	High impact polystyrene	PETP	Poly(ethylene terephthalate)
IR	Isoprene rubber	PF	Phenol-formaldehyde resin
IIR	Butyl rubber, isoprene-isobutylene copolymer	PI	Poly-trans-isoprene
		PIB	Polyisobutylene
MF	Melamin-formaldehyde resin	PMMA	Poly(methyl methacrylate)

321

Table A. 1. (continued)

PMP	Poly(4-methylpentene-1)	PVDF	Poly(vinylidene fluoride)
PαMS	Poly-α-methyl styrene	PVF	Poly(vinyl fluoride)
POM	Polyoxymethylene = poly-acetal = polyformaldehyde	PVFM	Poly(vinyl formal)
		PY	Unsaturated polyester resin
PP	Polypropylene	RF	Resorcinol-formaldehyde resin
PPO	Poly(2,6-dimethyl-1,4-phenylene oxide)	RP	Reinforced plastic
		SAN	Styrene-acrylonitrile copolymer
PS	Polystyrene		
PSU	Polysulfone	SB	Styrene-butadiene copolymer
PTFE	Poly(tetrafluoroethylene)	SBR	Styrene-butadiene rubber
PTMT	Poly(tetramethylene terephthalate)	SBS	Styrene-butadiene block copolymer
PUR	Polyurethane	SI	Silicone rubber
PVA(L)	Poly(vinyl alcohol)	SIR	Styrene-isoprene rubber
PVAC	Poly(vinyl acetate)	SMS	Styrene-α-methylstyrene copolymer
PVB	Poly(vinyl butyral)		
PVC	Poly(vinyl chloride)	UE	Polyurethane rubber
PVCA	Vinyl chloride-vinyl acetate copolymer	UF	Urea-formaldehyde resin
		UP	Unsaturated polyester resin
PVDC	Poly(vinylidene chloride)	UR	Polyurethane rubber

Table A. 2. List of Abbreviations not Referring to Polymer Names

CN	Center notched specimen (fracture mechanics)
COD	Crack opening displacement (fracture mechanics)
CT	Compact tensile specimen (fracture mechanics)
DCB	Double cantilever beam (fracture mechanics)
DT	Double torsion (fracture mechanics)
ESR	Electron spin resonance
IR	Infra red
NMR	Nuclear magnetic resonance
SANS	Small angle neutron scattering
SAXS	Small angle X-ray scattering
SEM	Scanning electron microscopy
SEN	Single edge notch specimen (fracture mechanics)
UV	Ultra violet
WAXS	Wide angle X-ray scattering
WOL	Wedge open loaded specimen (fracture mechanics)

Table A. 3. List of Symbols

Symbol	General Meaning (if not indicated otherwise in the section where the symbol appears)	Unit Used
C	Concentration of end groups	$mol\ m^{-3}$
C	Bending compliance	$m\ N^{-1}$
C_p	Specific heat	$J\ kg^{-1}\ K^{-1}$
D	Bond dissociation energy	$kJ\ mol^{-1}$
$\bar{D}$	Average spherulite diameter	μm
E	Young's modulus	$GN\ m^{-2}$
E'	Storage modulus	$GN\ m^{-2}$
E''	Loss modulus	$GN\ m^{-2}$
E^*	Complex modulus	$GN\ m^{-2}$
E_A	Activation energy	$kJ\ mol^{-1}$ or $kcal\ mol^{-1}$
E_c	Elastic modulus of crystalline regions	$GN\ m^{-2}$
E_k	Young's modulus of chains	$GN\ m^{-2}$
E_t	Creep modulus	$GN\ m^{-2}$
ΔF	Activation energy	$kJ\ mol^{-1}$
G	Shear modulus	$GN\ m^{-2}$
G'	Storage modulus	$N\ m^{-2}$
G''	Loss modulus	$N\ m^{-2}$
G^*	Free enthalpy of activation	$kcal\ mol^{-1}$
G_A	Energy release rate at crack arrest	$J\ m^{-2}$
G_c	Energy release rate corresponding to K_c	$J\ m^{-2}$
G_{c1}	Plain-strain energy release rate	$J\ m^{-2}$
G_{c2}	Plain-stress energy release rate (plastic component of G_c)	$J\ m^{-2}$
G_d	Dynamic energy release rate	$J\ m^{-2}$
G_{Ic}	Critical energy release rate	$J\ m^{-2}$
K	Compressive modulus	$GN\ m^{-2}$
K_o	Initial stress intensity factor	$MN\ m^{-3/2}$
K_c	Fracture toughness – the largest value of stress-intensity factor that exists prior to onset of rapid fracture	$MN\ m^{-3/2}$
K_{c1}	Plane strain fracture toughness	$MN\ m^{-3/2}$
K_{c2}	Plane stress fracture toughness	
K_I	Stress intensity factor	$MN\ m^{-3/2}$
K_m	Stress intenstiy factor below which no crazing occurs	$MN\ m^{-3/2}$
L	Contour length of a segment	nm or Å
L_o	End-to-end distance of a segment	nm or Å
M	Molecular weight	$(g\ mol^{-1})$
M_c	Chain molecular weight between crosslinks	$(g\ mol^{-1})$
M_n	Number average molecular weight	$(g\ mol^{-1})$
M_w	Weight average molecular weight	$(g\ mol^{-1})$
N	Number of chains between cross links	
N_F	Number of cycles to failure	
N_L	Avogadro number	$6.022 \cdot 10^{23}\ mol^{-1}$
Q_b	Amount of heat dissipated	kJ
R	Molar gas constant	$8.314\ J\ mol^{-1}\ K^{-1}$
S	Internal entropy	$kJ\ K^{-1}$
S_v	Volume of liquid absorbed/unit vol. of polymer	—

Table A. 3. (continued)

Symbol	General Meaning (if not indicated otherwise in the section where the symbol appears)	Unit Used
T	Absolute temperature	K
T_g	Glass transition temperature	°C
U	Internal energy	$kJ\ mol^{-1}$
U_o	Activation energy for bond scission	$kJ\ mol^{-1}$ or $kcal\ mol^{-1}$
a	Crack length	m
a_n	Impact strength	$kJ\ m^{-2}$
a_T	Shift factors of relaxation times	
f	Fractional unoccupied volume	cm^3
g	Spectroscopic splitting factor	
h	Planck's constant	$6.626 \cdot 10^{-34}\ Js$
k	Boltzmann's constant	$1.380 \cdot 10^{-23}\ JK^{-1}$
k_c	Rate of thermomechanical chain scission	s^{-1}
k_S	Notch sensitivity factor of strength	
k_T	Notch sensitivity factor of energy to break	
m_I	Spin quantum number	
n	Number of chain atoms	
Δn	Birefringence	
p_c	Craze yield stress	$MN\ m^{-2}$
q	Molecular cross-section	m^2
r	Chain end-to-end distance	nm or Å
r_p	Plastic zone size	μm
t_b	Time to break	s
α	Activation volume	$m^3\ mol^{-1}$ or nm^3
β	Activation volume for chain scission	$m^3\ mol^{-1}$ or nm^3
γ	Activation volume of fracture under tensile stress	$m^3\ mol^{-1}$ or nm^3
δ	Solubility parameter	$cal^{1/2}\ cm^{-3/2}$
$\tan \delta$	Loss tangent	—
ϵ	Strain	—
ϵ_c	Critical strain	—
ϵ_y	Yield strain	—
η	Viscosity (dynamic)	$Ns\ m^{-2}$
λ	Uniaxial extension ratio	—
λ	Heat conductivity	$W\ K^{-1}\ m^{-1}$
λ_s	Linear swelling ratio	—
μ	Coefficient of friction	—
ν	Poisson's ratio	—
ρ	Polymer density	$g \cdot cm^{-3}$
ρ_a	Radius of crack tip	μm
σ	Macroscopic stress	$GN\ m^{-2}$
σ_b	Breaking strength	$GN\ m^{-2}$
σ_c	Critical stress	$GN\ m^{-2}$
τ	Relaxation time	s
τ^*	Critical shear stress	$GN\ m^{-2}$
ψ	Axial chain stress	$GN\ m^{-2}$
ψ_c	Chain strength	$GN\ m^{-2}$
ω	Loading frequency	Hz
ω	Angular frequency	Hz

324

Table A. 4. Conversion Factors

Frequently found non-SI Units	SI-Units	Conversion
atmosphere, at	Nm^{-2} or Pascal, Pa	$1\ at = 98.07 \cdot 10^3\ N\ m^{-2}$
	bar	$1\ at = 0.9807\ bar$
calorie, cal	Joule, J	$1\ cal = 4.187\ J$
centipoise, cP	Pascal second, Pa $\cdot$ s	$1\ cP = 1.000 \cdot 10^{-3}\ Pa \cdot s$
dyn	Newton, N	$1\ dyn = 1 \cdot 10^{-5}\ N$
erg	Joule, J	$1\ erg = 10^{-7}\ J$
foot, ft	meter, m	$1\ ft = 3.048 \cdot 10^{-1}\ m$
foot pound-force, lbft	Joule, J	$1\ lb.ft = 1.356\ J$
inch, in	meter, m	$1\ in = 2.540 \cdot 10^{-2}\ m$
kilopond, kp	Newton, N	$1\ kp = 9.807\ N$
kilogram-force per square centimeter	NM^{-2} or Pascal, Pa	$1\ kgf\ cm^{-2} = 9.807 \cdot 10^4\ N\ m^{-2}$
mil (= 0.001 in)	micrometer, μm	$1\ mil = 25.4\ \mu m$
pound per square inch, PSI	Nm^{-2} or Pascal, Pa	$1\ PSI = 6.897 \cdot 10^3\ N\ m^{-2}$

Subject Index

Key words and page numbers set in *italics* refer to the headings.

Subject Index